Cognition and Exploratory Learning in the Digital Age

This book series focuses on the implications of digital technologies for educators and educational decision makers. Taking an interdisciplinary and visionary perspective on cognition and exploratory learning in the digital age, the series examines how digital technologies can or should be exploited to address the needs and propel the benefits of large-scale teaching, learning and assessment. It includes the newest technical advances and applications, and explicitly addresses digital technologies and educational innovations.
The series incorporates volumes stemming from the Cognition and Exploratory Learning in the Digital Age (CELDA) conference. Since 2004 the CELDA conference has focused on discussing and addressing the challenges pertaining to the evolution of the learning process, the role of pedagogical approaches and the progress of technological innovation, in the context of the digital age. Some of CELDA's main topics include: assessment of exploratory learning approaches and technologies, educational psychology, learning paradigms in academia and the corporate sector, student-centred learning and lifelong learning. In line with and in addition to these CELDA conference volumes, this series publishes monographs and edited volumes covering the following subject areas:

- K-12, Higher Education, Vocational Education, Business Education, Health Education
- Learning Design, Instructional Design
- Change Management, Technology Adoption, Learning Organization
- Data analytics
- Learning, Teaching, Assessment

The book series represents the most recent work in the area where cognitive psychology, information and communications technology and instructional design intersect.

More information about this series at http://www.springer.com/series/16424

Dirk Ifenthaler • Demetrios G. Sampson
Pedro Isaías
Editors

Balancing the Tension between Digital Technologies and Learning Sciences

Springer

Editors
Dirk Ifenthaler
Learning, Design and Technology
University of Mannheim
Mannheim, Baden-Württemberg, Germany

UNESCO Deputy Chair of Data Analytics
in Higher Education Learning and Teaching
Curtin University
Perth, Western Australia, Australia

Demetrios G. Sampson
Department of Digital Systems
University of Piraeus
Greece

Pedro Isaías
School of Information Systems &
Technology Management
UNSW Sydney
Sydney, New South Wales, Australia

ISSN 2662-5628 ISSN 2662-5636 (electronic)
Cognition and Exploratory Learning in the Digital Age
ISBN 978-3-030-65656-0 ISBN 978-3-030-65657-7 (eBook)
https://doi.org/10.1007/978-3-030-65657-7

This Springer imprint is published by the registered company Springer Nature Switzerland AG
The registered company address is: Gewerbestrasse 11, 6330 Cham, Switzerland

Preface

Educational technology is frequently associated with or even used interchangeably with Information Communication Technology (ICT), which indicates an oversimplified definition with a narrow focus on technology as a tool. What is often left out of the critical discourse in educational technology is the large body of knowledge from learning sciences. Still, the effectiveness of educational technology in improving learning and teaching remains the focus of rigorous research on educational technology (Mao, Ifenthaler, Fujimoto, Garavaglia, & Rossi, 2019).

The Cognition and Exploratory Learning in the Digital Age (CELDA; www.celda-conf.org) conference focuses on discussing and addressing the challenges pertaining to the evolution of learning processes, the role of pedagogical approaches, and the progress of technological innovation, in the context of the digital age. In each edition since 2004, CELDA gathers researchers and practitioners in an effort to cover both technological and pedagogical issues in ground-breaking studies. Some of CELDA's main topics include assessment of exploratory learning approaches and technologies, educational psychology, learning paradigms in academia and the corporate sector, and student-centered learning and lifelong learning. The 2019 edition of the CELDA conference received a wide array of papers examining the deployment of learning technologies, proposing pedagogical approaches and practices to address digital transformation, presenting case studies of specific technologies and contexts, and overall debating the contribution of learning technologies for the improvement of the learning process and the experience of students and for the development of key competences.

This edited volume *Balancing the Tension between Digital Technologies and Learning Sciences* provides a platform for the continuous conversation stemming from the CELDA conference series. It comprises three parts focusing different stakeholder groups and cases: Part I—Cases from Preschool and Primary School; Part II—Cases from Secondary School; and Part III—Cases from Higher Education and Further Education.

In Part I, the first chapter "The Enhancing of Numeracy Skills Through Pencil-and-Paper or Computerized Training for Kindergarteners" aims at promoting the use of educational tools during preschool trainings for the improvement of numerical skills (Maria Lidia Mascia, Mirian Agus, Maria Chiara Fastame, and Maria Pietronilla Penna, Chap. 1). The next chapter "Does Chess Training Affect Meta-Cognitive Processes and Academic Performance?" investigates the relationship between chess, general meta-cognitive abilities, and academic school skills using an experimental design (Carla Meloni and Rachele Fanari, Chap. 2). The concluding chapter in Part I is "Developing Computational Thinking in Early Childhood Education," which focuses on the development of young children's computational thinking using robotics activities taking into consideration individual cognitive differences (Kyriakoula Georgiou and Charoula Angeli, Chap. 3).

In Part II, the first chapter "Network Analytics of Collaborative Problem-Solving" describes a task-focused approach to network analysis of trace data from collaborative problem-solving in a digital learning environment (Simon Kerrigan, Shihui Feng, Rupa Vuthaluru, Dirk Ifenthaler, and David Gibson, Chap. 4). Next, "Experiences with Virtual Reality at Secondary Schools. Is There an Impact on Learning Success?" examines the effects of immersive virtual reality learning units on the learning success of secondary school students (Thomas Keller and Elke Brucker-Kley, Chap. 5). The following chapter "Pre-service Teachers' Adoption of a Makerspace" presents a study focusing on education majors in an instructional technology class in the United States learning about the concept of a makerspace (Junko Yamamoto, Chap. 6). The concluding chapter in Part II, "Relationship Between Learning Time and Dimensions of a Learning Organization," compares the IT sector and the education sector in terms of learning time and assessment of the individual dimensions of the learning organization (Vaclav Zubr, Chap. 7).

In Part III, the first chapter "Learning Analytics Dashboard Supporting Metacognition" discusses the designs and development of a learning analytics dashboard to support learners' metacognition (Li Chen, Min Lu, Yoshiko Goda, Atsushi Shimada, and Masanori Yamada, Chap. 8). The following chapter "Diversity as an Advantage: An Analysis of the Demand for Specialized and Social Competencies for STEM Graduates Using Machine Learning" asks which unique perspectives STEM graduates from underrepresented groups can bring to their future careers (Karin Maurer, Annika Hinze, Heidi Schuhbauer, and Patricia Brockmann, Chap. 9). Next, "Student Perceptions of Virtual Reality in Higher Education" explores student perceptions on the possible uses of virtual reality in their universities, as a way of easing their access to learning material (Tebogo John Matome and Mmaki Jantjies, Chap. 10). Then, "Open Distance Learning and Immersive Technologies: A Literature Analysis" aims to explore how new immersive technology can be used to enhance the experience of distance learning (Afika Ntaba and Mmaki Jantjies, Chap. 11). In their chapter "Technological, Organisational and Socio-Interactional Affordances In Simulation-Based Collaborative Learning," Kirsi Lainema, Timo Lainema, Kirsi Heinonen, and Raija Hämäläinen show how teams of learners employ the different types of affordances in their collaborative tasks (Chap. 12). Next, "Enhancement of Experiential Learning in Software Factory Project-Based Course" presents the design of a software factory course and student and teacher experiences as well as discusses the importance of reflective learning diaries and serious games (Muhammad Ovais Ahmad and Kari Liukkunen, Chap. 13). The following chapter "How to Generate Exercise Questions for Web-Based Investigative Learning" reflects on how to develop skills in question expansion for Web-based investigative learning (Rei Saito, Akihiro Kashihara, Yoshiki Sato, Miki Hagiwara, and Koichi Ota, Chap. 14). Then, "Collaborative Learning: Collegiate Pedagogy Utilizing Web Conferencing" reports on a study which was completed in two phases: the first employed systematic literature analysis of collegiate instructional use of web conferencing followed by a case study of a multicampus collaborative course utilizing web conferencing (Joan Ann Swanson, Susan L. Renes, and Anthony T. Strange, Chap. 15). The concluding chapter of Part III, "Interaction Effects of Teachers' Educational Policies for Seminars and Students' Learning Goal Orientation on Students' Learning-as-Duty Conception," examines interaction effects between teachers' educational polices and students' learning goal orientation on students' learning-as-duty conception (Mai Yokoyama and Kazuhisa Miwa, Chap. 16).

The contributions and outcomes collected in this edited volume are consistent with the overall goal of the Cognition and Exploratory Learning in the Digital Age book series (www.springer.com/series/16424) and further document the advances presented and published from previous editions of the CELDA conference. In their first publication, Spector, Ifenthaler, Isaias, Kinshuk, and Sampson (2010) approach the general developments and challenges of learning and instruction in the digital age. More specifically, the editors gathered contributions that examined cognitive approaches to learning and instruction, knowledge representation and mental models technology, facilitated tools and techniques, communications and methods, and integrative methods and online learning. In Ifenthaler, Kinshuk, Isaias, Sampson, and Spector (2011), the editors compiled research initiatives that emphasize multiple perspectives on problem-solving and learning in the context of the digital age by exploring related topics such as pedagogical usability issues in web-based learning objects, automated measurement of critical thinking for discussion forum participants, expanding global awareness with virtual collaboration, and simulation games as learning experience. In Isaias, Ifenthaler, Kinshuk, Sampson, and Spector (2012), the editors intended to assess the impact of web 3.0 in learning and instruction by focusing on student-centered learning, collaborative learning, and exploratory technologies and addressing educational precepts such as just-in-time learning, constructivism, and web 3.0's adoption in education. Following the tendency for the adoption of mobile devices in education, Sampson, Isaias, Ifenthaler, and Spector (2013) compiled the most relevant contributions pertaining to ubiquitous and mobile learning in the digital age and all its fundamental ramifications, such as formal and informal learning environments, social web technologies, virtual worlds and game-based learning, and location-based and context-aware environments. On a later publication Sampson, Ifenthaler, Spector, and Isaias (2014) emphasized the importance of digital systems for open access in the context of both formal and informal learning and gathered contributions that covered the theoretical and practical aspects of open access, as well as different methods and technologies used to support it. In Isaias, Spector, Ifenthaler, and Sampson (2015) the focus was placed on e-learning systems, which were scrutinized from different perspectives: exploratory learning technologies, e-learning social web design, learner communities through e-learning implementations, and collaborative and student-centered e-learning design. In the following year, Spector, Ifenthaler, Sampson, and Isaias (2016) gathered contributions about the competencies, challenges, and transformation that stem from the deployment of digital technologies. The publication introduces this subject, reflects about the changes in learning and instructional paradigms, debates assessments and analytics for teachers and decision makers, and examines the changing tools and environments teachers and learners must face. In Sampson, Ifenthaler, Spector, and Isaias (2018), digital technologies were explored from the perspective of their role as promoters of sustainable educational innovations for the enhancement of teaching, learning, and assessment in all educational levels. The research depicted in this publication addressed the importance of digital technologies in transforming the learning environment, enriching the student learning experiences, measuring and assessing teaching and learning, and cultivating student competences for the digital smart society. Then, Sampson, Spector, Ifenthaler, Isaias, and Sergis (2019) focused on the transformational potential that learning technologies have for large-scale teaching, learning, and assessment. The editors gathered the outcomes of research efforts featuring state-of-the-art case studies examining the innovative influence of learning technologies, such as Massive Open Online Courses and educational data analytics. Two additional volumes have been published exploring the conceptual and practical aspects of technologies that are used to support learning, with a multidisciplinary approach that encompasses all levels of education (Isaias, Sampson, & Ifenthaler, 2020a, 2020b).

The CELDA conferences (www.celda-conf.org) and related book series (www.springer.com/series/16424) strive to continue the critical discourse regarding the support of learning processes and learning outcomes through digital technologies. This edited volume adds to this continuous conversations and documents the advances of our field.

Mannheim, Germany — Dirk Ifenthaler
Perth, WA, Australia
Piraeus, Greece — Demetrios G. Sampson
Sydney, NSW, Australia — Pedro Isaias

References

Ifenthaler, D., Kinshuk, Isaias, P., Sampson, D. G., & Spector, J. M. (Eds.). (2011). *Multiple perspectives on problem solving and learning in the digital age*. New York, NY: Springer.

Isaias, P., Ifenthaler, D., Kinshuk, Sampson, D. G., & Spector, J. M. (Eds.). (2012). *Towards learning and instruction in Web 3.0. Advances in cognitive and educational psychology*. New York, NY: Springer.

Isaias, P., Sampson, D. G., & Ifenthaler, D. (Eds.). (2020a). *Online teaching and learning in higher education*. Cham: Springer.

Isaias, P., Sampson, D. G., & Ifenthaler, D. (Eds.). (2020b). *Technology supported innovations in school education*. Cham: Springer.
Isaias, P., Spector, J. M., Ifenthaler, D., & Sampson, D. G. (Eds.). (2015). *E-Learning systems, environments and approaches*. New York, NY: Springer.
Mao, J., Ifenthaler, D., Fujimoto, T., Garavaglia, A., & Rossi, P. G. (2019). National policies and educational technology: a synopsis of trends and perspectives from five countries. *TechTrends, 63*(3), 284–293. https://doi.org/10.1007/s11528-019-00396-0
Sampson, D. G., Ifenthaler, D., Spector, J. M., & Isaias, P. (Eds.). (2014). *Digital systems for open access to formal and informal learning*. New York, NY: Springer.
Sampson, D. G., Ifenthaler, D., Spector, J. M., & Isaias, P. (Eds.). (2018). *Digital technologies: Sustainable innovations for improving teaching and learning*. Cham: Springer.
Sampson, D. G., Isaias, P., Ifenthaler, D., & Spector, J. M. (Eds.). (2013). *Ubiquitous and mobile learning in the digital age*. New York, NY: Springer.
Sampson, D. G., Spector, J. M., Ifenthaler, D., Isaias, P., & Sergis, S. (Eds.). (2019). *Learning technologies for transforming teaching, learning and assessment at large scale*. Cham: Springer.
Spector, J. M., Ifenthaler, D., Isaias, P., Kinshuk, & Sampson, D. G. (Eds.). (2010). *Learning and instruction in the digital age*. New York, NY: Springer.
Spector, J. M., Ifenthaler, D., Sampson, D. G., & Isaias, P. (Eds.). (2016). *Competencies in teaching, learning and educational leadership in the digital age*. New York, NY: Springer.

Acknowledgments

We acknowledge the support of the International Association for Development of the Information Society (http://www.iadisportal.org/) and for its continuing sponsorship of the ***International Conference on Cognition and Exploratory Learning in Digital Age*** (CELDA; http://www. celda-conf.org). In addition, we wish to express our gratitude to all the members of the CELDA 2019 international program committees for their timely and thoughtful reviews of all papers submitted to the conference. We also would like to acknowledge the support of Springer Science + Business Media and Springer's education editor, Melissa James. Springer has been very supportive of CELDA for many years by supporting the publication of an edited volume based on expanded and edited versions of the best CELDA papers.

Mannheim, Germany — Dirk Ifenthaler
Perth, WA, Australia
Piraeus, Greece — Demetrios G. Sampson
Sydney, NSW, Australia — Pedro Isaias

Contents

Part I Cases from Pre-School and Primary School

1 **The Enhancing of Numeracy Skills Through Pencil-and-Paper or Computerized Training for Kindergarteners** 3
Maria Lidia Mascia, Mirian Agus, Maria Chiara Fastame, and Maria Pietronilla Penna

2 **Does Chess Training Affect Meta-Cognitive Processes and Academic Performance?** 19
Carla Meloni and Rachele Fanari

3 **Developing Computational Thinking in Early Childhood Education: A Focus on Algorithmic Thinking and the Role of Cognitive Differences and Scaffolding** 33
Kyriakoula Georgiou and Charoula Angeli

Part II Cases from Secondary School

4 **Network Analytics of Collaborative Problem-Solving** 53
Simon Kerrigan, Shihui Feng, Rupa Vuthaluru, Dirk Ifenthaler, and David Gibson

5 **Experiences with Virtual Reality at Secondary Schools: Is There an Impact on Learning Success?** 77
Thomas Keller and Elke Brucker-Kley

6 **Pre-service Teachers' Adoption of a Makerspace** 95
Junko Yamamoto

7 **Relationship Between Learning Time and Dimensions of a Learning Organization** 111
Vaclav Zubr

Part III Cases from Higher Education and Further Education

8 Learning Analytics Dashboard Supporting Metacognition 129
Li Chen, Min Lu, Yoshiko Goda, Atsushi Shimada, and Masanori Yamada

9 Diversity as an Advantage: An Analysis of the Demand for Specialized and Social Competencies for STEM Graduates Using Machine Learning 151
Karin Maurer, Annika Hinze, Heidi Schuhbauer, and Patricia Brockmann

10 Student Perceptions of Virtual Reality in Higher Education 167
Tebogo John Matome and Mmaki Jantjies

11 Open Distance Learning and Immersive Technologies: A Literature Analysis 183
Afika Ntaba and Mmaki Jantjies

12 Technological, Organisational and Socio-Interactional Affordances in Simulation-Based Collaborative Learning 199
Kirsi Lainema, Timo Lainema, Kirsi Heinonen, and Raija Hämäläinen

13 Enhancement of Experiential Learning in Software Factory Project-Based Course 217
Muhammad Ovais Ahmad and Kari Liukkunen

14 How to Generate Exercise Questions for Web-Based Investigative Learning 235
Rei Saito, Akihiro Kashihara, Yoshiki Sato, Miki Hagiwara, and Koichi Ota

15 Collaborative Learning: Collegiate Pedagogy Utilizing Web Conferencing 253
Joan Ann Swanson, Susan L. Renes, and Anthony T. Strange

16 Interaction Effects of Teachers' Educational Policies for Seminars and Students' Learning Goal Orientation on Students' Learning-as-Duty Conception 269
Mai Yokoyama and Kazuhisa Miwa

Index 285

Editors and Contributors

About the Editors

Dirk Ifenthaler (dirk@ifenthaler.info; www.ifenthaler.info) is Professor and Chair of Learning, Design and Technology at the University of Mannheim, Germany, and UNESCO Deputy Chair of Data Science in Higher Education Learning and Teaching at Curtin University, Australia. His previous roles include Professor and Director, Centre for Research in Digital Learning at Deakin University, Australia, Manager of Applied Research and Learning Analytics at Open Universities, Australia, and Professor for Applied Teaching and Learning Research at the University of Potsdam, Germany. He was a 2012 Fulbright Scholar-in-Residence at the Jeannine Rainbolt College of Education, at the University of Oklahoma, USA. Dirk's research focuses on the intersection of cognitive psychology, educational technology, data analytics, and organizational learning. His research outcomes include numerous coauthored books, book series, book chapters, journal articles, and international conference papers, as well as successful grant funding in Australia, Germany, and USA. He is the Editor-in-Chief of the *Technology, Knowledge and Learning*, Senior Editor of the *Journal of Applied Research in Higher Education*, and Deputy Editor-in-Chief of the *International Journal of Learning Analytics and Artificial Intelligence for Education*.

Demetrios G. Sampson (sampson@unipi.gr) is a Professor of Digital Systems for Learning and Education at the Department of Digital Systems, University of Piraeus, Greece, where he serves as Academic Staff since 2001, and a Professor of Learning Technologies at the School of Education, Curtin University, Australia, since 2015. He is the coauthor of 340 articles in scientific books, journals, and conferences and the editor of 12 books, 32 special issues, and 35 international conference proceedings. He has received ten times Best Paper Award in international conferences. He has been a Keynote/Invited Speaker/Lecturer in 85 international/national conferences and/or postgraduate programs around the world. He has been project director, principal investigator, and/or research consultant in 70 Research and Innovation projects with external funding at the range of 16 M€. He has supervised 155 honors and postgraduate students to successful completion. He has developed and delivered the first MOOC on the use of Educational Data Analytics by School Teachers (Analytics for the Classroom Teacher), offered by the edX platform which has attracted more than 12,000 participants from 160 countries around the world since October 2016. He served as Editor-in-Chief of one of the first open-access journals in educational technology, the *Educational Technology & Society* journal, 2003–2018. He has also served or serves as a member of the steering committee and/or advisory and/or editorial board of 25 international/national journals, in various leadership roles in 75 international conferences, and at the program committee of 525 international/national conferences. He is the recipient of the *IEEE Computer Society Distinguished Service Award* (July 2012) and named a *Golden Core Member* of IEEE Computer Society in recognition of his contribution to the field of Learning Technologies (January 2013). He is also the recipient of the *Golden Nikola Tesla Chain Award* of the International Society for Engineering Pedagogy (IGIP) for "International outstanding achievements in the field of Engineering Pedagogy" (September 2018).

Pedro Isaías (pedro.isaias@unsw.edu.au) is an associate professor at the Information Systems & Technology Management School of the University of New South Wales (UNSW—Sydney), Australia. Previously, he was an associate professor at the University of Queensland, Brisbane, Australia. Before moving to Australia, he was also an associate professor at the Universidade Aberta (Portuguese Open University) in Lisbon, Portugal, where he was responsible for several courses and director of the master's degree program in management/MBA. Pedro was director of the master's degree program in Electronic Commerce and Internet for 10 years. He is co-founder and president of IADIS—International Association for Development of the Information Society, a scientific nonprofit association. He holds a PhD in Information Management (in the specialty of information and decision systems) from the New University of Lisbon. Author of several books, book chapters, papers, and research reports, all in the information systems area, he has headed several conferences and workshops within the mentioned area. He has also been responsible for the scientific coordination of several EU-funded research projects. He is also a member of the editorial board of several journals and program committee member of several conferences and workshops. Currently, he conducts research activity related to MIS in general, and more specifically Learning Technologies, Data Analytics, Business Intelligence, Digital Transformation, e-Business, and WWW-related areas. He has more than 160 journal articles, book chapters, and books to his credit and has received over 2438 citations resulting in an h-index of 23 (Google Scholar, January 2020).

About the Authors

Mirian Agus (mirian.agus@unica.it) is a psychologist and received her PhD from the University of Barcelona. Since 2001 she has been working as statistical technician at the University of Cagliari. Her research interests include the study of research methodology, the construction of assessment instruments, and the conduction of statistical data analyses applied to behavioral sciences.

Muhammad Ovais Ahmad (ovais.ahmad@kau.se) is a Senior Lecturer at Karlstad University, Sweden. He received his PhD in Information Processing Science from the University of Oulu. Previously, he worked as European Master's in Software Engineering program coordinator at M3S research unit at the University of Oulu, Finland, and Visiting Professor of software engineering at Gdansk University of Technology, Poland. His research interests are software development methodologies, process assessment and improvement, software engineering curriculum and pedagogy, optimization techniques, technology, and digital transformations.

Charoula Angeli (cangeli@ucy.ac.cy) is Professor of Instructional Technology at the University of Cyprus and the Director of the Center of Teaching and Learning at the University of Cyprus. She completed her undergraduate and graduate studies at Indiana University-Bloomington, USA (BS in Computer Science, MS in Computer Science, and PhD in Instructional Systems Technology). Her research interests include the development of teachers' technological pedagogical content knowledge, the utilization of educational technologies in K-12, the design of computer-enhanced curricula, teacher training, online learning, and the design of learning environments for the development of computational thinking and critical thinking skills. She has published extensively in referred journals, presented her work in numerous conferences worldwide, and participated in various research projects funded by the European Union and other agencies.

Patricia Brockmann (patricia.brockmann@th-nuernberg.de) is a full professor at the Nuremberg Institute of Technology in Germany. She teaches information systems and databases. Her research topics include global software engineering and bionic computation.

Elke Brucker-Kley (brck@zhaw.ch) is an information scientist and Senior Research Associate at Zurich University of Applied Science, School of Management and Law, since 2011. Her focus is on interdisciplinary projects dealing with the human-centric design of innovative solutions and the social impact of technologies. Her main research interest currently lies in the technology-enabled visualization of future scenarios for human-machine interaction and on communication with embodied conversational agents in immersive virtual reality. After completing her studies in information science at the University of Konstanz (MSc 1995), she gained practical experience in various industries in the fields of technology and knowledge management.

Li Chen (chenli@mark-lab.net) is a doctoral student at the Graduate School of Human-Environment Studies, Kyushu University, Japan. She received her master's degree in education from Kyushu University in 2018. She worked as a chemistry and science teacher at Shanghai Aurora Foreign Language Middle School in China from 2000 to 2015. Her research interests include learning analytics, self-regulated learning, and collaborative problem-solving in STEM education.

Rachele Fanari (rfanari@unica.it) is Associate Professor in Developmental and Educational Psychology at the Department of Education, Psychology and Philosophy, University of Cagliari (Italy). She is lecturer of "Developmental Psychology" (Bachelor's degree course in Psychology) and of "Specific Learning Disorder" (Master's degree course in Psychology). Her main interests are language and math development and processing. Her current research interests are focused on the following areas: Specific Learning Disorders; written and oral language acquisition both in typical and atypical populations; relationship between language, first reading, math skills, and cognitive abilities; semantic ability development; relationship between semantic competence and lexical development in preschool children.

Maria Chiara Fastame (chiara.fastame@unica.it) is Associate Professor in Developmental Psychology at the University of Cagliari, Italy. Her research interests include the study of the development and enhancement of visuo-spatial abilities in kindergartners and students attending the primary and secondary schools. A further research topic concerns the investigation of the role played by visuo-spatial abilities for the development of numeracy skills.

Shihui Feng (shihuife@hku.hk) is an assistant professor at the Faculty of Education at the University of Hong Kong. Her research interest is on studying social behavior, complex networks, and collaborative learning. In particular, she is interested in studying the social effects on individual and collective behavior with self-administrated data and large-scale online data from a network perspective.

Kyriakoula Georgiou (odsgeorgiou@gmail.com) is a math teacher assigned by the Ministry of Education of Cyprus at the Secondary Education. She has worked as a special Scientist at the University of Cyprus and at the Open University of Cyprus. She completed her undergraduate and graduate studies at the University of Crete, University of Patras, and University of Cyprus (BSc in Mathematics, 2003, MSc in Science, 2006, and PhD in Educational Technology, 2020). Her research interests include the application of educational technologies in preK-12 and the design of learning environments for the development of computational thinking. She has presented her work in various conferences worldwide.

David Gibson (david.c.gibson@curtin.edu.au) is Director of Learning Futures at Curtin University in Australia and UNESCO Chair of Data Science in Higher Education Learning and Teaching. His research interests are collaborative problem-solving, complex systems analysis and modeling, data science in higher education learning and teaching, digital media learning, dynamic modeling, education, educational systems change, educational technology, game-based learning, leadership, leadership and policy studies, learning sciences, symbolic regression, and systems thinking.

Yoshiko Goda (ygoda@kumamoto-u.ac.jp) is an associate professor at the Research Center for Instructional Systems, Kumamoto University, Japan. She used to be a director of the International Board of Standards for Training, Performance, and Instruction (IBSTPI) from 2015 to 2019. She received her PhD (Science Education) from the Florida Institute of Technology in 2004 with the partial support of a Fulbright Scholarship. Her current research interests include self-regulated e-learning, online education program evaluation, technology-enhanced language learning, and innovative communities for global education.

Miki Hagiwara (miki.hagiwara@uec.ac.jp) received her BS and MS degrees in Engineering from the University of Electro-Communications in 2018 and 2020, respectively. She had engaged in the research on support for web investigative learning by recommending learning questions for question expansion and presenting question keywords adapting learners' skills.

Raija Hämäläinen (raija.h.hamalainen@jyu.fi) works in the field of technology-enhanced learning at the Center for Research for Learning and Teaching at the University of Jyväskylä, Finland. Belonging to the elite or the top 3% in the world, the JYU research center is among the leading European research groups in learning and teaching. Hämäläinen's research interests include collaboration and creativity at the technology-enhanced learning settings, workplace learning, and teacher-student interaction. She has participated in one publication that has defined the research area of collaboration scripts, and it has become one of the standard references in such an area. Hämäläinen has been the principal investigator/leader of several research projects related to technology-enhanced learning. She has designed a long-term research line that focuses on designing and investigating new learning environments for future educational efforts. Hämäläinen's work has been based on active international collaboration. In practice, her international collaboration efforts include visits to international research groups and joint high-level publishing. Hämäläinen's recent publications include articles in *Computers & Education*, the *British Journal of Educational Technology*, and *The Internet and Higher Education*. She is an associate editor of the *Educational Research Review* (5-Year Impact Factor: 8.815).

Kirsi Heinonen (kirsi.m.heinonen@jyu.fi) is a PhD student at the Faculty of Education and Psychology, University of Jyväskylä. Her research interests are in the field of information and communication technology (ICT) competences, technology-enhanced teaching and learning, and computer-supported collaborative learning.

Annika Hinze (hinze@waikato.ac.nz) is an associate professor and head of the School of Computing and Mathematical Sciences at the University of Waikato in Hamilton, New Zealand. She teaches information systems and databases. Her research topics include event notification systems, alerting in IOT, cultural heritage, and women in computer science.

Dirk Ifenthaler (dirk@ifenthaler.info; www.ifenthaler.info) is Professor and Chair of Learning, Design and Technology at the University of Mannheim, Germany, and UNESCO Deputy Chair of Data Science in Higher Education Learning and Teaching at Curtin University, Australia. Dirk's research focuses on the intersection of cognitive psychology, educational technology, data analytics, and organizational learning. He is the Editor-in-Chief of the *Technology, Knowledge and Learning*, Senior Editor of the *Journal of Applied Research in Higher Education*, and Deputy Editor-in-Chief of the *International Journal of Learning Analytics and Artificial Intelligence for Education*.

Mmaki Jantjies (mjantjies@uwc.ac.za) is an Associate Professor in Information Systems at the University of the Western Cape, an Aspen New Voices fellow, and a member of the South African Young Academy of Science.

Akihiro Kashihara (akihiro.kashihara@inf.uec.ac.jp) received his BS, and MS degrees in Computer Science and System Engineering from Tokushima University in 1987 and 1989, respectively, and his PhD in Information and Computer Science from Osaka University in 1992. He was an assistant/associate professor of the Institute of Scientific and Industrial Research in Osaka University from 1992 to 2004. He is now a professor of Graduate School of Informatics and Engineering at The University of Electro-Communications. Since 1990, he has been engaged in knowledge engineering and knowledge-based systems especially in designing intelligent and adaptive learning environments. His main research interests lie in designing of learning models, robot in learning, metacognition in hyperspace, knowledge externalization, cognitive loading, and scaffolding with cognitive tools. He has been active in both local and international academic societies, which include executive committee member of the *International Journal of Artificial Intelligence in Education* (IJAIED), executive committee member of Asia-Pacific Society for Computers in Education (APSCE), and the president of Japanese Society for Information and Systems in Education (JSiSE). He has also served as Local Steering Committee Vice Chair in ICALT2007, LOC Chair in ICCE2014, senior program committee in AIED2007-2015, etc. in several international conferences.

Thomas Keller (kell@zhaw.ch) studied electrical engineering at the Swiss Federal Institute of Technology (ETH) Zurich between 1987 and 1993. From 1993 he was responsible for setting up and realizing a hardware-in-the-loop (HIL) simulator for the propulsion system of rail vehicles at ABB Switzerland. At the same time, he successfully completed a degree program in information technology at Zurich University, where he also obtained his PhD. Thomas Keller is an expert in the definition of complex tool systems, business process automation, and emerging technologies. Since 2003, he has been a senior lecturer at the ZHAW School of Management and Law, and since 2007 as a professor of information technology. From 2008 till end of 2015 he acted as the head of the Institute which he built up. His field of expertise is business process management, modeling and automation, the application of new technologies, e.g., mixed reality and blockchain, in entrepreneurial contexts, and scenarios about our future digital lives.

Simon Kerrigan (simon.kerrigan@curtin.edu.au) is an IT Project manager at Curtin University, focused on the management of learning platforms. His education background is in Information Systems Management, and his key research activities revolve around the management of data integrity and validating statistical findings against observational behavior.

Kirsi Lainema (kirsi.k.lainema@jyu.fi) studies interaction and organizing and the use and effects of technology in digital environments. Her research covers topics and contexts such as managerial meetings, municipal government, networked businesses, strategic practices, hospitals, and digital learning environments. Lainema's interests lie particularly in studying collaborative learning and interactional processes at work. Currently, Lainema works in the Well@digiwork research group.

Timo Lainema (timo.lainema@utu.fi) holds a PhD in economics and business administration from Turku School of Economics with an emphasis on Information Management. His research interests are learning through simulation gaming, knowledge sharing in virtual working contexts, and decision making in time-intensive environments.

Kari Liukkunen (kari.liukkunen@oulu.fi) has over 30 years of management and R&D experience in industry and academy. He has worked at the University of Oulu since 2001. Now he is working as an Adjunct Professor and the head of the Software Factory laboratory at M3S research group. He has worked as a Visiting Professor in Fraunhofer IESE, Germany. He was also nominated as a Visiting Professor to Nanking Institute of Technology in 2019. He has published over 60 papers in international conferences and journals and served as chair and committee member in organizing numerous international conferences. His main research interests are service design, SW development processes, and educational technology.

Min Lu (lu@artsci.kyushu-u.ac.jp) is an assistant professor at the Faculty of Arts and Science, Kyushu University. He received his master's degree on Cartology and Geographic Information System in 2011 from the School of Earth and Space Science, Peking University, and PhD in Socio-Cultural Environmental Studies in 2014 from the Graduate School of Frontier Sciences, the University of Tokyo, and used to be a project researcher at the Center for Spatial Information Science, the University of Tokyo, before joining Kyushu University. His research interests include human-centered mobile mapping, human-computer interaction, learning support environments, and prototype developments at these fields.

Maria Lidia Mascia (marialidia.mascia@unica.it) is a psychologist and received her PhD at the University of Pavia. Since 2011 she is a contract professor in Experimental Psychology at the University of Cagliari. She is mainly involved in educational technology investigation, in motivational and self-regulation processes, and in academic drop-out prevention.

Tebogo John Matome (3366229@myuwc.ac.za) works at the University of the Western Cape, with a focus on the use of Immersive Technology in the Education field.

Karin Maurer (karinmaurer92@gmail.com) completed her bachelor's and master's degrees in information systems at the Nuremberg Institute of Technology in Germany. She currently works as an IT consultant.

Carla Meloni (cmeloni@unica.it) is a researcher in psychology at the Department of Education, Psychology, and Philosophy, University of Cagliari (Italy). She is a lecturer in General Psychology for the Degree in Philosophy and lecturer for the Laboratory on Methods and Techniques of Research in Psychology, for the degree in Psychology at the University of Cagliari. Her current research interests include cognitive and meta-cognitive processes involved in mathematics skills in children aged between 4 and 12 years. She has carried out a series of studies and researches on general domain precursors (working memory and executive functions) and specific domain precursors of mathematical skills (mechanisms of quantification, comparison, serialization, mental calculation strategies, mental number line). Another topic of interest is the effect of chess play, as a strategic game, on developing children meta-cognitive ability and mathematical problem-solving skills.

Kazuhisa Miwa (miwa@is.nagoya-u.ac.jp) received the BS degree in Applied Physics from Nagoya University in 1984. He received his PhD degree in Information Science from Nagoya University in 1989. He stayed in CMU as a visiting scholar from 1991 to 1992. He is currently a professor of Nagoya University. He was the president of Japanese cognitive science society from 2017 to 2018. His research interests include cognitive science and artificial intelligence, especially interested in fundamental studies on higher order cognition, such as creative thinking, insight problem-solving, scientific discovery, AI-based learning support systems, and human factors of automobile driving.

Afika Ntaba (3428062@myuwc.ac.za) is a first year master's student at the University of the Western Cape. She holds a Honors degree in Information Systems and a Diploma in Immersive Technology. She is an immersive technology enthusiast and has worked on various virtual reality training applications with the University of the Western Cape.

Koichi Ota (ota@jaist.ac.jp) received his BS, MS, and PhD degrees in Information and Communication Engineering from the university of Electro-Communications in 2005, 2007, and 2011, respectively. He is now an assistant professor at the Research Center for Advanced Computing Infrastructure in Japan Advanced Institute of Science and Technology. Since 2004, he has been engaged in research on web-based self-directed learning support. His main research interests lie in designing of learning models, metacognition in hyperspace, and scaffolding with cognitive tools.

Maria Pietronilla Penna (penna@unica.it) is Full Professor in Experimental Psychology at the University of Cagliari. She is mainly involved in the investigation and assessment of brain functions and cognitive processes in life span. Her research interests include the study of the factors underpinning well-being in adulthood, the educational technology investigation, and the motivational and self-regulation processes in academic drop-out prevention.

Susan L. Renes (slrenes@alaska.edu) is Emeritus Professor at the University of Alaska Fairbanks. Her research focuses on factors related to Indigenous and rural student recruitment and retention in higher education. Dr. Renes presented Expanding Access for Indigenous Students to the Behavioral Health Workforce through Distance Education at the International Congress of Arctic Social Sciences in Akureyri, Iceland, in 2011. In 2013, she received the Elders Award given by Alaska Native Elders to acknowledge those whose deeds and practices promote cultural understanding and balance in all things for a better world community. In 2013, she chaired the roundtable Indigenous Studies in Alaska: Interdisciplinary Initiatives in Research, Service and Pedagogy for the Native American and Indigenous Studies Association (NAISA) in Saskatoon, Saskatchewan. In 2014, Dr. Renes authored Amplifying Indigenous Voices. In 2016 she presented Teaching in Higher Education: You Can't Call it an Adventure if You Know How it's Going to Turn Out at the Lilly International Conference on Teaching and Learning in Oxford, Ohio; and served on a panel addressing Reclaiming Indigenous Spaces In Higher Education: Examining Recruitment, Retention, And Mentoring of Alaska Native Students at the University of Alaska Fairbanks for the National Indian Education Association in Anchorage, Alaska.

Rei Saito (rei.saito@uec.ac.jp) received his BS degrees in Engineering from the University of Electro-Communications in 2019. He has engaged in the research on scaffolding for web-based investigative learning by exercise question.

Yoshiki Sato (yoshiki.sato@uec.ac.jp) received his BS and MS degrees in Engineering from the University of Electro-Communication in 2018 and 2020, respectively. He had engaged in the research on diagnosis of self-directed learning process in web-based investigative learning from 2017 to 2020.

Heidi Schuhbauer (heidi.schuhbauer@th-nuernberg.de) is a full professor at the Nuremberg Institute of Technology in Germany. She teaches information systems, knowledge management, and electronic government. Her special research interests are social network analysis and knowledge maps.

Atsushi Shimada (atsushi@ait.kyushu-u.ac.jp) received his ME and DE degrees from Kyushu University in 2007. He is a professor of Faculty of Information Science and Electrical Engineering, Kyushu University. He also worked as a JST-PRESTO researcher since 2015 to 2019. His current research interests include learning analytics, pattern recognition, media processing, and image processing. He received MIRU Interactive Presentation Award (2011, 2017), MIRU Demonstration Award (2015), Background Models Challenge (2012, secured first place), PRMU research award (2013), SBM-RGBD Challenge (2017, secured first place), ITS Symposium Best Poster Award (2018), JST PRESTO Interest Poster Award (2019), IPSJ/IEEE-Computer Society Young Computer Researcher Award (2019), and CELDA Best Paper Award (2019).

Anthony T. Strange (stranget@wbu.edu) earned his doctorate in Education from the University of Southern California. He has worked in a number of settings including High School Counseling, Assistant and Associate Professor, Chair of Counseling Department and Chair of Graduate Studies at the University of Alaska Fairbanks. He has been invited to present at local, state, and national level. He is a Past President of the Alaska School Counseling Association and consultant to school districts. He is currently a faculty member at Wayland Baptist University. Dr. Strange is a member of the South Carolina Counseling Association and serves on the Executive Governing Board.

Joan Ann Swanson (Swansonj@franklinpierce.edu) is the Director of Education at Franklin Pierce University in New Hampshire. She earned her PhD in Educational Psychology and Methodology from the State University of New York, University at Albany. She also has an MS in Educational Counseling from McDaniel College and a BS in Education and Psychology from Indiana Wesleyan University. Dr. Swanson holds teaching certification for Nursery through 12th grade and has experience both as a school teacher and college professor. Dr. Swanson's research interests include the study of instruction and learning, with a particular focus on technology use by emerging adults and adolescents. Dr. Swanson has presented her research both nationally and internationally and recently authored a book chapter in which she presented her theory, "Technology as Skin."

Rupa Vuthaluru (r.vuthaluru@curtin.edu.au) is a Physicist and Data Scientist at Curtin University with current research activities focusing on modeling, predictive analytics, and natural language processing techniques to enhance innovation in learning and teaching.

Masanori Yamada (mark@mark-lab.net) is Associate Professor in the Faculty of arts and science, and Learning Analytics Center at Kyushu University. He is engaged in learning analytics research, in particular, the relationship between learning behaviors and psychological factors such as self-regulated learning. He worked for NTT Comware Corporation where he was engaged in the research on the technology related to moving images from 2000 to 2003. He received his MA and PhD in Human System Science from Tokyo Institute of Technology in 2005 and 2008, respectively. He is a recipient of several best paper awards such as ICWL 2015 and CELDA 2016.

Junko Yamamoto (junko.yamamoto@sru.edu) grew up in Kyoto, Japan. She earned her Doctorate of Education from Duquesne University in Pittsburgh, PA, USA. She has taught instructional technology and world language teaching methodology at Slippery Rock University of Pennsylvania. She is interested in human factors related to technology, such as innovation, creativity, critical thinking, and digital citizenship. She advocates world language education as global collaboration is increasingly required.

Mai Yokoyama (yokoyama.mai@c.mbox.nagoya-u.ac.jp) received her BS degree in System Information Science from Future University Hakodate in 2013. She received her MA degree from Hokkaido University in 2015. She is currently a doctoral student of Nagoya University. Her research interests include educational psychology and educational technology.

Vaclav Zubr (vaclav.zubr@uhk.cz) is a postdoc at the Department of Management, Faculty of Informatics and Management, University of Hradec Kralove. His research activities concentrate on the area of organizational culture, learning organization, and knowledge management. The title of his dissertation of 2018 was "Learning as a Part of Strategy." He has published 3 papers in journals with IF, 10 papers in journals with SJR, and has recorded 24 citations in WoS (h-index 2). He has worked on a number of projects, e.g., ESF CZ.1.07/2.2.00/28.0104 "Interdisciplinary, informative-cognitive, linguistic and modular development of study," 2012–2015. He has been a co-investigator of the specific research projects solved at the University of Hradec Kralove and focused on the same topic for the last 4 years.

Part I
Cases from Pre-School and Primary School

Chapter 1
The Enhancing of Numeracy Skills Through Pencil-and-Paper or Computerized Training for Kindergarteners

Maria Lidia Mascia, Mirian Agus, Maria Chiara Fastame, and Maria Pietronilla Penna

1.1 Introduction

In recent decades, considerable attention has been paid to the diagnosis of learning disabilities (e.g. dyslexia, dyscalculia). Despite this, the difference between learning 'difficulty' and 'disability' is not always fully understood. From a practical viewpoint, it is fundamental to understand the differences as well as the analogies between those two concepts, because thanks to this distinction, it would be possible for teachers to better decide how to act, in order to promote the development and learning of any pupil. By learning disability, we mean any conditions with a neurological basis that are marked by substantial deficits in acquiring certain scholastic or academic skills, particularly those associated with written or expressive language. Learning disabilities may include learning problems resulting from perceptual disabilities, brain injury, and minimal brain dysfunction. However, disabilities resulting from visual impairment or hearing loss, intellectual disability; emotional disturbance; or environmental, cultural, or economic factors should not be included in this definition (VandenBos, 2015). A growing consensus among researchers is that children with Mathematical learning disabilities (MLD) show fundamental deficiencies in numerical understanding and domain-general functioning such as language and spatial skills compared with average achieving (AA) children (Geary, 2004). By 'difficulty' we mean a non-pathological and non-innate condition which does not meet the clinical criteria for the disability and that can be modified with appropriate targeted interventions. Learning difficulty is mainly due to environmental factors such as the lack of/poor education, emotional difficulties or

M. L. Mascia (✉) · M. Agus · M. C. Fastame · M. P. Penna
Department of Pedagogy, Psychology, Philosophy, University of Cagliari, Cagliari, Italy
e-mail: marialidia.mascia@unica.it; mirian.agus@unica.it; chiara.fastame@unica.it; penna@unica.it

D. Ifenthaler et al. (eds.), *Balancing the Tension between Digital Technologies and Learning Sciences*, Cognition and Exploratory Learning in the Digital Age,
https://doi.org/10.1007/978-3-030-65657-7_1

environmental or family problems (Fletcher, Lyon, Fuchs, & Barnes, 2018). A body of studies state that both cognitive and environmental factors may be associated with children's early mathematical learning (Passolunghi, Cargnelutti, & Pastore, 2014). The label 'learning disabilities' is usually restricted to a small group of students with persistent problems, whereas the label 'learning difficulties' describe the experiences of a larger group of students who do not respond to classroom programmes appropriately (Elkins, 2002).

Since different theories about the construction of number concept agree that the period between 2 and 8 years is crucial for the construction of number concept, the development of numerical knowledge should be supported since the kindergarten age (Bonny & Lourenco, 2013). If children leave kindergarten with poor numerical competence, they will find themselves with a disadvantage at the beginning of primary school, and will have greater difficulty in reaching (and perhaps never will) schoolmates who begin school with good numerical competence (Jordan, Kaplan, Ramineni, & Locuniak, 2009). The authors conclude their work by suggesting that supporting children in the development of their numerical skills already at preschool level and in the subsequent educational degrees is very important, especially for children coming from more disadvantaged socio-economic situations. In this case, schools should propose programmes fostering the development of numerical intelligence.

This chapter aims at encouraging teachers and educators to pay attention to those aspects, by systematically carrying out activities that would allow them to identify any learning difficulties among kindergarteners. After identifying those difficulties, teachers and educators should adopt adequate cognitive enhancement tools. This aspect is fundamental because the effect of developmental factors on numeracy competence (i.e. the ability to use numerical information to perform different daily life activities, such as counting, seriation and number comparison tasks) already begins in early infancy and continues along the whole life span. Moreover, this effect can predict academic achievements (Aunola, Leskinen, Lerkkanen, & Nurmi, 2004; Stock, Desoete, & Roeyers, 2010; Watts, Duncan, Siegler, & Davis-Kean, 2014) as well as also mental health (Fastame, Manca, Penna, Lucangeli, & Hitchcott, 2019).

Which kind of interventions are more useful to improve mathematical learning at kindergarten? Which cognitive abilities should be enhanced in order to help the development of mathematical achievement in life? Those issues will be discussed along this chapter.

1.1.1 Theoretical Frame

A wide literature has shown that different variables are fundamental cognitive predictors of later arithmetic achievement (Cornu, Schiltz, Pazouki, & Martin, 2019).

One of the main variables related to the mathematical learning is the working memory, particularly its visuo-spatial component. Many studies show that the working memory is an important predictor for mathematical learning (Bull, Espy, & Wiebe, 2008; Miller, Müller, Giesbrecht, Carpendale, & Kerns, 2013; Passolunghi & Lanfranchi, 2012). Strong correlations between the visuo-spatial component of working memory and mathematics have been found among preschool's children (Bull et al., 2008; Kroesbergen, van Luit, Naglieri, Taddei, & Franchi, 2010; Kyttälä, Aunio, Lehto, Van Luit, & Hautamäki, 2003) and among pupils attending the first years of school (Lee & Bull, 2016; Toll, Kroesbergen, & Van Luit, 2016). Today, working memory deficits have been found during the screening of children with mathematical learning disabilities (Praet & Desoete, 2019). Furthermore, children with higher working memory would perform better in mathematical field (Lee & Bull, 2016). This research highlights that the relation between basic numerical skills (i.e. counting sequence knowledge, number symbols recognition, or number sense) (Dehaene, 2011; Wei, Li, & Su, 2020; Zhang et al., 2020) and advanced mathematical achievement (symbolic skills such as counting and number recognition) (Landerl, Bevan, & Butterworth, 2004) can be associated with visuo-spatial processing (Sella, Sader, Lolliot, & Cohen Kadosh, 2016). Therefore, visuo-spatial training is considered as a successful approach to provide young children with a sound foundation for later mathematical learning (Cornu et al., 2019). Recently, some studies have found evidence for positive effects of visuo-spatial training applied to children's mathematical performance (Allen, Higgins, & Adams, 2019; Cheng & Mix, 2014; Yang, Chung, & McBride, 2019). These aspects stress on the need to apply this approach before formal schooling age (Cornu et al., 2019). However, other numerous factors such as motivation, metacognition and attention, could influence mathematical learning. These processes are fundamental in order to perform written and mental calculation, as well as to solve math problems in daily life (e.g. paying a bill) (Lucangeli, Iannitti, & Vettore, 2007). A recent study carried out among 198 Afro–American children attending kindergarten has shown that three aspects of cognition (fluid intelligence, executive functioning, and crystallized intelligence) could predict the later achievement of math and reading skills (Blankson, Gudmundson, & Kondeh, 2019).

A longitudinal study has shown how early numeracy assessed in kindergarten could predict children's mathematical performance in the first grade, after controlling for the effects of age, gender, and parents' education. Results from this study underline that the achievement of counting and relational skills before formal schooling would predict the achievement of basic arithmetical skills and overall mathematical performance during the first year of school, above and beyond the effects of demographic factors (Aunio & Niemivirta, 2010).

Other studies have also highlighted the importance of demographic factors in the mathematical achievement, such as gender influence (Casey, Dearing, Dulaney, Heyman, & Springer, 2014; Casey, Erkut, Ceder, & Young, 2008; Gunderson, Ramirez, Levine, & Beilock, 2012; Hyde, Lindberg, Linn, Ellis, & Williams, 2008).

1.1.2 Pencil-and-Paper or Computerized Trainings

Many studies have proved that psychoeducational programmes would contribute to in the improvement of cognitive functions in childhood (Passolunghi & Costa, 2014; Penna & Stara, 2010; Penna, Stara, & Bonfiglio, 2002; Ramani, Siegler, & Hitti, 2012). A series of follow-up studies reports that specific metacognitive and cognitive (e.g. visuo-spatial attention and working memory) psychoeducational trainings in formal education can be an useful tool for the empowerment of learning in the classroom (e.g. Fastame & Callai, 2015).

A wide body of studies shows the efficacy of pre-literacy psychoeducational interventions in improving early numeracy skills (e.g. classification, seriation, counting) of pre-schoolers (e.g. Agus et al., 2015).

Many authors state that psychoeducational trainings have shown some advantages and that the same specific training can be settled and presented in two different formats: computerized training or pencil-and-paper training (Penna et al., 2002). A further trend of research (Chen, Lin, Wei, Liu, & Wuang, 2013) has highlighted that in order to get cognitive empowerment cognitive interventions based on the use of new technologies would be more effective than the traditional pencil-and-paper method. This is because videogame-like activities can enhance both cognitive function efficiency and pupil motivation. Computerized trainings can therefore provide for intensive and individualized training for children having learning difficulties (Hellstrand, Korhonen, Linnanmäki, & Aunio, 2020). Through computer games, children can be more motivated through a more entertaining context (Hellstrand et al., 2020). Another research (Fastame & Manca, 2020) has recently shown that computerized training is effective in order to foster the empowerment of spatial comprehension, mental imagery, and processing speed among during the second year of primary school.

Nevertheless, pencil-and-paper mode, could promote another important aspect such as novelty aspect. Novelty could be identified in the presence of a new teacher, who could be specialized in the promotion of training activities (Agus et al., 2015; Slavin, 2013).

In the Italian context, there is evidence for the effectiveness of combined pencil-and-paper and computer-assisted interventions, including both visuospatial and numeracy tasks, for the empowerment of mathematical skills in children attending kindergartens (Agus et al., 2015). In one of their studies, Mascia et al. (2015) trained one group of 5-year-old children with a computer-assisted mathematical programme, and a further group with the same computerized intervention, combined with a pencil-and-paper numeracy programme. A third group did not receive any specific training (i.e. control group). After the end of training, and after comparing the two groups with the control group, both trained groups seemed to gain an advantage from the psychoeducational interventions. However, no significant differences in terms of numeracy efficiency between the two trained groups were found. These results could also depend on the use of computers in kindergarten, which are not so used in Italian schools.

This study mainly aimed at examining the effect of the presentation modality (i.e. computer-assisted versus pencil-and-paper) and the combination modality (i.e. pencil-and-paper visuospatial training and pencil-and-paper mathematical or pencil-and-paper visuospatial training and computerized mathematical training) on the empowerment of numeracy skills in 5-year-old children, at post-test time and at follow-up time. Early hypothesis seems to show that pupils in the experimental combined groups would obtain higher scores in the assessment and in the follow-up time of numerical abilities than pupils in the control group.

1.2 Method

1.2.1 Participants, Materials and Procedure

Seventy-three pupils (38 males, 52.1%; mean age 63.3 ± 4.5 months) attending the last year of kindergarten in Italian schools (Sardinian area) were divided into five groups: the control group ($n = 16$) and four experimental groups (pencil-and-paper mathematical training: $n = 14$; computerized mathematical training: $n = 15$; combined pencil-and-paper visuospatial training and pencil-and-paper mathematical training $n = 14$; combined pencil-and-paper visuospatial training and computerized mathematical training $n = 14$). The activities aiming at enriching numerical knowledge were developed collectively during 10 weekly meetings; each meeting lasted approximately lasted 1 h. A follow-up was carried out after 3 months. The mathematical psychoeducational training consisted of some activities developed by Lucangeli and her colleagues (Lucangeli, Poli, & Molin, 2003, 2010) in 'Sviluppare l'intelligenza numerica I' and 'L'intelligenza numerica I'. The visuo spatial training consisted of some activities developed in pencil-and-paper form by Lucangeli et al. in 'Conosco le forme' (Lucangeli, Mammarella, Todeschini, Miele, & Cornoldi, 2009) for the development and enhancement of manipulative skills. More specifically, thanks to these activities, pupils would be able to discover, the shape and name of the geometric figures, the concepts of side, angle, spatial orientation and size by means of a funny operational path. In the control group, pupils just performed the curricular activities proposed by their teachers. Assessment was carried out by the presentation (at pre-test, post-test and follow-up sessions) of Raven's Coloured Progressive Matrices (CPM) (Belacchi, Scalisi, Cannoni, & Cornoldi, 2008; Raven, 1958) and the BIN numerical intelligence scale (Molin, Poli, & Lucangeli, 2007). These tests aim at achieving a measure of the pupils' fluid intelligence and numerical knowledge. The BIN test is used to investigate four principal areas: lexical, semantic, pre-syntactic, and counting. Each area is evaluated by means of specific activities, such as reading and writing. More specifically, the lexical numeric knowledge allows teachers to appraise reading and writing skills of Arabic numbers. The semantic numerical knowledge allows teachers to appraise the ability to associate numerical sizes, dots, and Arabic digits. The pre-syntactical numeric knowledge assesses the ability to connect numbers to their number representation. Finally, the counting scale assesses the ability to declaim the number words sequence forward and backward.

1.2.2 Findings

The above-mentioned five groups showed similar behavioural patterns during pre-test assessment regarding both fluid intelligence and numerical knowledge (lexical area: $F_{(4;68)} = 1.412$ $p = 0.239$; semantic area: $F_{(4;68)} = 0.352$ $p = 0.842$; counting area: $F_{(4;68)} = 1.105$ $p = 0.361$; pre-syntactic area: $F_{(4;68)} = 0.330$ $p = 0.857$; CPM $F_{(4;68)} = 0.347$ $p = 0.845$).

In order to assess the effect of trainings on numerical abilities, the gain score was computed in relation to post-test [(Post-test score − Pre-test score)/Pre-test score] and follow-up [(Follow-up score − Pre-test score)/Pre-test score] conditions, respectively.

In order to measure the effects of training activities, the linear mixed effects models (LME) by restricted maximum likelihood estimators was applied (Bates, Mächler, Bolker, & Walker, 2015; Kwok et al., 2008; Muth et al., 2016). These models allow the authors to overcome the problems concerning the reduced sample size in repeated measures, missing values and unbalanced designs. Thanks to these models, reliable esteems of parameters, accounting for random and fixed effects were obtained. Coefficients that might vary from cluster to cluster (and in this case, from participant to participant) are defined as random coefficients, and their mean (fixed expected value) is defined as fixed coefficients. Indeed, in this work, the evaluation of the effects of each training in relation to time (repeated measures) was needed; therefore it was necessary to bookkeep the dependency of the data. This dependency was accounted by allowing the intercept to vary from pupil to pupil. By using this approach, each pupil might have a high or low overall gain (average score over time) and the residuals (error terms) might be computed as the deviation from the pupil's mean score. This method seizes the dependency among repeated measures, thus overcoming the limitations related to the classical repeated measures analysis of variance.

The statistical analyses were carried out by using the Jamovi software (version 1.1.9, https://www.jamovi.org/) (AA.VV., 2019); in Jamovi the random effects were performed by the *lme4* R package (Bates et al., 2015) (https://cran.r-project.org/web/packages/lme4/index.html).

In the LME model the pupils are the cluster variable that is used to assess the intercept random effect; the age was used as a covariate. The fixed effects were then assessed in relation to the following variables:

- BIN scale (lexical, semantic, pre-syntactic, counting);
- training (control group; pencil-and-paper mathematical training; computerized mathematical training; combined pencil-and-paper visuo-spatial training and pencil-and-paper mathematical training; combined pencil-and-paper visuospatial training and computerized mathematical training);
- gender (male, female).

The model highlighted a marginal R-squared of 0.160 [the variance is due to the fixed effects, referring to the total expected variance of the gain (dependent variable)] and a Conditional R-squared of 0.428 [the variance is due to both the random and the fixed effects]. Then, the overall model (considering both random and fixed effects) showed a good proportion of variance of the gain.

The random components highlighted a 0.319 Intraclass Correlation Coefficient (ICC), thus assessing the correlation among observations in the same participant (pupil) (Tables 1.1 and 1.2).

In order to compare groups' performances, the Bonferroni's post hoc analyses were carried out. The Bonferroni's correction was applied in order to identify which groups' means were significantly different from the other means. A pairwise comparison of the means has therefore been carried out. In Table 1.3, the significant data were showed. The results highlighted that the female gain score in the scale 2 (semantic) is higher in the combined *visuospatial and pencil and paper math training* than in the *control group* for the scale 1 (lexical), for the scale 3(pre-syntactic). These results are consistent with some studies in the literature (e.g. Casey et al., 2014), showing that an association between spatial skill and mathematics is more robust in girls' performances, compared with males. Figure 1.1 illustrates the general effects, and highlights the gain obtained in relation to the trainings. Figure 1.2 focuses on the specific significant effects observed in the statistical analyses.

Table 1.1 Fixed effect omnibus tests

	F	Num df	Den df	p
Training	1.103	4	61.4	0.363
Age	2.256	1	63.2	0.138
Scale	6.314	3	446.7	$< 0.001^{**}$
Gender	0.296	1	61.7	0.589
Training * Scale	2.278	12	446.7	0.008^{**}
Training * Gender	1.189	4	61.4	0.325
Scale * Gender	4.439	3	446.7	0.004^{**}
Training * Scale * Gender	2.236	12	446.7	0.010^{**}

Note $^{*}p < 0.05$; $^{**}p < 0.01$

Table 1.2 Fixed effects parameter estimates—significant effects

	Effect	Names	Estimate	SE	95% CI lower	95% CI upper	df	t	p
1	Training * Scale	Training3 – (Training1, Training2) * Scale1 – Scale2	−2.267	0.627	−3.498	−1.037	446.7	−3.612	< 0.001**
2	Training * Scale	Training3 – (Training1, Training2) * Scale2 – Scale3	1.511	0.627	0.281	2.741	446.7	2.407	0.016*
3	Training * Scale	Control – (Training1, Training2, Training3, Training4) * Scale3 – Scale4	−1.095	0.536	−2.148	−0.043	446.7	−2.041	0.042*
4	Training * Scale * Gender	Training3 – (Training 1, Training2) * Scale1 – Scale2 * Gender f – Gender m	−2.994	1.255	−5.455	−0.533	446.7	−2.385	0.018*
5	Training * Scale * Gender	Training3 – (Training 1, Training2) * Scale2 – Scale3 * Gender f – Gender m	3.406	1.255	0.946	5.867	446.7	2.713	0.007**
6	Training * Scale * Gender	Training4 – (Training 1, Training2, Training3) * Scale3 – Scale4 * Gender f – Gender m	3.236	1.237	0.811	5.661	446.7	2.615	0.009**

Note: *$p < 0.05$; **$p < 0.01$; Training 1 = pencil-and-paper mathematical training; Training 2 = computerized mathematical training; Training 3 = combined pencil-and-paper visuospatial training and pencil-and-paper mathematical training; Training 4 = combined pencil-and-paper visuospatial training and computerized mathematical training; Control = control group; Scale 1 = lexical; Scale 2 = semantic; Scale 3 = pre-syntactic; Scale 4 = counting

Table 1.3 Post hoc comparisons for Training * Scale * Gender

Training	Scale	Gender	Vs	Training	Scale	Gender	Difference	SE	t	df	p
Training1	1	m	–	Training3	2	f	−4.282	1.029	−4.160	155	0.041*
Training1	1	f	–	Training3	2	f	−4.307	1.045	−4.124	159	0.047*
Training1	2	m	–	Training3	2	f	−4.469	1.029	−4.342	155	0.020*
Training1	2	f	–	Training3	2	f	−4.288	1.045	−4.105	159	0.050*
Training1	3	m	–	Training3	2	f	−4.384	1.029	−4.260	155	0.028*
Training2	1	f	–	Training3	2	f	−4.612	1.014	−4.547	150	0.009*
Training4	1	m	–	Training4	4	m	−2.586	0.638	−4.052	446	0.047*
Training4	1	m	–	Training3	2	f	−4.528	0.973	−4.655	153	0.005*
Training4	2	m	–	Training3	2	f	−4.472	0.973	−4.598	153	0.007*
Training3	1	f	–	Training3	2	f	−4.008	0.760	−5.275	446	< 0.001**
Training3	2	f	–	Training2	3	f	4.196	1.014	4.136	150	0.046*
Training3	2	f	–	Training4	4	f	4.755	1.137	4.181	156	0.038*
Training3	2	f	–	Control	3	f	4.180	0.983	4.251	144	0.030*
Control	1	m	–	Training3	2	f	−4.480	1.023	−4.381	149	0.017*
Control	1	f	–	Training3	2	f	−4.419	0.983	−4.493	144	0.011*
Control	3	m	–	Training3	2	f	−4.397	1.023	−4.300	149	0.024*

Note: *$p < 0.05$; **$p < 0.01$; Training 1 = pencil-and-paper mathematical training; Training 2 = computerized mathematical training; Training 3 = combined pencil-and-paper visuospatial training and pencil-and-paper mathematical training; Training 4 = combined pencil-and-paper visuospatial training and computerized mathematical training; Control = control group; Scale 1 = lexical; Scale 2 = semantic; Scale 3 = pre-syntactic; Scale 4 = counting

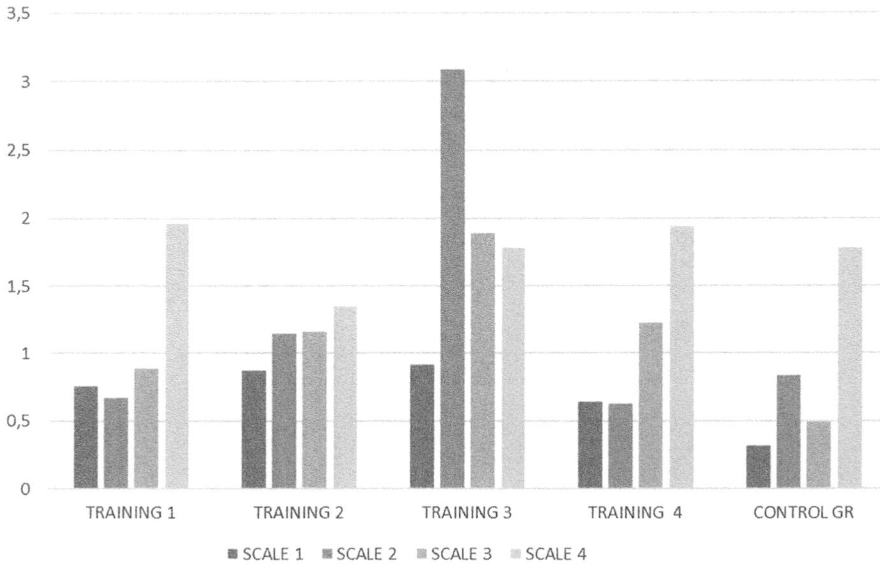

Fig. 1.1 General effects plots. Note: $*p < 0.05$; $**p < 0.01$; Training 1 = pencil-and-paper mathematical training; Training 2 = computerized mathematical training; Training 3 = combined pencil-and-paper visuospatial training and pencil-and-paper mathematical training; Training 4 = combined pencil-and-paper visuospatial training and computerized mathematical training; Control = control group; Scale 1 = lexical; Scale 2 = semantic; Scale 3 = pre-syntactic; Scale 4 = counting

1.3 Discussion

Our data show the efficacy of psychoeducational intervention that can be carried out both by pencil-and-paper and computerized formats (Agus, Mascia, Fastame, & Penna, 2016; Mascia, Agus, Fastame, & Addis, 2016; Mascia, Fastame, Agus, & Penna, 2019). The combination between visuo-spatial and mathematical contents has been very useful to empower children skills. These results have been confirmed during the follow-up time. Our results have shown some differences in gender variable. Hutchison, Lyons, and Ansari (2019) stated that although investigation about gender differences in basic numerical skills are not so common in literature, the majority of basic numerical tasks show some similarities in terms of gender. Furthermore, male advantages in foundational numerical skills are the exception rather than the rule. In relation to visuo-spatial abilities, a recent meta-analysis has shown a male advantage in visuospatial working memory, especially when performing complex span tasks, as well as in mental rotation (Voyer, Voyer, & Saint-Aubin, 2017; Wang, 2020). However, our study has highlighted interesting results concerning girls' achievement after the combined math and visuospatial training. These findings might be consistent with some other studies (Anderson et al., 2008; Carr & Davis, 2001; Carr & Jessup, 1997; Carr, Jessup, & Fuller, 1999; Casey et al., 2008).

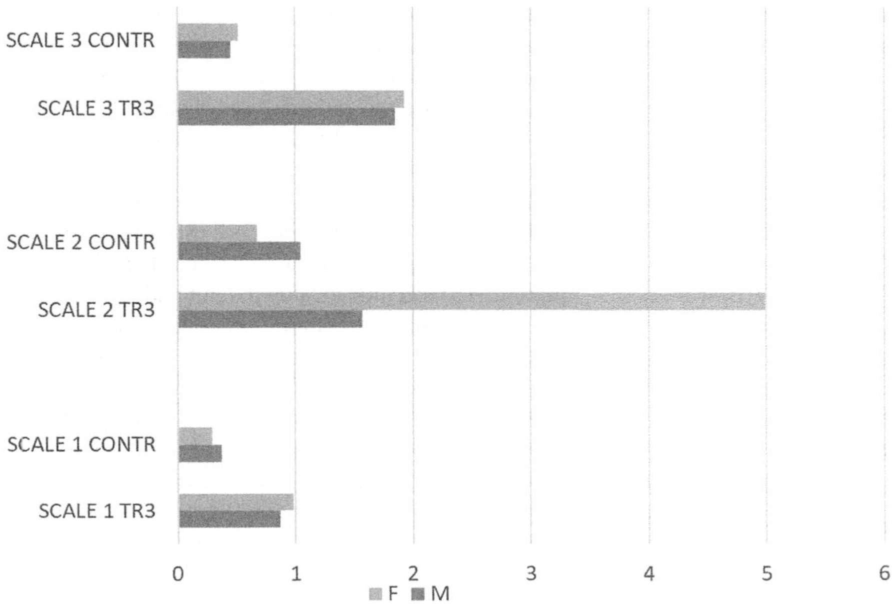

Fig. 1.2 Plot of significant effects (regarding the interaction among Training * Scale * Gender). Note: $^*p < 0.05$; $^{**}p < 0.01$; Training 1 = pencil-and-paper mathematical training; Training 2 = computerized mathematical training; Training 3 = combined pencil-and-paper visuospatial training and pencil-and-paper mathematical training; Training 4 = combined pencil-and-paper visuospatial training and computerized mathematical training; Control = control group; Scale 1 = lexical; Scale 2 = semantic; Scale 3 = pre-syntactic; Scale 4 = counting

According to those studies, potential gender differences in mathematical skills might just be found when a detailed learning process is analysed, and not when overall test scores in mathematical ability are carried out. Indeed, these authors have found that young girls seem to prefer applying concrete manipulation during math problem solving. As far as these aspects are concerned, we might suppose that these concrete manipulations may have been supported by the combined visuospatial and math training proposed in this study. Furthermore, the interesting female advantage in the performance for some scales might be related also to the young age of these pupils. In this context, the potential effects of adults' gender expectancies and stereotypes regarding mathematical achievements could still be limited.

1.4 Conclusion

Children with low numerical skills at preschool are at high risk for low mathematic achievement over the early elementary school grades (Barnes et al., 2016; Chu, VanMarle, & Geary, 2015; Martin, Cirino, Sharp, & Barnes, 2014; Zhang et al.,

2020) and in life in general. Consequently, in the future those children might face a lack of work opportunities, as well as a reduction in personal independence while performing tasks in everyday life (Benavides-Varela et al., 2020). Therefore, in those cases, it will be fundamental to find a way to enhance numerical abilities. First of all, in order to plan specific interventions, cognitive variables related to mathematical skills will have to be found, especially with working memory (Passolunghi et al., 2014). Moreover, mixing pencil and paper trainings with computerized trainings will be also very important. Pencil and paper trainings are essential to further cognitive development as a body of a growing literature states (Wollscheid, Sjaastad, & Tømte, 2016). Kersey and James (2013) has indeed shown that handwriting practice can activate particular areas of children's brains more than other forms of fine motor manipulation tasks (Wollscheid et al., 2016). On the other side, nowadays computerized training are very common, as well as useful especially for some specific features like high mobility, multiple sensory inputs, high-quality graphics and feedback speed. Those features can also support one-to-one learner-centred interactive training (Ifenthaler & Schweinbenz, 2013; Lee & Choi, 2020). Therefore, it is highly important to carry out trainings aiming at enhancing those variables since kindergarten age, in order to support the development of mathematical learning. Results from our study have allowed us to reflect on the importance of sustaining mathematical learning with combined and long-term training in kindergarten learning programmes. This kind of training will foster kindergarteners' interest for this subject, and it will motivate them, by raising their attention, and by uplifting their self-esteem (Lee & Choi, 2020).

Acknowledgements The authors would like to thank the schools and the children who participated in the study.

References

AA.VV. (2019). The jamovi project (version 1.0). Retrieved from https://www.jamovi.org

Agus, M., Mascia, M. L., Fastame, M. C., Melis, V., Pilloni, M. C., & Penna, M. P. (2015). The measurement of enhancement in mathematical abilities as a result of joint cognitive trainings in numerical and visual-spatial skills: A preliminary study. *Journal of Physics: Conference Series, 588*(1), 012041. https://doi.org/10.1088/1742-6596/588/1/012041

Agus, M., Mascia, M. L., Fastame, M. C., & Penna, P. M. (2016). Difficoltà matematiche: percorsi di potenziamento delle abilità numeriche e visuo-spaziali. *Psicologia dell'educazione, 2*, 29–37.

Allen, K., Higgins, S., & Adams, J. (2019). The relationship between visuospatial working memory and mathematical performance in school-aged children: A systematic review. *Educational Psychology Review, 31*, 509–531.

Anderson, K. L., Casey, B., Thompson, W. L., Burrage, M. S., Pezaris, E., & Kosslyn, S. M. (2008). Performance on middle school geometry problems with geometry clues matched to three different cognitive styles. *Mind, Brain, and Education, 2*(4), 188–197.

Aunio, P., & Niemivirta, M. (2010). Predicting children's mathematical performance in grade one by early numeracy. *Learning and Individual Differences, 20*(5), 427–435. https://doi.org/10.1016/j.lindif.2010.06.003

Aunola, K., Leskinen, E., Lerkkanen, M.-K., & Nurmi, J.-E. (2004). Developmental dynamics of math performance from preschool to grade 2. *Journal of Educational Psychology, 96*, 699–713. https://doi.org/10.1037/0022-0663.96.4.699

Barnes, M. A., Klein, A., Swank, P., Starkey, P., McCandliss, B., Flynn, K., … Roberts, G. (2016). Effects of tutorial interventions in mathematics and attention for low-performing preschool children. *Journal of Research on Educational Effectiveness, 9*(4), 577–606. https://doi.org/10.1080/19345747.2016.1191575

Bates, D., Mächler, M., Bolker, B., & Walker, S. (2015). Fitting linear mixed-effects models using lme4. *Journal of Statistical Software, 67*(1), 1–48. https://doi.org/10.18637/jss.v067.i01

Belacchi, C., Scalisi, T. G., Cannoni, E., & Cornoldi, C. (2008). *CPM-Coloured Progressive Matrices Standardizzazione Italiana [CPM-Coloured Progressive Matrices Italian Standardization]*. Firenze: Giunti OS Organizzazioni Speciali.

Benavides-Varela, S., Callegher, C. Z., Fagiolini, B., Leo, I., Altoè, G., & Lucangeli, D. (2020). Effectiveness of digital-based interventions for children with mathematical learning difficulties: A meta-analysis. *Computers & Education, 157*, 103953.

Blankson, A. N., Gudmundson, J. A., & Kondeh, M. (2019). Cognitive predictors of kindergarten achievement in African American children. *Journal of Educational Psychology, 111*(7), 1273–1283. https://doi.org/10.1037/edu0000346

Bonny, J. W., & Lourenco, S. F. (2013). The approximate number system and its relation to early math achievement: Evidence from the preschool years. *Journal of Experimental Child Psychology, 114*(3), 375–388. https://doi.org/10.1016/j.jecp.2012.09.015

Bull, R., Espy, K. A., & Wiebe, S. A. (2008). Short-term memory, working memory, and executive functioning in preschoolers: Longitudinal predictors of mathematical achievement at age 7 years. *Developmental Neuropsychology, 33*(3), 205–228. https://doi.org/10.1080/87565640801982312

Carr, M., & Davis, H. (2001). Gender differences in arithmetic strategy use: A function of skill and preference. *Contemporary Educational Psychology, 26*(3), 330–347. https://doi.org/10.1006/ceps.2000.1059

Carr, M., & Jessup, D. L. (1997). Gender differences in first-grade mathematics strategy use: Social and metacognitive influences. *Journal of Educational Psychology, 89*(2), 318–328. https://doi.org/10.1037/0022-0663.89.2.318

Carr, M., Jessup, D. L., & Fuller, D. (1999). Gender differences in first-grade mathematics strategy use: Parent and teacher contributions. *Journal for Research in Mathematics Education, 30*(1), 20. https://doi.org/10.2307/749628

Casey, B., Dearing, E., Dulaney, A., Heyman, M., & Springer, R. (2014). Young girls' spatial and arithmetic performance: The mediating role of maternal supportive interactions during joint spatial problem solving. *Early Childhood Research Quarterly, 29*(4), 636–648. https://doi.org/10.1016/j.ecresq.2014.07.005

Casey, B., Erkut, S., Ceder, I., & Young, J. M. (2008). Use of a storytelling context to improve girls' and boys' geometry skills in kindergarten. *Journal of Applied Developmental Psychology, 29*(1), 29–48. https://doi.org/10.1016/j.appdev.2007.10.005

Chen, Y.-N., Lin, C.-K., Wei, T.-S., Liu, C.-H., & Wuang, Y.-P. (2013). The effectiveness of multimedia visual perceptual training groups for the preschool children with developmental delay. *Research in Developmental Disabilities, 34*(12), 4447–4454. https://doi.org/10.1016/j.ridd.2013.09.023

Cheng, Y.-L., & Mix, K. S. (2014). Spatial training improves children's mathematics ability. *Journal of Cognition and Development, 15*(1), 2–11. https://doi.org/10.1080/15248372.2012.725186

Chu, F. W., VanMarle, K., & Geary, D. C. (2015). Early numerical foundations of young children's mathematical development. *Journal of Experimental Child Psychology, 132*, 205–212. https://doi.org/10.1016/j.jecp.2015.01.006

Cornu, V., Schiltz, C., Pazouki, T., & Martin, R. (2019). Training early visuo-spatial abilities: A controlled classroom-based intervention study. *Applied Developmental Science, 23*(1), 1–21. https://doi.org/10.1080/10888691.2016.1276835

Dehaene, S. (2011). *The number sense: How the mind creates mathematics* (2nd ed.). New York, NY: Oxford University Press.

Elkins, J. (2002). Learning difficulties/disabilities in literacy. *Australian Journal of Language and Literacy, 25*(3), 11.

Fastame, M. C., & Callai, D. (2015). Empowering visuo-spatial ability in primary school: Results from a follow-up study. *Educational Psychology in Practice, 31*(1), 86–98. https://doi.org/10.1080/02667363.2014.989315

Fastame, M. C., & Manca, C. (2020). The enhancement of visuospatial abilities through pencil-and-paper or computer-aided training: What is more effective in 7-year-old pupils? *International Journal of School and Educational Psychology*, 1–9.

Fastame, M. C., Manca, C., Penna, M. P., Lucangeli, D., & Hitchcott, P. K. (2019). Numeracy skills and self-reported mental health in people aging well. *Psychiatric Quarterly, 90*, 629. https://doi.org/10.1007/s11126-019-09655-y

Fletcher, J. M., Lyon, G. R., Fuchs, L. S., & Barnes, M. A. (2018). *Learning disabilities: From identification to intervention*. New York, NY: Guilford Publications.

Geary, D. C. (2004). Mathematics and learning disabilities. *Journal of Learning Disabilities, 37*, 4–15. https://doi.org/10.1177/00222194040370010201

Gunderson, E. A., Ramirez, G., Levine, S. C., & Beilock, S. L. (2012). The role of parents and teachers in the development of gender-related math attitudes. *Sex Roles, 66*(3–4), 153–166. https://doi.org/10.1007/s11199-011-9996-2

Hellstrand, H., Korhonen, J., Linnanmäki, K., & Aunio, P. (2020). The number race–computer-assisted intervention for mathematically low-performing first graders. *European Journal of Special Needs Education, 35*(1), 85–99.

Hutchison, J. E., Lyons, I. M., & Ansari, D. (2019). More similar than different: Gender differences in children's basic numerical skills are the exception not the rule. *Child Development, 90*(1), e66–e79.

Hyde, J. S., Lindberg, S. M., Linn, M. C., Ellis, A. B., & Williams, C. C. (2008). Gender similarities characterize math performance. *Science, 321*(5888), 494–495. https://doi.org/10.1126/science.1160364

Ifenthaler, D., & Schweinbenz, V. (2013). The acceptance of tablet-PCs in classroom instruction: The teachers' perspectives. *Computers in Human Behavior, 29*(3), 525–534.

Jordan, N. C., Kaplan, D., Ramineni, C., & Locuniak, M. N. (2009). Early math matters: Kindergarten number competence and later mathematics outcomes. *Developmental Psychology, 45*(3), 850.

Kersey, A. J., & James, K. H. (2013). Brain activation patterns resulting from learning letter forms through active self-production and passive observation in young children. *Frontiers in Psychology, 4*, 567.

Kroesbergen, E. H., van Luit, J. E. H., Naglieri, J. A., Taddei, S., & Franchi, E. (2010). Pass processes and early mathematics skills in Dutch and Italian kindergarteners. *Journal of Psychoeducational Assessment, 28*(6), 585–593. https://doi.org/10.1177/0734282909356054

Kwok, O.-M., Underhill, A. T., Berry, J. W., Luo, W., Elliott, T. R., & Yoon, M. (2008). Analyzing longitudinal data with multilevel models: An example with individuals living with lower extremity intra-articular fractures. *Rehabilitation Psychology, 53*(3), 370–386. https://doi.org/10.1037/a0012765

Kyttälä, M., Aunio, P., Lehto, J. E., Van Luit, J., & Hautamäki, J. (2003). Visuospatial working memory and early numeracy. *Educational and Child Psychology, 20*(3), 65–76.

Landerl, K., Bevan, A., & Butterworth, B. (2004). Developmental dyscalculia and basic numerical capacities: A study of 8–9 year old students. *Cognition, 93*, 99–125. https://doi.org/10.1016/j.cognition.2003.11.004

Lee, K., & Bull, R. (2016). Developmental changes in working memory, updating, and math achievement. *Journal of Educational Psychology, 108*(6), 869–882. https://doi.org/10.1037/edu0000090

Lee, H. K. & Choi, A. (2020). Enhancing early numeracy skills with a tablet-based math game intervention: A study in Tanzania. Educational Technology Research and Development. https://doi.org/10.1007/s11423-020-09808-y.

Lucangeli, D., Iannitti, A., & Vettore, M. (2007). *Lo sviluppo dell'intelligenza numerica*. Roma: Carocci.

Lucangeli, D., Mammarella, I. C., Todeschini, M., Miele, G., & Cornoldi, C. (2009). *Conosco le forme*. Firenze: Giunti.

Lucangeli, D., Poli, S., & Molin, A. (2003). *L'intelligenza numerica*. Trento: Erickson.

Lucangeli, D., Poli, S., & Molin, A. (2010). *Sviluppare l'intelligenza numerica I*. Trento: Erickson.

Martin, R. B., Cirino, P. T., Sharp, C., & Barnes, M. (2014). Number and counting skills in kindergarten as predictors of grade 1 mathematical skills. *Learning and Individual Differences, 34*, 12–23. https://doi.org/10.1016/j.lindif.2014.05.006

Mascia, M. L., Agus, M., Fastame, M. C., & Addis, A. (2016). Enhancement in mathematical abilities: A system approach. In G. Minati, M. R. Abram, & E. Pessa (Eds.), *Towards a post-bertalanffy systemics SE—25* (pp. 243–249). Basel: Springer. https://doi.org/10.1007/978-3-319-24391-7_25

Mascia, M. L., Agus, M., Fastame, M. C., Penna, M. P., Sale, E., & Pessa, E. (2015). The development and empowerment of mathematical abilities: The impact of pencil and paper and computerised interventions for preschool children. *Cognition and exploratory learning in the digital age (CELDA 2015)* (pp. 59–68). Retrieved from http://files.eric.ed.gov/fulltext/ED562093.pdf#page=76

Mascia, M. L., Fastame, M. C., Agus, M., & Penna, M. P. (2019). Numeracy skills empowerment from preschool. *16th international conference on cognition and exploratory learning in digital age, CELDA 2019* (pp. 425–428).

Miller, M. R., Müller, U., Giesbrecht, G. F., Carpendale, J. I., & Kerns, K. A. (2013). The contribution of executive function and social understanding to preschoolers' letter and math skills. *Cognitive Development, 28*(4), 331–349. https://doi.org/10.1016/j.cogdev.2012.10.005

Molin, A., Poli, S., & Lucangeli, D. (2007). *Batteria Intelligenza Numerica 4–6 [Battery for numerical intelligence]*. Trento: Erickson.

Muth, C., Bales, K. L., Hinde, K., Maninger, N., Mendoza, S. P., & Ferrer, E. (2016). Alternative models for small samples in psychological research. *Educational and Psychological Measurement, 76*(1), 64–87. https://doi.org/10.1177/0013164415580432

Passolunghi, M. C., Cargnelutti, E., & Pastore, M. (2014). The contribution of general cognitive abilities and approximate number system to early mathematics. *British Journal of Educational Psychology, 84*(4), 631–649.

Passolunghi, M. C., & Costa, H. M. (2014). Working memory and early numeracy training in preschool children. *Child Neuropsychology, 7049*(November), 1–18. https://doi.org/10.1080/09297049.2014.971726

Passolunghi, M. C., & Lanfranchi, S. (2012). Domain-specific and domain-general precursors of mathematical achievement: A longitudinal study from kindergarten to first grade. *British Journal of Educational Psychology, 82*(1), 42–63. https://doi.org/10.1111/j.2044-8279.2011.02039.x

Penna, M. P., & Stara, V. (2010). Opinions on computers, and efficacy of a computer-based learning: A pilot study. *Education and Information Technologies, 15*(3), 181–204. https://doi.org/10.1007/s10639-009-9104-1

Penna, M. P., Stara, V., & Bonfiglio, N. (2002). A systemic proposal on the use of a new technology as a learning tool in school context. In *Emergence in complex, cognitive, social, and biological systems* (pp. 153–157). https://doi.org/10.1007/978-1-4615-0753-6_13.

Praet, M., & Desoete, A. (2019). A pilot study about the effect and sustainability of early interventions for children with early mathematical difficulties in kindergarten. *Learning Disabilities, 17*(1), 29–40.

Ramani, G. B., Siegler, R. S., & Hitti, A. (2012). Taking it to the classroom: Number board games as a small group learning activity. *Journal of Educational Psychology, 104*(3), 661–672. https://doi.org/10.1037/a0028995

Raven, J. C. (1958). *Guide to using the coloured progressive matrices*. Oxford: H. K. Lewis & Co..

Sella, F., Sader, E., Lolliot, S., & Cohen Kadosh, R. (2016). Basic and advanced numerical performances relate to mathematical expertise but are fully mediated by visuospatial skills. *Journal of Experimental Psychology: Learning, Memory, and Cognition, 42*(9), 1458–1472. https://doi.org/10.1037/xlm0000249

Slavin, R. E. (2013). Effective programmes in reading and mathematics: Lessons from the best evidence encyclopaedia 1. *School Effectiveness and School Improvement, 24*(4), 383–391.

Stock, P., Desoete, A., & Roeyers, H. (2010). Detecting children with arithmetic disabilities from kindergarten: Evidence from a 3-year longitudinal study on the role of preparatory arithmetic abilities. *Journal of Learning Disabilities, 43*(3), 250–268. https://doi.org/10.1177/0022219409345011

Toll, S. W. M., Kroesbergen, E. H., & Van Luit, J. E. H. (2016). Visual working memory and number sense: Testing the double deficit hypothesis in mathematics. *British Journal of Educational Psychology, 86*(3), 429–445. https://doi.org/10.1111/bjep.12116

VandenBos, G. R. (2015). *APA dictionary of psychology* (2nd ed.). Washington, DC: American Psychological Association.

Voyer, D., Voyer, S. D., & Saint-Aubin, J. (2017). Sex differences in visual-spatial working memory: A meta-analysis. *Psychonomic Bulletin & Review, 24*(2), 307–334. https://doi.org/10.3758/s13423-016-1085-7

Wang, L. (2020). Mediation relationships among gender, spatial ability, math anxiety, and math achievement. *Educational Psychology Review, 32*(1), 1–15.

Watts, T. W., Duncan, G. J., Siegler, R. S., & Davis-Kean, P. E. (2014). What's past is prologue: Relations between early mathematics knowledge and high school achievement. *Educational Researcher, 43*(7), 352–360.

Wei, W., Li, Y., & Su, H. Y. (2020). Predicting the growth patterns in early mathematics achievement from cognitive and environmental factors among Chinese kindergarten children. *Learning and Individual Differences, 79*, 101841.

Wollscheid, S., Sjaastad, J., & Tømte, C. (2016). The impact of digital devices vs. pen (cil) and paper on primary school students' writing skills–A research review. *Computers & Education, 95*, 19–35.

Yang, X., Chung, K. K. H., & McBride, C. (2019). Longitudinal contributions of executive functioning and visual-spatial skills to mathematics learning in young Chinese children. *Educational Psychology, 39*(5), 678–704.

Zhang, X., Räsänen, P., Koponen, T., Aunola, K., Lerkkanen, M. K., & Nurmi, J. E. (2020). Early cognitive precursors of children's mathematics learning disability and persistent low achievement: A 5-year longitudinal study. *Child Development, 91*(1), 7–27.

Chapter 2
Does Chess Training Affect Meta-Cognitive Processes and Academic Performance?

Carla Meloni and Rachele Fanari

2.1 Introduction

2.1.1 Cognitive Abilities and Chess

An interesting topic in the cognitive psychology field is the one investigating how specific skills and strategies used in strategic games can be transferred to other learning areas to improve students' outcome. Among the strategic games, chess has perhaps been the most studied and many researchers considered, for instance, how the skills that make a chess player a good player and the skills which differentiate a chess master from a naive player could be related to general learning outcome, making of chess a sort of model environment for research in problem-solving and expertise. De Groot (1966), in his seminal work, compared experts and novices players, considering the abilities to remember pieces' position on a chessboard and to evaluate the best moves during a game: expert players remembered more accurately the position of the pieces on the chessboard compare to novices, and experts were also more skilled than novices in choosing the best moves. Holding and Reynolds (1982) wondered if specific knowledge of the positions of pieces on the chessboard was what that made "expert" an expert player. They compared experts and novices player considering the same abilities investigated by De Groot in the study mentioned above, but testing memory of pieces' position and ability to find the best moves in an experimental set in which the pieces were arrange by random pattern. Results showed that both experts and novices did not differ in the random patter recall test, but that if the participants were asked to evaluate the best moves even starting from a random configuration, the expert players indicated better quality

C. Meloni (✉) · R. Fanari
Department of Education, Psychology, and Philosophy, University of Cagliari, Cagliari, Italy
e-mail: cmeloni@unica.it; rfanari@unica.it

D. Ifenthaler et al. (eds.), *Balancing the Tension between Digital Technologies and Learning Sciences*, Cognition and Exploratory Learning in the Digital Age,
https://doi.org/10.1007/978-3-030-65657-7_2

moves, even if they did not have a previous specific schemes, so the authors emphasized that the strategies developed by expert chess players are what make the difference.

The studies in the field of the expertise leaded to the construction of computational models of thought. One of these is the *long-term working memory model* by Ericsson and Kintsch (1995), which considered the cognitive abilities implied in complex cognitive tasks, as chess, as different from those involved in standard tasks. By Ericsson and Kintsch, the experts' knowledge of pieces configurations on a chessboard is mediated by structures that allows the production of mental representations more general than the actual configuration scheme of the chessboard and should also include planning general skills, thus assuming a relationship between specific domain and general domain skills.

In a recent meta-analysis, Sala, Foley, and Gobet (2017) examined studies investigating the cognitive performance of chess players and of non-chess players of different ages. The results showed that chess players' general cognitive ability was better than age-matched participants even when the education level was controlled for, suggesting a relationship between cognitive ability and chess skill and that chess activity requires domain-general cognitive abilities. Burgoyne et al., 2016 in a meta-analysis identified which could be these general cognitive abilities and reported statistically significant correlations between chess skill and four broad measures of cognitive ability: fluid intelligence (the ability to solve new problems); processing speed (the efficiency of basic mental operations, as measured in reaction-time tasks); Working Memory (to retain, manipulate, and recall information) and comprehension knowledge (the ability to use knowledge acquired through experience).

The positive correlation between cognitive ability and chess, though, does not tell us anything certain about transfer issue. The basic idea under the chess instruction recommendation is to train general cognitive abilities through chess, general abilities that later could be transferred to other domains (see Sala & Gobet, 2016 for a discussion).

The idea that skills acquired in a specific domain such as chess can be transferred to other specific domains, such as those related to school learning or other general domain cognitive skills, has promoted, in several European countries, projects involving the introduction of chess instruction in primary school. Many schools offer chess as an optional subject, while for some schools chess teaching is a part of the standard school program; this also happens following the favorable opinion of the European Parliament itself that promotes the chess game as an important educational tool (Binev, Attard-Montalto, Deva, Mauro, & Takkula, 2011).

In particular, the hypothesis that chess can influence skills in learning mathematics and so favoring students' choice of STEM (Science, Technology, Engineering and Mathematics) educational careers is of special interest, given that the job market demands always graduates in STEM subjects.

Several studies and meta-analysis investigated chess instruction in order to verify if the skills acquired in this specific field can improve academic performance, for instance in mathematics and reading, and may lead to an improvement even in general domain of cognitive skills, but the scientific literature does not present strong evidence that specific skills acquired in the chess practice can be transferred to other domains (see Sala & Gobet, 2016 for a discussion).

2.1.2 The Problem of Transfer

Transfer is a process that occurs when skills acquired in a given domain are transferred to another specific or general domain, but the exact nature of the transfer process is not yet entirely clear. Transfer is a central issue in cognitive psychology because it is a manifestation of how humans acquire and process information. It is customary to distinguish between near and far transfer. Near transfer is the generalization of a set of skills across two (or more) domains tightly related to each other. Far transfer occurs when a set of skills generalizes across two (or more) domains that are only loosely related to each other (e.g., mathematics and Latin). So, far transfer indicates the transfer of skills across domains that are not, or very weakly, related to each other. The distinction between near and far transfer relies on the overlap between the source and target domains. In other words, the definition of the type of transfer is directly related to the extent to which the domains share common features. The more the shared features, the nearer the transfer. That means that while near transfer is predicted to occur often, far transfer is supposed to be rare. Substantial research into learning, skill acquisition, and expertise has corroborated the theory (Ritchie, Bates, & Deary, 2015; Sala & Gobet, 2017).

In 1901, Thorndike and Woodworth, in their seminal work, formulated the hypothesis that transfer depends on the number of features shared between two domains. More recently, Anderson (1990) stated that transfer is a function of the degree of overlap of the cognitive elements present in two tasks, an idea suggesting that the transfer from one specific task to another is often limited. Sternberg (2000) suggested a different approach to the transfer issue: transferable abilities are those constituting the basis of intelligence (general abilities as the verbal or visuospatial abilities) that can be applied in different domains but that, being innate, cannot be increased through practice.

Some experimental evidence (e.g., Ericsson & Charness, 1994) has shown that the higher the level of expertise in a given specific domain, the more the transfer is limited. Generic learning skills (learning strategies, problem-solving methods, and reasoning techniques), on the other hand, are useful for more domains, but their teaching seems to have immediate, but not long-term benefits (Grotzer & Perkins, 2000).

Regarding the potential of transferring of skills acquired in chess playing to other domains, Gobet and Campitelli (2006), in a critical review, emphasized that the results of the works done on the topic, even if they seem to support a possible transfer of abilities, are often weak and contradictory due to methodological problems. Based on their review, the empirical evidence suggests that chess players tend to be smarter than non-chess players, and that, at least with children, there is a correlation between chess skills and general intelligence even if, quite surprisingly, a direct link between chess and visuospatial skills has not been identified. However, these results could be explained mainly by sample selection processes: more intelligent people are more likely to choose, and to excel, in intellectual activities such as chess.

In a recent meta-analysis, Sala, Foley, and Gobet (2017) investigated the effects of chess programs both on cognitive abilities and school performances in primary school children. The authors concluded that the effects of chess training are more evident, even if moderate, on math performance and general cognitive skills than reading skills and that at least 25–30 h of chess training seem to be required to have positive effects.

2.1.3 Chess and Math Skills

In the literature, it is often stated that chess playing improves math skills because chess practice has some elements in common with the domain of mathematics and promotes skills independent from the chess-specific context, such as the ability to understand the existence of a problem and reasoning skills.

Trinchero (2013) investigated chess training effects on math skills in children between 8 and 10 years old. The author proposed to children a training that combined sessions of chess game in the classroom, in primary school, with a chess computer assisted training (CAT), observing a significant increase in math problem-solving skills. The author observed how chess increased attention and concentration, promoting the identification of the most relevant information needed to plan and implement the most useful strategies to win the game. Attention, concentration and planning allow an analysis of the chessboard configuration and help to create a mental representation of the effects of the possible moves. All these abilities could be also the basis for mathematical problem-solving and so their increment, due to chess training, may bring to better math abilities.

In the same vein are also the results obtained by Sala, Gorini, and Pravettoni (2015) who studied the effects of combined chess training, classroom, and CAT, observing a positive effect on mathematical problem-solving ability. The authors speak of a virtuous circle generated by playing chess: chess is an amusing and rewarding activity, this encourages children to play more so improving attention and planning skills.

Trinchero and Sala (2016) to deeper investigate the role of heuristics in chess game, compared two experimental chess training groups and one control group in mathematical problem-solving ability. The first experimental had a combination of chess training in classroom and chess CAT, the second experimental group, in addition, had a chess master teaching them heuristics to solve chess problems, the control group did not play chess nor had any supplemental instruction about general cognitive strategies or heuristics. The group that had received the teaching of heuristics, significantly improved their skills in mathematical problem-solving compared to the other two groups. The authors consider that chess involves, in addition to the pure mathematical-geometric elements (geometric space represented by the chessboard, movements of the pieces according to the rules of geometry, number of moves) the ability to reflect on their own thinking. Thus, chess could favor the increase of meta-cognitive skills which could affect mathematical problems-solving ability.

2.1.4 Chess, Math and Meta-Cognitive Abilities

Understanding whether teaching chess problem-solving heuristics helps children also to solve mathematical problems is a question of interest not only for the field of education, but also for the general psychological issue of transfer of skills.

The literature investigated the relationship between chess instruction and math problem-solving skills often hypothesized that the improvement in math ability stemmed from the fact that chess practice improved meta-cognitive skills, especially planning and self-regulation (e.g., Sala et al., 2015; Trinchero, 2013; Trinchero & Sala, 2016). But there are practically no studies that have directly investigated the relationship between chess, general meta-cognitive abilities, and specific academic school skills (e.g., reading and math abilities) in the same experimental design.

To our knowledge, there is only one study that focused not on general meta-cognitive skills but rather on meta-cognitive skills specifically related to math performance. The study, published in 2012 by Kazemi, Yektayar, and Abad found a positive effect of the use of chess programs on math meta-cognitive abilities. The authors concluded that chess instruction is a way to develop higher-order thinking skills useful for math problem-solving and that the meta-cognitive abilities boosted by chess practice can be successfully transferred into mathematics domain.

Very little is known on chess training's influence on general meta-cognitive skills, skills that children can use in other domains, such motivation to study, organization of personal work, strategic elaboration of the learning material, flexibility of the modality of studying, ability to concentrate, anxiety and attitude toward school, and the knowledge and use of more or less functional study strategies. The aim of the work here described was to fill this gap and to explore the link between chess training and both general meta-cognitive study abilities and verbal and math academic skills in primary school children. We compared two groups of children—one group participating in a chess training and a control group—in their ability to solve math problems, to comprehend and recall a written text, and in their approach to studying and using study strategies.

Following the literature on the relationship between chess playing and academic skills, we expected chess training to influence mathematical problem-solving abilities, but not verbal skills. As we already told, there is no literature on the effect of chess training on general meta-cognitive abilities of the approach to studying and on the knowledge and use of more or less functional strategies of studying. Therefore, with this research we tried to can cast light on the debated issue of the potential to improve general meta-cognitive abilities, useful in various domains, through the training of a specific skill.

2.2 The Experimental Study: Method

2.2.1 Participants

Eighty-five typically developing children were recruited from a public primary school in Cagliari, Italy. Both the school and the children's parents agreed to let the children take part in the research study and signed informed consent forms.

Forty-eight children were randomly assigned to a chess training group (mean age = 9.27 years and SD = 0.84; 24 males and 24 females), and 37 were randomly assigned to a control group (mean age = 9.25 years and SD = 0.76; 17 males and 20 females). At the start of the study, all the children were chess novices. The participants came from different classes in which the same teachers evenly rotated; thus, any teacher effects were controlled for and the teaching provided to the children was the same even if the children came from different classes. After conducting the random assignment to the experimental and control groups, the teachers were asked if they believed, based on their daily experience with the children, that there were differences between the two groups related to academic performance or differences in attitudes toward school/learning. The teachers noted that the two randomly selected groups were comparable with respect to these variables.

2.2.2 Procedure

The children in the experimental group participated in a chess program during the school year. Chess lessons were held by a chess master once per week from November to May during school hours. Following Sala et al.'s (2017) meta-analysis results, a 30-h program was chosen. The control group followed a sport training: specifically, an introduction to basketball.

At the end of the training, the children were presented with a test battery aimed to assess their meta-cognitive skills (approach to studying, knowledge and actual use of functional and dysfunctional strategies) and were tested on their ability to solve mathematical problems and on their level of text comprehension and text recall.

2.2.3 Assessment Tools

The tools used to assess children's meta-cognitive abilities were taken from the AMOS 8-15 Skills and Motivation Study Battery (De Beni, Moè, & Cornoldi, 2003). The battery is composed of seven questionnaires indicating different aspects of meta-cognitive abilities involved in academic performance. The three questionnaires we used in this work were: the Questionnaire of Approach to Studying (QAS), the Questionnaire of Study Strategies 1 (QS1), a questionnaire measuring the effectiveness of the study strategies know by the children and the Questionnaire of Study Strategies 2 (QS2), a questionnaire evaluating the children's actual use of the study strategies.

The QAS questionnaire on the approach to studying investigates seven different dimensions (part A: study motivation; part B: personal work organization; part D: strategic information processing; part E: study flexibility; part N: concentration; part U: anxiety; part V: attitude toward school), for each dimension, seven different statements are presented to the child and he/she must indicate with a cross how true each written statement is to him/her (1 = not true, 2 = enough true, and 3 = very true). The QAS allows for a total score for the approach to studying ability as well as a single score for each dimension.

The second questionnaire used, QS1, identifies the children's beliefs on the effectiveness of functional and dysfunctional strategies that can be used while studying. In particular, the QS1 measures 32 studying strategies (example item: "Thinking about what you already known about the topic you are studying"), and the child is asked to read them carefully and evaluate how much these strategies, according to him/her, are useful for studying, giving each strategy a rating from 1 to 4 (1 = not useful, 2 = not very useful, 3 = useful, and 4 = very useful). The third questionnaire, QS2, detects the child's actual use of the same strategies proposed by the QS1 questionnaire. In the QS2, therefore, 32 studying strategies are proposed (example item: "If you do not understand a part of the text, read it again") and the child is asked to think about their approach to studying and to indicate how often he/she uses the activity with a rating from 1 to 4 (1 = I never use it, 2 = I use it sometimes, 3 = I use it often, and 4 = I always use it). These two last questionnaires (QS1 and QS2) allow a summary index of strategic coherence to be calculated that reflects the correspondence between utility judgments and the estimation of the use of the same strategies by children.

The tools used to assess children's school performance were the "Mathematical Problem-Solving" (SPM) test (Lucangeli, Tressoldi, & Cedron, 2003) used to test mathematical problem-solving and a test taken from the AMOS 8-15 Battery (De Beni et al., 2003) called the "Studying Test", used to evaluate the level of text comprehension and text recall.

In the SPM, the child is presented with some mathematical problems with different difficulty levels depending on the level of schooling. The SPM evaluates the following skills: problem understanding (understanding the information present in the problem and their relationships), problem representation (the representation of information through a scheme able to integrate problem information), problem categorization (ability to identify among a series of alternatives the problem that has the same deep structure), problem-solving planning, problem-solving procedure, and self-assessment of the correctness of the used procedure.

Moreover, to evaluate the ability to understand, store, and recall information, the Studying Test was used. The test asks the children to study a written text for 30 min. After 30 min, they are involved in other activities for 10 min and then questions about the text are presented to evaluate three indexes: the ability to select the main aspects of the text (asking the child to choose a suitable title), the ability to identify specific information (open questions), and the ability to recognize true/false information with respect to the studied text (multiple-choice test).

2.2.4 Results

For the data analysis, the scores obtained in each test were calculated following the indications provided. Since the participants in the study were of different ages and school classes, the raw scores were transformed into Z points to compare children's scores, following the test norms.

We conducted six multivariate analyses of variance (MANOVAs). All the MANOVAs had the factor "training" as the independent variable with two levels: chess training and control.

The first MANOVA was carried out to analyze the effects of training on the dimensions of the approach to studying investigated by the QAS; the second, third, and fourth MANOVA were carried out to analyze, respectively: the effects of training on children's beliefs on the effectiveness of functional and dysfunctional study strategies (QS1 questionnaire), the effects of training on the actual use of functional and dysfunctional study strategies (QS2 questionnaire), and the index of strategic coherence in the usage of study strategies; the fifth MANOVA was carried out to analyze the effects of training on children's ability to understand and memorize a text, evaluated in the Studying Test. Finally, with the sixth MANOVA, we aimed to analyze the effects of training on children's ability to solve mathematical problems, evaluated through the SPM. Univariate tests were performed where necessary.

The first MANOVA did not show a significant effect of the chess training on the approach to studying dimensions investigated by the QAS: Wilks/'Lambda = 0.928, F (7, 77) = 0.86, $p = 0.54$, $\eta^2 = 0.072$. The children who participated in the chess training seem to approach the study in the same way as the control group.

The MANOVAs for the QS1 and QS2 questionnaires and for the index of strategic coherence were separately calculated: Wilks' Lambda (QS1) = 0.951, F (2, 82) = 2.09, $p = 0.13$, $\eta^2 = 0.049$; Wilks' Lambda (QS2) = 0.994, F (2, 82) = 0.23, $p = 0.79$, $\eta^2 = 0.006$; Wilks' Lambda (strategic coherence) = 0.979, F (2, 82) = 0.89, $p = 0.42$, $\eta^2 = 0.021$. The results show no significant difference between the experimental and control groups relative to the children's beliefs about the effectiveness of study strategies on the degree of their actual usage and on the degree of coherence with which they actually use study strategies they consider most effective.

From the fifth MANOVA, no effects of chess training on children's ability to understand and memorize a text (evaluated by the Studying Test) emerged: Wilks' Lambda = 0.954, F (3, 81) = 1.31, $p = 0.28$, $\eta^2 = 0.046$.

The sixth MANOVA, which compared the two groups of children on their ability to solve mathematical problems (SPM test), instead highlighted a difference between the experimental and control groups: Wilks' Lambda = 0.801, F (6, 78) = 3.23, $p = 0.01$, $\eta^2 = 0.199$. This last result has been examined in detail through

Table 2.1 Analysis of variance for the control and chess training groups' scores in SPM battery sub-tests ($n = 85$)

	Control group Mean	Control group SD	Chess training group Mean	Chess training group SD	F	p-value
Problem comprehension	−0.09	1.66	0.01	1.83	0.07	0.79
Problem representation	−0.11	1.34	0.41	1.06	3.91	0.05*
Problem categorization	−0.30	0.95	0.37	0.83	11.91	0.001*
Problem-solving planning	0.25	1.05	0.26	0.83	0.00	0.98
Problem-solving execution	−0.59	0.98	−0.60	1.28	0.00	0.99
Self-evaluation	−0.41	0.65	−0.15	0.55	4.05	0.05*

Mean score, SD, F, and p-value (***$p < 0.001$, **$p < 0.01$, *$p < 0.05$)

a series of individual ANOVAs, one for each of the single dimensions investigated by the SPM battery.

The ANOVA findings indicate that chess training seems to primarily influence the ability to create a mental representation. This ability is measured by the SPM battery with a test in which the child must choose between a series of more or less abstract graphic representations (vignettes or diagrams) of the problem to be solved.

Another dimension in which the children in the experimental group exhibited significantly better performance than those in the control group is that of categorization, which investigates children's ability to extend their knowledge on the solution to a given problem to other similar problems. Finally, the children in the chess group achieved better results in the self-assessment dimension, showing a greater ability to objectively assess their problem-solving performance (Table 2.1).

2.3 Study Discussion and General Conclusion

As already pointed out, the present job market requires graduates in Science, Technology, Engineering, and Mathematics (STEM) subjects and it is well known that promoting children' mathematical abilities since primary school is crucial to increment the number of future high education STEM students. Educational researchers have explored various approaches to improve the efficacy of mathematical teaching, and teaching chess in schools is one of these approaches. Chess instruction has been recently proposed as an educational tool able to enhance children's cognitive and academic abilities and it has recently become part of the school curriculum in several countries.

Several studies have been carried out to demonstrate the benefits of chess instruction, especially for children's mathematical abilities. According to these studies, chess instruction may increase children's mathematical skills, because playing chess helps to shape children's way of thinking in a way that is particularly effective when the child must face mathematical problems (Trinchero & Sala, 2016).

Chess instruction seems to increase children's mathematical abilities, but some doubts remain on the efficacy of such practice. In fact, to evaluate the specific effect of chess instruction an ideal experiment is difficult to conduct, and many studies lack a proper experimental design (see Sala & Gobet, 2016 for review). A common methodological problem is that in most studies the experimental groups are already formed, for instance with children attending a chess club vs. children not attending, without the possibility of assigning participants randomly to different conditions.

To overcome this bias, in our work all the children were initially selected among chess novices and the participants were assigned to the experimental group played chess or to the control group in a random fashion.

The aim of this work was to observe the effects of chess training on general meta-cognitive abilities and on skills closely related to school performance (math and reading abilities). Two groups of children were compared: an experimental group that participated in a chess training and a control group that participated in a sports program.

For the study trial in which participants were asked to understand and remember written text informational content, no significant difference was found between the chess and control groups. This is an expected result: The researchers who have investigated the relationship between chess and verbal skills have not found relationships between these two abilities, the explanation probably lies in the fact that these two types of skills do not share common elements (see Sala & Gobet, 2016 for further discussion). However, it must be emphasized that in our work we have considered the abilities of understanding and retaining information that have more in common with chess from the point of view of the underlying cognitive abilities than the simple reading skills investigated in other studies; despite this, however, we did not find significant effects.

Significant differences between the two groups of children have emerged, instead, in the SPM test score assessing the ability to solve mathematical problems, particularly in the dimensions of problem representation, problem categorization, and self-evaluation. Chess training children are more capable of organizing information by creating a coherent representation of a problem and more able to categorize problems; they even demonstrate a greater capacity to extend their knowledge of a problem's structure to other similar problems to be able to solve them faster. The meta-analyses previously considered show that the domain of mathematics is the one that benefits most from chess training, leading most authors to think that these two areas involve common cognitive abilities (e.g., Sala & Gobet, 2016). The literature results suggest that playing chess allows children to develop skills that can be de-contextualized, such as problem-solving skills and the ability of identifying quantitative relationships—abilities that can be transferred to the domain of mathematics. Our findings, demonstrating a difference between the experimental and control groups only in the math problem-solving domain, confirm this hypothesis.

Furthermore, our results showed that the chess players have a better capacity, respect to the children in the control group, to self-evaluate their school performance. This is in line with Aciego, García, and Betancort (2012) findings that chess practice improved not only cognitive skills but also the socio-emotional sphere, especially the ability to self-evaluate.

In our data, as for the general meta-cognitive abilities involved in learning, no significant differences emerged between the two groups.

Some authors (Bart, 2014; Kazemi, Yektayar, & Abad, 2012; Sala et al., 2015; Trinchero, 2013) have proposed that chess can boost mathematical abilities not only due to the features in common between chess and mathematics but also to the general heuristics that chess players use during games. We did find an improvement of math ability. If the math improvement is due to a meta-cognitive improvement, as the above cited literature suggest, we would have had to find differences between the two groups also in their meta-cognitive skills. But this was not the case.

In our data, no significant differences emerged between the two groups in the different dimensions related to the QAS test assessing the general way in which they approach studying and some related sub-dimensions (e.g., motivation, strategic elaboration of the study material, ability to concentrate, and attitude toward schooling). No differences emerged between the two groups regarding children's beliefs about the effectiveness and the actual usage of study strategies. The studies that have explored the benefits of chess training on cognitive skills and school performance are scarce in the literature. To our knowledge, only one study, by Kazemi et al. (2012), has specifically considered meta-cognitive abilities. In their study, the authors found that the effects of chess training were closely related to meta-cognitive abilities linked to math problem-solving. However, the skills considered in our study are quite different from the ones explored by Kazemi and colleagues: They tested meta-cognitive skills specifically involved in math problem-solving, while we tested general meta-cognitive abilities applicable to any kind of subject of study. One potential explanation for our findings is that the dimensions tested in our study (e.g., motivation, attitude toward schooling, knowledge and usage of study strategies) are quite different from the skills acquired through chess practice, such as elaborating game plans. They require more than merely a simple transfer, but what Mestre (2005) defined as a far transfer, that is, a transfer between areas that are far from each other, and much more difficult to gain.

Although we have investigated only meta-cognitive abilities, our data seem to go in the direction of the studies that have investigated more general cognitive abilities, for example, Scholz et al. (2008), who did not find an effect of training with chess on focused attention. A meta-analysis by Burgoyne et al. (2016) considered 19 studies that related cognitive abilities to chess skills and found a positive correlation between general cognitive abilities and chess practice that seemed, however, to be mediated by age and chess skill level. In particular, the younger and more inexperienced participants were, the greater the correlation with cognitive abilities. But it must be noted that in the meta-analysis, the percentage of explained variance of cognitive abilities on chess performance is on average 6%, a fairly low value. In the already cited meta-analysis done by Sala and Gobet (2016), the authors underlined how in the works considered, the effect size is not large enough to strongly support the hypothesis that the improvement in cognitive abilities is due to the chess itself. Moreover, the authors highlighted that most studies considered did not consider the placebo effect: very often the control groups were not involved in other activities that could enhance their cognitive abilities. Therefore, it is difficult to determine whether the (already small) effect found was due strictly to playing chess or merely to being involved in a stimulating activity. The results of Sala and Gobet's (2016) meta-analysis regarding cognitive abilities therefore seem to support the difficulty in the transfer of chess-related abilities to general cognition. In sum, our findings confirm that chess practice can be useful for primary school children to enhance their mathematical problem-solving abilities, learning-related, and self-evaluation abilities, but that chess practice alone seems to be of relatively scarce use to improve more general and transferable meta-cognitive skills. It is possible that math-specific meta-cognitive skills and general meta-skills are in fact less related than one might think and needed to be stimulated differently.

We think, in this regard, that is useful to think into the perspective of meta-cognitive teaching/learning model and that explicit teaching of meta-cognitive heuristics applied to chess could lead to an increase in general meta-cognitive abilities. The main characteristic of meta-cognitive teaching is to favor strategic thinking, favoring, so, the acquisition of those transversal skills allowing to adapt to different and new situations, beyond the structured learning contexts (Borkowski, Carr, Rellinger, & Pressley, 1990).

The suggestion, worth of further study, is, so, to teach chess at school with an explicitly meta-cognitive didactics, leading the children to reflect on their cognitive processes (e.g., attention and memorization) and on the strategies to be applied and adjusted during the games. This idea is sustained by studies (e.g., Trinchero & Sala, 2016) showing that combining chess practice with explicit teaching of chess-problem-solving heuristics has proved to be more effective than chess rules teaching alone to improve meta-cognitive skills.

The operational suggestion emerging from this work is that insert chess in the school curriculum can be very useful, primarily to enhance math abilities and also, in the perspective of a meta-cognitive teaching, to make children a "good strategy user" (Borkowski & Muthukrishna, 1992).

Our work has the merit of having used a sample taken from a primary school where the children were randomly assigned to an experimental group that played chess and to a control group. Often the works in this area use samples drawn from populations already selected a priori, such as children who are part of real chess clubs. In line with literature recommendations, the time extent of the training was 30 h, and the control group was engaged in an alternative activity.

However, there are several limitations to note. The main limitation is that the participants did not undergo a pre-test to assess whether differences existed before the treatment, given that we had to limit the number of testing sessions to meet a school demand. To overcome this limitation, we used a large sample, randomly drawn from the whole school population and not pre-selected. We tested classes in which the same teachers were regularly involved and asked the teachers, in advance, to evaluate whether the cognitive and academic levels of the children randomly assigned to the two groups were comparable. Nonetheless, we cannot exclude the possibility that one of the groups would start higher than the other in some of the dimensions considered. This is a serious limitation, even if the lack of a rigorous test-retest methodology is often observed in this type of work (see Sala & Gobet, 2016 for further discussion).

Our findings suggest that the topic of the transfer of skills gained through chess practice to the academic domain is worth further investigation. Future research should be undertaken using a pre-posttest experimental design and a longitudinal approach to investigate the effects of chess practice over time.

Future research should compare chess simple practice with training that integrates chess rules teaching with heuristics' teaching to control if the effects of training on both meta-cognitive skills in mathematics and general meta-cognitive skills can favor the desired transfer from the chess-specific domain to a more general domain.

Acknowledgments The authors would like to thank the principal of the school where the data collection took place. Special thanks go to the children, families, and teachers who participated in the study.

References

Aciego, R., García, L., & Betancort, M. (2012). The benefits of chess for the intellectual and social-emotional enrichment in schoolchildren. *The Spanish Journal of Psychology, 15*(2), 551–559.

Anderson, J. R. (1990). *Cognitive psychology and its implication* (3rd ed.). New York: Freeman.

Bart, W. M. (2014). On the effect of chess training on scholastic achievement. *Frontiers in Psychology, 5*, 762.

Binev, S., Attard-Montalto, J., Deva, N., Mauro, M., & Takkula, H. (2011). Declaration of the European Parliament, 0050/2011.

Borkowski, J. G., Carr, M., Rellinger, E., & Pressley, M. (1990). Self-regulated cognition: Interdependence of metacognition, attributions, and self-esteem. In *Dimensions of thinking and cognitive instruction* (pp. 53–92). Hillsdale, NJ: Erlbaum.

Borkowski, J., & Muthukrishna, N. (1992). Moving metacognition into the classroom: "Working models" and effective strategy teaching. In M. Pressley, K. R. Harris, & J. T. Guthrie (Eds.), *Promoting academic competence and literacy in school* (pp. 477–501). Cambridge: Academic Press.

Burgoyne, A. P., Sala, G., Gobet, F., Macnamara, B. N., Campitelli, G., & Hambrick, D. Z. (2016). The relationship between cognitive ability and chess skill: A comprehensive meta-analysis. *Intelligence, 59*, 72–83.

De Beni, R., Moè, A., & Cornoldi, C. (2003). *AMOS 8-15. Abilità e motivazione allo studio: Prove di valutazione e orientamento* (Vol. 30). Trento, IT: Edizioni Erickson.

De Groot, A. D. (1966). Perception and memory versus thought: Some old ideas and recent findings. In B. Kleinmuntz (Ed.), *Problem solving: Research, methods and theory*. New York: Wiley.

Ericsson, K. A., & Charness, N. (1994). Expert performance: Its structure and acquisition. *American Psychologist, 49*(8), 725.

Ericsson, K. A., & Kintsch, W. (1995). Long-term working memory. *Psychological Review, 102*(2), 211.

Gobet, F., & Campitelli, G. (2006). Educational benefits of chess instruction: A critical review. In *Chess and education: Selected essays from the Koltanowski conference* (pp. 124–143). Chess Program at the University of Texas at Dallas, Dallas, TX.

Grotzer, T. A., & Perkins, D. N. (2000). Teaching intelligence: A performance conception. In R. J. Sternberg (Ed.), *Handbook of intelligence* (pp. 492–515). Cambridge, UK: Cambridge University Press. https://doi.org/10.1017/CBO9780511807947.023

Holding, D. H., & Reynolds, R. I. (1982). Recall or evaluation of chess positions as determinants of chess skill. *Memory & Cognition, 10*(3), 237–242.

Kazemi, F., Yektayar, M., & Abad, A. M. B. (2012). Investigation the impact of chess play on developing meta-cognitive ability and math problem-solving power of students at different levels of education. *Procedia-Social and Behavioral Sciences, 32*, 372–379.

Lucangeli, D., Tressoldi, P. E., & Cendron, M. (2003). *SPM-Test delle abilità di soluzione dei problemi matematici*. Trento, IT: Edizioni Erickson.

Mestre, J. P. (2005). *Transfer of learning from a modern multidisciplinary perspective*. Greenwich, CT: Information Age.

Ritchie, S. J., Bates, T. C., & Deary, I. J. (2015). Is education associated with improvements in general cognitive ability, or in specific skills? *Developmental Psychology, 51*(5), 573.

Sala, G., Burgoyne, A., Macnamara, B. N., Hanbrick, Z., Campitelli, G., & Gobet, F. (2017). Checking the "academic selection" argument. Chess players outperform non-chess players in cognitive skills related to intelligence. *Intelligence, 61*, 130–139. https://doi.org/10.1016/j.intell.2017.01.013

Sala, G., Foley, J. P., & Gobet, F. (2017). The effects of chess instruction on pupils' cognitive and academic skills: State of the art and theoretical challenges. *Frontiers in Psychology, 8*, 238.

Sala, G., & Gobet, F. (2016). Do the benefits of chess instruction transfer to academic and cognitive skills? A meta-analysis. *Educational Research Review, 18*, 46–57.

Sala, G., & Gobet, F. (2017). Does chess instruction improve mathematical problem-solving ability? Two experimental studies with an active control group. *Learning & Behavior, 45*(4), 414–421.

Sala, G., Gorini, A., & Pravettoni, G. (2015). Mathematical problem-solving abilities and chess: an experimental study on young pupils. *Sage Open*. https://doi.org/10.1177/2158244015596050.

Scholz, M., Niesch, H., Steffen, O., Ernst, B., Loeffler, M., Witruk, E., et al. (2008). Impact of chess training on mathematics performance and concentration ability of children with learning disabilities. *International Journal of Special Education, 23*, 138–148.

Sternberg, R. J. (Ed.). (2000). *Handbook of intelligence*. Cambridge, UK: Cambridge University Press.

Thorndike, E. L., & Woodworth, R. S. (1901). The influence of improvement in one mental function upon the efficiency of other functions. *Psychological Review, 9*, 374–382.

Trinchero, R. (2013). Can chess training improve Pisa scores in mathematics? An experiment in Italian primary schools. Kasparov Chess Foundation Europe. Retrieved from www.kcfe.eu/sites/default/files/Trinchero_KCFE.pdf

Trinchero, R., & Sala, G. (2016). Chess training and mathematical problem-solving: The role of teaching heuristics in transfer of learning. *Eurasia Journal of Mathematics, Science & Technology Education, 12*(3), 655–668.

Chapter 3
Developing Computational Thinking in Early Childhood Education: A Focus on Algorithmic Thinking and the Role of Cognitive Differences and Scaffolding

Kyriakoula Georgiou and Charoula Angeli

3.1 Introduction

Computational thinking is defined as "*the process of recognizing aspects of computation in the world that surrounds us, and applying tools and techniques from computer science to understand and reason about both natural and artificial systems and processes*" (Furber, 2012, p. 29). A large body of literature emphasizes the importance of teaching computational thinking in educational settings in order to better prepare students to cope with the challenges of the twenty-first century (Berry, 2011; Bers, Flannery, Kazakoff, & Sullivan, 2014; Ching, Hsu, & Baldwin, 2018; Kazakoff, Sullivan, & Bers, 2013; Lye & Koh, 2014; Yadav, Hong, & Stephenson, 2016).

Despite the fact that many studies have been undertaken during the last decade in order to examine the teaching of computational thinking skills, most of them were conducted within the context of higher education settings (e.g., Astrachan, Hambrusch, Peckham, & Settle, 2009; Howland, Good, & Nicholson, 2009; Kazimoglu, Kiernan, Bacon, & Mackinnon, 2012). Very few studies investigated the development of computational thinking skills in educational settings with younger students (Angeli & Valanides, 2020; Bers et al., 2014), and, therefore, the field urgently needs more research toward this direction in order to remedy for the gap in the literature and better inform the efforts of curriculum developers who are striving toward integrating computational thinking in school curricula (Román-González, Pérez-González, & Jiménez-Fernández, 2017).

In the educational technology literature, there is a major body of research that shows consistently the effects of cognitive type, and in particular, field dependence/

K. Georgiou · C. Angeli (✉)
Department of Education, University of Cyprus, Nicosia, Cyprus
e-mail: cangeli@ucy.ac.cy

D. Ifenthaler et al. (eds.), *Balancing the Tension between Digital Technologies and Learning Sciences*, Cognition and Exploratory Learning in the Digital Age,
https://doi.org/10.1007/978-3-030-65657-7_3

independence (FDI) on young students' ability to process and complete cognitive tasks (Angeli & Valanides, 2004a, 2004b, 2013; Angeli & Valanides, 2013; Chen & Macredie, 2004; Evans, Richardson, & Waring, 2013). FDI reflects the ways in which individuals perceive and process information from their surrounding environment (Evans et al., 2013; Witkin, Moore, Goodenough, & Cox, 1977). Witkin et al. (1977) conceptualized FDI as a bipolar construct with two distinct modes of perception, namely, Field Dependence (FD) and Field Independence (FI). FI learners are characterized as analytical and visually perceptive, whereas FD learners are referred to as global and not visually perceptive (Hall, 2000).

Scaffolding is also another important variable to consider since current research evidence strongly indicates that it plays a significant role in young learners' cognitive performance especially when children learn with technology tools (Azevedo & Hadwin, 2005). The concept of scaffolding is grounded within the socio-cultural theory of Vygotsky (1978), and it is conceptualized as cognitive support provided to learners in order to enable them to complete a task that could not otherwise complete by themselves. Scaffolding provision is of great importance especially for young students, because in its absence they may not be able to complete a task, and, thus, fail to develop important cognitive skills (Belland, 2014; Chen, Kao, & Sheu, 2003; Mashburn, Justice, Downer, & Pianta, 2009; Myhill & Warren, 2005; Rodgers, 2005; Van de Pol, Volman, & Beishuizen, 2010; Van Merriënboer, Kirschner, & Kester, 2003).

To this end, the study herein set out to investigate the effects of scaffolding on pre-primary education children's computational thinking with the context of learning with robotics activities, and the extent to which cognitive type affected their performance both as a main effect and in interaction with the cognitive scaffolds.

3.2 Theoretical Background

Computational thinking is a fundamental concept of computer science arising from its basic principles and practices, and drawing on methods from various disciplinary contexts (Fluck et al., 2016; Sengupta, Kinnebrew, Basu, Biswas, & Clark, 2013; Wing, 2006). Despite the fact that currently there is not one unanimous definition of computational thinking, after a systematic examination of what is known in the literature, Grover and Pea (2018) concluded that researchers have come to accept that computational thinking is a thought process that utilizes among other elements algorithmic thinking. Algorithmic thinking is a problem-solving skill related to devising a step-by-step solution to a problem, and it involves putting actions in the correct sequence.

During the last decade, the research community has embraced educational robotics with genuine enthusiasm as an approach for teaching computational thinking to pre-primary education students (Bers, 2008; Bers et al., 2014; Kazakoff & Bers, 2012). Educators use educational robotics in order to engage young learners in active and playful learning activities through building and programming tangible robotic devices (Angeli & Valanides, 2020).

Angeli and Valanides (2020) reported statistically significant differences in the initial and final evaluation of young learners' computational thinking, reporting learning benefits for 50 children. Analytically, they studied the effects of learning with a floor robot on the development of children's computational thinking using two different scaffolding techniques. The results showed a statistically significant interaction effect between gender and scaffolding technique reporting that boys had better learning outcomes during learning with kinesthetic activities, while girls had better learning outcomes when they collaborated with others to solve problems.

Papadakis, Kalogiannakis, and Zaranis (2016) conducted a study with 43 children who worked with ScratchJr to create various artifacts. They concluded that activities with ScratchJr could be easily integrated in early childhood education curricula and that ScratchJr could be an important teaching tool for developing young children's computational thinking skills.

Along the same line of reasoning, Bers et al. (2014) investigated the use of innovative new technology in early elementary school. In particular, they focused on computer programming and robotics with the goal of understanding what was developmentally appropriate for young children. Within this context, they developed and piloted an innovative programming environment called CHERP (Creative Hybrid Environment for Robotic Programming), a hybrid tangible/graphical computer language, in eight classrooms and numerous afterschool programs, reaching approximately 240 children. The researchers concluded that the curriculum helped children discover aspects of educational robotics and programming while developing their computational thinking skills.

Scaffolding becomes important when examining young children's learning with technological tools and educational robotics (Angeli & Valanides, 2020). The construct of scaffolding is directly associated with the socio-cultural theory of Vygotsky (1978), and the notion of "zone of proximal development," which is defined as "*the distance between the actual development level as determined by independent problem solving and the level of potential development as determined through problem solving under adult guidance or in collaboration with more capable peers*" (Vygotsky, 1978, p. 86). Scaffolding can take the form of tools and strategies, computer tutors, more capable peers, or even animated pedagogical agents assisting learners to comprehend and conceptualize meanings emerging during teaching that are far beyond their mental and cognitive capabilities (Graesser, Wiemer-Hastings, Wiemer-Hastings,, & Kruez, 2000). It is reported that in computer-based learning environments that lack scaffolding, learners exhibit low ability in regulating their learning and fail to acquire deep concept understanding (Azevedo & Cromley, 2004; Hill & Hannafin, 1997; Land & Greene, 2000).

Furthermore, studies in the field of educational technology also show that learning with scaffolds differentially affects student learning when individual cognitive differences exist. Research shows that cognitive type is an important variable to consider when children are engaged in technology-enhanced learning (Angeli &

Valanides, 2013; Jonassen, 1992; Riding, 1997, 2002). Learners' cognitive type directly influences the way individuals think, perceive, remember, analyze, organize and represent information from their environment (Burnett, 2010; Morgan, 1997; Witkin et al., 1977). Field Dependence/Independence (FDI) is the most extensively studied cognitive type in the educational technology literature (Angeli & Valanides, 2013; Jonassen, 1992; Pithers, 2002). Witkin et al. (1977) conceptualized FDI as a bipolar construct with two distinct modes of perception, namely, FD and FI (Morgan, 1997). The core difference between FI and FD learners is the way they perceive and process visual clues and complex representations (Angeli & Valanides, 2004a, 2004b; Witkin et al., 1977; Canelos, Taylor, & Gates, 1980; Davis, 1991; Morgan, 1997; Saracho, 2000; Snowman & Biehler, 1993).

Succinctly, according to Witkin et al. (1977), FDI is the "*the extent to which the person perceives part of a field as discrete from the surrounding field as a whole, rather than embedded in the field; or the extent to which the organization of the prevailing field determines perception of its components*" (p. 6–7). FI learners have the ability of extracting information from a context and creating a mental model of a problem before solving it (Angeli & Valanides, 2004a; Dufresne & Turcotte, 1997; Jonassen, 1992; Pithers, 2002; Saracho, 1989, 2000). They are characterized as analytical and visually perceptive exhibiting an analytical approach to problem solving (Saracho & Spodek, 1981; Zhang & Sternberg, 2009). At the other end of the dipole, FD learners function less autonomously and they depend on a more knowledgeable other to guide their problem solving and decision making (Hergovich, 2003).

Even though cognitive type has not yet been extensively studied for the age group of pre-primary students, there is preliminary evidence showing the importance of it on young learners' cognitive performance. For example, Guisande, Tinajero, Cadaveira, and Páramo (2012) investigated the relationship of visuospatial abilities and FDI for 149 children aged between 8 and 11 years old. Their findings reported that FI children outperformed the FD children on verbal working memory tasks, and complex cognitive tasks. Similarly, Nicolaou and Xistouri (2011) reported statistically significant findings in favor of FI learners in terms of their ability to pose and solve problems.

To this end, the authors investigated the effectiveness of two scaffolding techniques on FD and FI children's computational thinking during problem-solving activities with the Bee-Bot and hypothesized that scaffolding and FDI will both play a significant role on children's problem-solving performance. The research questions are stated as follows:

1. Are there any statistically significant differences between FD and FI learners' computational thinking?
2. Are there any statistically significant differences on the effects of the scaffolding techniques on learners' computational thinking?
3. Does scaffolding differentially affect FD and FI learners' computational thinking?

3.3 Method

3.3.1 Participants

One hundred and eighty children (82 females and 98 males) aged between 5 and 6 years old from nine public pre-primary schools in a Southern European country participated in the study. Parents/guardians signed consent forms for each child prior to participating in the research study. Children who took part in the study had no special needs or previous experience with the programmable floor robot that was used for the purposes of this research, or educational robotics in general. It is also worth mentioning that due to the fact that students' cognitive type had to be screened through the administration of a research instrument, the total number of students who participated in the study was greater than 180, because of the difficulty in identifying students with type FI. In particular, of the total number of 425 students who were screened for FDI, the authors were able to randomly select and analyze research data from 180 of them.

Participants were first classified as FD or FI learners based on their scores on the Children's Embedded Figures Test, and, subsequently, FD and FI learners were randomly assigned to two experimental groups and a control group. In order to eliminate selection bias, participants were randomized into groups with the block randomization method.

3.3.2 Research Instruments

3.3.2.1 Children's Embedded Figures Test

The Children's Embedded Figures Test (CEFT) (Karp & Konstadt, 1971) was administered individually to assess participants' FDI. The CEFT has an internal reliability Cronbach's $\alpha = 0.87$ and is specially designed to identify the cognitive type of children aged from 5 to 9 years old. It is often applied in neuropsychological assessments as an indication of perceptive ability and the ability to break down an organized visual field in order to extract a part from the whole (Guisande et al., 2012). The CEFT includes 38 complex figures composed of smaller and simpler figures like a triangle (△) and a small house (⌂). Participants are instructed to discover a simpler figure embedded in a complex figure as shown in Fig. 3.1. Each participant has 30 seconds at his disposal to identify each simpler shape. The total administration time for the test is 20 min. One point is given for each shape correctly recognized, and the maximum score on the test is 20 points.

Fig. 3.1 An example from the CEFT test

Fig. 3.2 The floor robot Bee-Bot

3.3.3 Research Materials

3.3.3.1 Bee-Bot

The Bee-Bot, a programmable floor robot (see Fig. 3.2), was used in the study. The Bee-Bot is suitable for children aged between 3 and 8 years old (Highfield, 2010; Janka, 2008; Misirli & Komis, 2014). It consists of seven keys that direct Bee-Bot to move forward and backward, to turn left and right by 90°, to clear its memory, to pause and to execute a sequence of commands. The Bee-Bot can store a maximum of 40 commands in its memory.

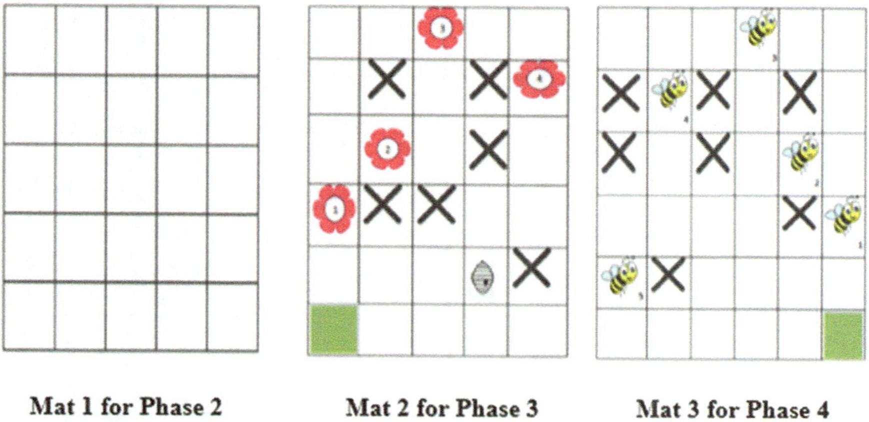

Fig. 3.3 The three Bee-Bot mats used in the research

3.3.3.2 Bee-Bot Mats

The Bee-Bot mats are surfaces made of durable plastic. Each surface is organized into squares of 15 cm × 15 cm because the Bee-Bot can move in 15 cm increments. The researchers designed and created three different mats that were used in different research phases of the study (see Fig. 3.3). The Bee-Bot mat for Phase 2 was a plain surface that included only squares of 15 cm × 15 cm and it was used to teach the basic commands of the Bee-Bot. The Bee-Bot mat that was designed for Phase 3 included four numbered flowers, a green square, a beehive, and several X marks. The green square represented the starting point of the Bee-Bot's journey to gather the pollen from the flowers. The route of the Bee-Bot included the stations from one flower to another and ended at the Bee-Bot's destination that was the beehive. The X mark denoted a square that the Bee-Bot could not pass through. Lastly, the third mat for Phase 4 included as routes Bee-Bot's friends invited to a party. The children were instructed to develop algorithms so the Bee-Bot could execute all paths described in the problem-solving tasks. At the beginning, these algorithms constituted simple sequences of commands, but gradually they were developed into more complex and longer sequences of commands.

3.3.3.3 Scaffolding

The authors used two scaffolding techniques, namely Scaffold A and Scaffold B, to facilitate children's efforts in constructing algorithms as external representations to depict Bee-Bot's movements on the mats. Scaffold A, shown in Fig. 3.4, included a model of the real mat on an A4 piece of paper and small laminated cards with the Bee-Bot commands. Children were able to use the cards and place them on the model mat to think about the algorithm without programming the Bee-Bot. After creating the algorithm, they tested it on the real mat by programming Bee-Bot.

Fig. 3.4 Scaffold A

Fig. 3.5 Scaffold B

Scaffold B, shown in Fig. 3.5, used small laminated cards with the Bee-Bot commands and a board that the cards could be placed on. The idea was to place the cards in a sequence to create an algorithm. After placing the cards in sequence, the children were able to test the algorithm by programming the Bee-Bot and testing the correctness of the algorithm on the real mat.

The difference between Scaffold A and B was that Scaffold A used a more concrete approach to scaffold children's efforts, because it enabled them to model and simulate the algorithm first on a smaller mat and then test the algorithm on the real mat. Scaffold B used a more abstract approach where students needed to depict visually the algorithm using only the cards without simulating it first on a smaller mat.

3.3.3.4 Problem-Solving Tasks

Learning by solving problems is regarded an appropriate strategy for teaching computational thinking skills (Aho, 2012; Wing, 2006). Accordingly, children's interactions with a robot is considered an important approach for teaching children how to think computationally (Flowers & Gossett, 2002; Williams, Ma, Prejean, Ford, & Lai, 2007). In the context of this research study, participants were engaged in a series of problem-solving tasks to program the Bee-Bot to move from one point on the mat to another. More specifically, children were asked to solve three problem-solving tasks, of increased difficulty and complexity, with the Bee-Bot.

The first problem-solving scenario introduced children to the Bee-Bot and its programming commands. The researcher engaged the children in a sequence of systematic problem-solving activities. Specifically, the scenario consisted of thirteen subtasks, and one subtask at a time was read out loud by the researcher, before proceeding to the next one. The subtasks guided the children to program the Bee-Bot to move out of the beehive and collect pollen from the flowers before returning home (beehive). The subtasks were specifically designed to include sequences of actions involving combinations of all four directional keys. Simple sequences included tasks that made use of the keys "move forward" or "move backward". More complex sequences included tasks that used the commands "move forward" and "move backward" in different combinations. Advanced sequences involved the commands "turn right" or "turn left" in different combinations with the other two directional keys (i.e., move forward, move backward). The length of a sequence varied from one to four commands. The second problem-solving task consisted of five subtasks, and the third task comprised of five subtasks. The subtasks were designed and presented to each child individually at increasing levels of complexity (Armoni & Gal-Ezer, 2014). Children were allowed 20 min for each problem-solving task.

3.3.4 Research Procedures

All research procedures were implemented for each participant individually since large group instruction often conceals the individual needs of learners. In addition, the provision of scaffolding tools is particularly effective in personalized learning environments (McNeill & Krajcik, 2009; Puntambekar & Kolodner, 2005; Saye & Brush, 2002; Tabak, 2004). Analytically, the research procedures comprised of four research phases. During the first phase, the researchers administered the CEFT to each child individually. Participants were classified into FD and FI learners by using the median split method that is commonly recommended in FDI research (Angeli & Valanides, 2004b). This method divides the upper half of the FDI scores as FI and the lower half as FD. The median for the FDI scores, in the study reported herein was 13 points with a range from 6 to 25 points. Children were first classified as FD or FI learners, and, subsequently, each group of FD and FI learners was randomly assigned into the two experimental groups (Scaffold A and Scaffold B), and the control group (No Scaffold).

The second phase allowed children to explore Bee-Bot and get familiar with its commands. During the third phase the two scaffolding techniques were systematically used to facilitate children's development of computational thinking. Lastly, during the fourth phase the scaffolds were withdrawn, and a post assessment was performed to assess children's computational thinking skills.

3.3.5 The Assessment of Computational Thinking

There is a dearth of suitable research instruments and techniques for assessing young children's computational thinking (Angeli & Valanides, 2020; Papadakis et al., 2016). Standardized pre and post tests are not applicable to every activity and are not suitable for young children (Giannakos, Papavlasopoulou, & Sharma, 2020). Interaction analysis is ideal for full class and verbal interactions (Jordan & Henderson, 1995), therefore, not applicable to individual learning environments with technological tools. Eye tracking was reported as not suitable for small children, because they can easily remove the eye tracking device, need extra time to get familiarized with the device, and they show low tolerance to the device's temperature increase (Giannakos et al., 2020). For these reasons, researchers prefer to devise their own rubrics in assessing computational thinking skills, and this is what the authors herein also chose to do (Sherman & Martin, 2015).

3.3.5.1 Computational Thinking Scoring Rubric

The researchers developed inductively, based on children's answers (i.e., sequences of commands), a rubric for assessing computational thinking. The authors first wrote down all the attempts that children made for all problem-solving tasks. For example, X's performance on Subtask2 in Phase 3 was recorded as follows:

X's first attempt (unsuccessful): MOVE BACKWARD-TURN RIGHT
X's second attempt (unsuccessful): MOVE BACKWARD-TURN RIGHT-MOVE FORWARD
X's third attempt (successful): MOVE BACKWARD-TURN RIGHT-TURN RIGHT-MOVE FORWARD

Then, the researchers collected all possible answers from all one hundred and eighty students for Subtask2 and created a table. Table 3.1 shows the number of possible attempts children made to solve the task. If for example, the maximum number of attempts made to find the correct answer was three, then the maximum score is three for finding the correct answer during the first attempt, two if it took two attempts and one if it took three attempts.

Table 3.1 Subtask2: Measurement of computational thinking

Code	Description	Score
3	Attempt 1: Success	3
2	Attempt 2: Success	2
1	Attempt 3: Success	1

Table 3.2 Descriptive statistics of children's computational thinking in Phase 3 and Phase 4 for each scaffold and cognitive type

Phase 3			
	Mean	**SD**	***N***
Scaffold A			
FD	234.93	48.31	30
FI	249.43	9.8	30
Total	241.18	35.33	60
Scaffold B			
FD	224.40	38.48	30
FI	231.83	27.44	30
Total	228.12	33.35	60
No scaffold (Control group)			
FD	163.00	55.31	30
FI	179.33	44.20	30
Total	171.17	50.32	60

Phase 4			
	Mean	**SD**	**N**
Scaffold A			
FD	165.77	63.95	30
FI	207.17	34.57	30
Total	186.47	55.08	60
Scaffold B			
FD	159.13	59.88	30
FI	201.63	46.36	30
Total	180.38	57.26	60
No scaffold (Control group)			
FD	185.77	56.69	30
FI	203.87	43.30	30
Total	194.82	50.84	60

3.4 Results

Table 3.2 presents children's descriptive statistics in regards to their performance on the problem-solving tasks in Phases 3 and 4. The descriptive statistics in Table 3.2 show that FI participants had better performance on all problem-solving tasks in both phases.

A 2 × 3 analysis of variance was performed and revealed that type of scaffold (F (2, 174) = 52.60, $p < 0.01$) was found to be statistically significant. Post-hoc comparisons showed that both experimental groups outperformed the control group. In addition, the analysis revealed that there were no statistically significant differences between the FI and FD participants.

A second 2 × 3 analysis of variance was performed for the fourth research phase during which the scaffolds were withdrawn. It was found that only FDI was a significant main effect (F (1, 174) = 19.38, $p < 0.01$) in favor of the FI children.

3.5 Discussion

In general, FI children outperformed FD children on all problem-solving tasks. This finding is in line with previous studies that documented the better problem-solving performance of FI learners in comparison with FD learners (e.g., Angeli & Valanides, 2004a, 2013; Angeli & Valanides, 2013; Chen & Macredie, 2004). This can be justified by the fact that FI learners are able to create internal representations of the various aspects of a problem and use them in a procedural way to guide their problem-solving process (Angeli & Valanides, 2020).On the other hand, FD learners perform better when they are guided by external reference cues (Dufresne & Turcotte, 1997). As a result, FI and FD learners need different kinds of scaffolds to assist them during problem solving (Canelos et al., 1980; Chen & Macredie, 2004; Davis, 1991; Morgan, 1997; Saracho, 2000). The quantitative results of the third research phase showed that scaffolding benefited FD learners, because the external scaffolds provided them with visualizations of the sequences of the commands facilitating this way the formation of the internal representation of the algorithm (Pape & Tchoshanov, 2001).

The present study also shows that children in the control group had a statistically significant lower performance than all other children because of the absence of scaffolding tools in the control group. This finding further strengthens the argument about the importance of scaffolding during children's learning with technology tools. Interestingly, FI and FD learners assigned to the control group had statistically significant differences from their counterparts in the experimental groups indicating that all learners, irrespective of cognitive type can benefit from scaffolding during learning in a new context (Evans et al., 2013). During Phase 4, when the scaffolds were removed, FI learners outperformed all other learners, signifying the effects of cognitive type in the absence of cognitive scaffolding.

All in all, the findings of the study provide empirical evidence for the importance of field type and scaffolding on young children's computational thinking during learning with educational robotics. The findings of this study strongly indicate that both FI and FD learners can benefit from scaffolding in order to succeed in problem-solving tasks with robotics. Consequently, teachers need to consider learners' cognitive type, adapt learning to the basis of children's cognitive differences (NAEYC, 2012) and provide appropriate scaffolds in order to ensure that all learners can equally learn with robotics (Thomas, Ratcliffe, Woodbury, & Jarman, 2002). The instructional tasks and scaffolding tools being designed and used in the current research can be potentially useful in various ways since they aid in the design of curriculum materials as well, they constitute applicable knowledge for teachers.

3.6 Future Research Directions

Our analyses provide critical insight into the association between the trend of field type attrition with robotics activities and the development of computational thinking. Nonetheless, some limitations should also be considered. The findings obtained during the last research phase, when the scaffolding tools were withdrawn, reported that the differences on the scores on the assessment of the computational thinking among the experimental and control groups were not statistically significant. This result can be attributed to various reasons. Firstly, the duration and the number of the lessons proved to be inadequate to enable the transfer of knowledge as other researchers concur (e.g., Bers et al., 2014). In regards of children's computational thinking development, it has been established by researchers that developing cognitive skills in young children requires sustained and immersive effort (Bers et al., 2014). Lastly, to trigger the augmentation of the pedagogical gains of the scaffolding is essential that the scaffolding to gradually fade out (Van de Pol et al., 2010). Therefore, future research effort should focus on expanding the duration of the interventions.

In addition, future research directions can benefit from investigations taking place in the authentic context of a school classroom. In a classroom context, it can be easily examined the extent to which group-based activities in collaboration with other school children can benefit the development of computational thinking skills, and whether in this context field type makes a difference. In addition, it will be of utmost importance to examine when a scaffold can be removed in order to maximize learning benefits for all children.

References

Aho, A. V. (2012). Computation and computational thinking. *Computer Journal, 55*(7), 832–835.

Angeli, C., & Valanides, N. (2004a). Examining the effects of text-only and text-visual instructional materials on the achievement of field-dependent and field independent learners during problem-solving with modeling software. *Educational Technology Research and Development, 52*(4), 23–36.

Angeli, C., & Valanides, N. (2004b). The effect of electronic scaffolding for technology integration on perceived task effort and confidence of primary student teachers. *Journal of Research on Technology in Education, 37*(1), 29–43.

Angeli, C., & Valanides, N. (2013). Using educational data mining methods to assess field-dependent and field-independent learners' complex problem solving. *Educational Technology Research & Development, 61*(3), 521–548.

Angeli, C., & Valanides, N. (2020). Developing young children's computational thinking with educational robotics: An interaction effect between gender and scaffolding strategy. *Computers in Human Behavior, 105*, 105954. Retrieved August 10, 2019, from https://www.sciencedirect.com/science/article/pii/S0747563219301104?via%3Dihub

Armoni, M., & Gal-Ezer, J. (2014). Early computing education: Why? what? when? who? *ACM Inroads, 5*(4), 54–59.

Astrachan, O., Hambrusch, S., Peckham, J., & Settle, A. (2009). The present and future of computational thinking. *ACM SIGCSE Bulletin, 41*(1), 549–550.

Azevedo, R., & Cromley, J. G. (2004). Does training on self-regulated learning facilitate students' learning with hypermedia? *Journal of Educational Psychology, 96*(3), 523.

Azevedo, R., & Hadwin, A. F. (2005). Scaffolding self-regulated learning and metacognition–implications for the design of computer-based scaffolds. *Instructional Science, 33*(5), 367–379.

Belland, B. R. (2014). Scaffolding: Definition, current debates, and future directions. In *Handbook of research on educational communications and technology* (pp. 505–518). New York, NY: Springer.

Berry, D. M. (2011). The computational turn: Thinking about the digital humanities. *Culture Machine, 12*, 1–22.

Bers, M. U. (2008). *Blocks, robots and computers: Learning about technology in early childhood*. New York: Teacher's College Press.

Bers, M. U., Flannery, L., Kazakoff, E. R., & Sullivan, A. (2014). Computational thinking and tinkering: Exploration of an early childhood robotics curriculum. *Computers & Education, 72*, 145–157.

Burnett, W. C. (2010). Cognitive style: A meta-analysis of the instructional implications for various integrated computer enhanced learning environments. Doctoral dissertation, Indiana University of Pennsylvania.

Canelos, J., Taylor, W. D., & Gates, R. B. (1980). The effects of three levels of visual stimulus complexity on the information processing of field-dependents and field-independents when acquiring information for performance on three types of instructional objectives. *Journal of Instructional Psychology, 7*(2), 65–70.

Chen, Y. S., Kao, T. C., & Sheu, J. P. (2003). A mobile learning system for scaffolding bird watching learning. *Journal of Computer Assisted Learning, 19*(3), 347–359.

Chen, S. Y., & Macredie, R. D. (2004). Cognitive modeling of student learning in web-based instructional programs. *International Journal of Human-Computer Interaction, 17*(3), 375–402.

Ching, Y. H., Hsu, Y. C., & Baldwin, S. (2018). Developing computational thinking with educational technologies for young learners. *TechTrends, 62*(6), 563–573.

Davis, J. K. (1991). Educational implications of field dependence independence. In S. Wapner & J. Demick (Eds.), *Field dependence independence: Cognitive style across the life span* (pp. 149–176). Mahwah, NJ: Lawrence Erlbaum Associates.

Dufresne, A., & Turcotte, S. (1997). Cognitive style and its implications for navigation strategies. In B. du Boulay & R. Mizoguchi (Eds.), Artificial intelligence in education knowledge and media learning system. *Proceedings of the 8th World Conference on Artificial Intelligence in Education*, (pp. 287–293). Amsterdam, Netherlands.

Evans, C., Richardson, J. T., & Waring, M. (2013). Field independence: Reviewing the evidence. *British Journal of Educational Psychology, 83*(2), 210–224.

Flowers, T. R., & Gossett, K. A. (2002). Teaching problem solving, computing, and information technology with robots. *Journal of Computing Sciences in Colleges, 17*, 45–55.

Fluck, A., Webb, M., Cox, M., Angeli, C., Malyn-Smith, J., Voogt, J., & Zagami, J. (2016). Arguing for computer science in the school curriculum. *Educational Technology and Society, 19*(3), 38–46.

Furber, S. (2012). *Shut down or restart? The way forward for computing in UK schools*. London: The Royal Society.

Giannakos, M. N., Papavlasopoulou, S., & Sharma, K. (2020). Monitoring children's learning through wearable eye-tracking: The case of a making-based coding activity. *IEEE Pervasive Computing, 19*(1), 10–21.

Graesser, A., Wiemer-Hastings, K., Wiemer-Hastings, P., Kruez, R., & The Tutoring Research Group. (2000). AutoTutor: A simulation of a human tutor. *Journal of Cognitive Systems Research, 1*, 35–51.

Grover, S., & Pea, R. (2018). Computational thinking: A competency whose time has come. In S. Sentance, E. Barendsen, & S. Carsten (Eds.), *Computer science education: Perspectives on teaching and learning in school* (pp. 19–38). New York: Bloomsbury.

Guisande, M. A., Tinajero, C., Cadaveira, F., & Páramo, M. F. (2012). Attention and visuospatial abilities: A neuropsychological approach in field-dependent and field-independent schoolchildren. *Studia Psychologica, 54*(2), 83.

Hall, J. K. (2000). *Field dependence–independence and computer-based instruction in geography*. Unpublished doctoral dissertation. Blacksburg, VA: Virginia Polytechnic Institute and State University.

Hergovich, A. (2003). Field dependence, suggestibility and belief in paranormal phenomena. *Personality and Individual Differences, 34*(2), 195–209.

Highfield, K. (2010). Robotic toys as a catalyst for mathematical problem solving. *Australian Primary Mathematics Classroom, 15*(2), 22–27.

Hill, J. R., & Hannafin, M. J. (1997). Cognitive strategies and learning from the world wide web. *Educational Technology Research and Development, 45*(4), 37–64.

Howland, K., Good, J., & Nicholson, K. (2009). *Language-based support for computational thinking*. In IEEE symposium on visual languages and human-centric computing, (pp. 147–150). Corvallis, USA.

Janka, P. (2008). Using a programmable toy at preschool age: Why and how. In *Teaching with robotics: didactic approaches and experiences. Workshop of International Conference on Simulation, Modeling and Programming Autonomous Robots* (pp. 112–121).

Jonassen, D. H. (1992). What are cognitive tools? In D. H. Jonassen (Ed.), *Cognitive tools for learning* (pp. 1–6). Berlin: Springer.

Jordan, B., & Henderson, A. (1995). Interaction analysis: Foundations and practice. *The Journal of the Learning Sciences, 4*(1), 39–103.

Karp, S. A., & Konstadt, M. (1971). *Children's embedded figures test*. Washington, WA: Consulting Psychologists Press.

Kazakoff, E., & Bers, M. (2012). Programming in a robotics context in the kindergarten classroom: The impact on sequencing skills. *Journal of Educational Multimedia and Hypermedia, 21*(4), 371–391.

Kazakoff, E. R., Sullivan, A., & Bers, M. U. (2013). The effect of a classroom-based intensive robotics and programming workshop on sequencing ability in early childhood. *Early Childhood Education Journal, 41*(4), 245–255.

Kazimoglu, C., Kiernan, M., Bacon, L., & MacKinnon, L. (2012). Learning programming at the computational thinking level via digital game-play. *Procedia Computer Science, 9*, 522–531.

Land, S. M., & Greene, B. A. (2000). Project-based learning with the world wide web: A qualitative study of resource integration. *Educational Technology Research and Development, 48*(1), 45–66.

Lye, S. Y., & Koh, J. H. L. (2014). Review on teaching and learning of computational thinking through programming: What is next for K-12? *Computers in Human Behavior, 41*, 51–61.

Mashburn, A. J., Justice, L. M., Downer, J. T., & Pianta, R. C. (2009). Peer effects on children's language achievement during pre-kindergarten. *Child Development, 80*(3), 686–702.

McNeill, K. L., & Krajcik, J. (2009). Synergy between teacher practices and curricular scaffolds to support students in using domain-specific and domain-general knowledge in writing arguments to explain phenomena. *The Journal of the Learning Sciences, 18*(3), 416–460.

Misirli, A., & Komis, V. (2014). Robotics and programming concepts in early childhood education: A conceptual framework for designing educational scenarios. In *Research on e-learning and ICT in education* (pp. 99–118). New York: Springer.

Morgan, H. (1997). *Cognitive styles and classroom learning*. Westport, CT: Pragerger Publisher.

Myhill, D., & Warren, P. (2005). Scaffolds or straitjackets? Critical moments in classroom discourse. *Educational Review, 57*(1), 55–69.
National Association for the Education of Young Children (NAEYC). (2012). Technology and interactive media as tools in early childhood programs serving children from birth through age 8. Retrieved from http://www.naeyc.org/files/naeyc/file/positions/PS_technology_WEB2.pdf
Nicolaou, A. A., & Xistouri, X. (2011). Field dependence/independence cognitive style and problem posing: An investigation with sixth grade students. *Educational Psychology, 31*(5), 611–627.
Papadakis, S., Kalogiannakis, M., & Zaranis, N. (2016). Developing fundamental programming concepts and computational thinking with ScratchJr in preschool education: A case study. *International Journal of Mobile Learning and Organisation, 10*(3), 187–202.
Pape, S. J., & Tchoshanov, M. A. (2001). The role of representation (s) in developing mathematical understanding. *Theory Into Practice, 40*(2), 118–127.
Pithers, R. T. (2002). Cognitive learning style: A review of the field dependent-field independent approach. *Journal of Vocational Education & Training, 54*(1), 117–132.
Puntambekar, S., & Kolodner, J. L. (2005). Toward implementing distributed scaffolding: Helping students learn science from design. *Journal of Research in Science Teaching, 42*(2), 185–217.
Riding, R. J. (1997). On the nature of cognitive style. *Educational Psychology, 17*(1–2), 29–49.
Riding, R. J. (2002). *School learning and cognitive style*. London: David Fulton Publishers.
Rodgers, E. M. (2005). Interactions that scaffold reading performance. *Journal of Literacy Research, 36*, 501–532.
Román-González, M., Pérez-González, J. C., & Jiménez-Fernández, C. (2017). Which cognitive abilities underlie computational thinking? Criterion validity of the computational thinking test. *Computers in Human Behavior, 72*, 678–691.
Saracho, O. N. (1989). Cognitive style: individual differences. *Early Child Development and Care, 53*(1), 75–81.
Saracho, O. N. (2000). A framework for effective classroom teaching: Matching teachers' and students' cognitive styles. In R. J. Riding & S. G. Raynor (Eds.), *International perspectives on individual differences* (pp. 297–314). Stamford, CT: Ablex.
Saracho, O. N., & Spodek, B. (1981). Teachers' cognitive styles: Educational implications. *The Educational Forum, 45*(2), 153–159.
Saye, J. W., & Brush, T. (2002). Scaffolding critical reasoning about history and social issues in multimedia-supported learning environments. *Educational Technology Research and Development, 50*(3), 77–96.
Sengupta, P., Kinnebrew, J. S., Basu, S., Biswas, G., & Clark, D. (2013). Integrating computational thinking with K-12 science education using agent-based computation: A theoretical framework. *Education and Information Technologies, 18*(2), 351–380.
Sherman, M., & Martin, F. (2015). The assessment of mobile computational thinking. *Journal of Computing Sciences in Colleges, 30*(6), 53–59.
Snowman, J., & Biehler, R. (1993). *Psychology applied to teaching* (7th ed.). Boston, MA: Houghton Mifflin Company.
Tabak, I. (2004). Synergy: A complement to emerging patterns of distributed scaffolding. *Journal of the Learning Sciences, 13*(3), 305–335.
Thomas, L., Ratcliffe, M., Woodbury, J., & Jarman, E. (2002). Learning styles and performance in the introductory programming sequence. *ACM SIGCSE Bulletin, 34*(1), 33–37.
Van de Pol, J., Volman, M., & Beishuizen, J. (2010). Scaffolding in teacher-student interaction: A decade of research. *Educational Psychology Review, 22*, 271–296.
Van Merriënboer, J. J., Kirschner, P. A., & Kester, L. (2003). Taking the load off a learner's mind: Instructional design for complex learning. *Educational Psychologist, 38*(1), 5–13.
Vygotsky, L. S. (1978). *Mind in society*. Cambridge, MA: Harvard University Press.
Williams, D. C., Ma, Y., Prejean, L., Ford, M. J., & Lai, G. (2007). Acquisition of physics content knowledge and scientific inquiry skills in a robotics summer camp. *Journal of Research on Technology in Education, 40*(2), 201–216.

Wing, J. (2006). Computational thinking. *Communications of the ACM, 49*(3), 33–36.
Witkin, H. A., Moore, C. A., Goodenough, D. R., & Cox, P. W. (1977). Field-dependent and field-independent cognitive styles and their educational implications. *ETS Research Bulletin Series, 1975*(2), 1–64.
Yadav, A., Hong, H., & Stephenson, C. (2016). Computational thinking for all: Pedagogical approaches to embedding 21st century problem solving in K-12 classrooms. *TechTrends, 60*(6), 565–568.
Zhang, L. F., & Sternberg, R. J. (2009). Revisiting the value issue in intellectual styles. In L. F. Zhang & R. J. Sternberg (Eds.), *Perspectives on the nature of intellectual styles* (pp. 63–85). New York, NY: Springer.

Part II
Cases from Secondary School

Chapter 4
Network Analytics of Collaborative Problem-Solving

Simon Kerrigan, Shihui Feng, Rupa Vuthaluru, Dirk Ifenthaler, and David Gibson

4.1 Introduction

Problems vary in terms of their structure. Jonassen (1997) classifies problems on a continuum from well-structured to ill-structured. Well-structured problems have a well-defined initial state, a known goal state or solution, and a constrained set of known procedures for solving a class of problems. In contrast, the solutions to ill-structured problems are neither predictable nor convergent because they often possess aspects that are unknown. Additionally, they possess multiple solutions or solution strategies or often no solutions at all (Funke, 2012). Jonassen (2011) reiterates that the structure of a problem often overlaps with complexity: Ill-structured problems tend to be more complex, especially those emerging from everyday practice, whereas most well-structured problems tend to be less complex. The complexity of a problem is determined by the number of functions, or variables it involves; the degree of connectivity among these variables; the type of functional relationships between these properties; and the stability of the properties of the problem over time (Funke, 1991). Simple problems are composed of few variables, while

S. Kerrigan · R. Vuthaluru · D. Gibson
Curtin University, Perth, Australia
e-mail: simon.kerrigan@curtin.edu.au; r.vuthaluru@curtin.edu.au; david.c.gibson@curtin.edu.au

S. Feng
University of Hong Kong, Hong Kong, People's Republic of China
e-mail: shihuife@hku.hk

D. Ifenthaler (✉)
Learning, Design and Technology, University of Mannheim, Mannheim, Baden-Württemberg, Germany
e-mail: dirk@ifenthaler.info

D. Ifenthaler et al. (eds.), *Balancing the Tension between Digital Technologies and Learning Sciences*, Cognition and Exploratory Learning in the Digital Age,
https://doi.org/10.1007/978-3-030-65657-7_4

ill-structured problems may include many variables that may interact in unpredictable ways. When the conditions of a problem change, a person must continuously adapt his or her understanding of the problem while searching for new solutions, because the old solutions may no longer be viable. Static problems are those in which the factors are stable over time while ill-structured problems tend to be more dynamic (Seel, Ifenthaler, & Pirnay-Dummer, 2009). Hence, in order to successfully solve complex and ill-structured problems, the person involved in problem-solving must be able to view and simulate the dynamic problem system in its entirety imagining the events that would take place if a particular action were to be performed (Eseryel, Ifenthaler, & Ge, 2013). It has been argued convincingly that all games serve as situated problem-solving environments, in which players are immersed in a culture and way of thinking (Eseryel, Ge, Ifenthaler, & Law, 2011; Gee, 2003).

Collaboration is an essential part in most working environments because it encompasses different views, multiple skills, diverse experiences, analytical judgments, and rich knowledge. Common characteristics of definitions of a collaborative team include at least two involved individuals, common objectives, shared responsibility and interdependence as well as optimal performance (Ifenthaler, 2014). Empirical research shows that through the use of combined resources, teams can successfully handle problems that otherwise would be too complex for a single individual (Badke-Schaub, Neumann, & Lauche, 2011; Cannon-Bowers & Salas, 2001). Digital learning environments, and especially games, designed for team performance, often are characterised by integrated, media-rich contexts with multiple layers of interaction with peers as well as computational resources, which provides a foundation for authentic performance of individual and team-based problem-solving processes with attendant opportunities for unobtrusive observation and documentation of strategies, tools, communications, intentional actions and artefacts (Clarke-Midura, Code, Dede, Mayrath, & Zap, 2012).

A *network* (or graph) is constructed from a set of *vertices* whose relationships are represented by *edges*. Basics of graph theory are necessary to describe the properties of such a network (Diestel, 2000). Various measures from network or graph theory have been applied to assess individual as well as team problem representations and, in addition, to track the development of problem-solving over time (Clariana, 2010). Appropriate structural measures include (a) *number of vertices*, (b) *number of edges*, (c) *connectedness*, (d) *ruggedness*, (e) *diameter*, (f) *number of cycles*, or (g) *average degree of vertices* (Ifenthaler, 2010b).

Both, problem-solving and collaboration are regarded as an essential part of twenty-first century skills (Griffin, McGaw, & Care, 2012). In this article, we define network measures of personal learning, collaboration and problem-solving and integrate them into a analysis based on a series of network states of team behaviour evolving during collaborative problem-solving. A case-study illustrates a semester-long collaborative problem-solving task where six teams of three students were engaged, leading to high performing and low performing teams being identified.

4.2 Dimensions of Personal Learning, Collaboration and Problem-Solving

The domain model of a learning assessment is a conceptual representation of the key indicators that experts 'might see people say, do, or make as evidence, and situations and activities that evoke it—in short, the elements of assessment arguments' (Mislevy, 2011, p. 13). For the analysis discussed here, the dimensions of the domain model are personal learning, collaboration and problem solving, which have been defined along with evidence indicators (Gibson, Irving, & Seifert, 2018). We will refer to these as the theory-based *evidence targets*.

The evidence targets represent potential observations of the data stream from a digital learning space, which are claimed as partial indicators of student performance relatable to the domain model. For example, a student working on a team might co-write and co-edit a statement or document with one or more other students—an observable action with exact and highly detailed traces in the digital record—which serves as partial evidence of establishing and maintaining team organisation and taking appropriate action to solve the problem.

Personal learning: acquisition of knowledge (e.g. new insights, capacities for thinking, acting and employing skills) that is evidenced for outside observers as well as an individual's own reflection and metacognition (Friedrichs & Gibson, 2003). *Evidence targets*:

PL1: Sharing experience

PL2: Expressing and examining diverse concepts

PL3: Articulating, applying and building understanding

PL4: Communicating new powers and creations

Collaboration: coordinated group activity resulting from continuous attempts to construct and maintain a shared conception of a problem (Roschelle & Teasley, 1995). *Evidence targets:*

C1: Establishing and maintaining shared understanding

C2: Taking appropriate action to solve the problem

C3: Establishing and maintaining team organisation

Problem solving: cognitive processing directed at achieving a goal when no solution method is obvious (Mayer & Wittrock, 1996). *Evidence targets:*

PS1: Exploring and understanding

PS2: Representing and formulating

PS3: Planning and executing

PS4: Monitoring and reflecting

Mapping from a measure such as 'network density of the degree to which a student works with others' (labelled (a) in Table 4.1) to evidence targets such as 'taking appropriate action' (C2) or 'maintaining team organisation' (C3) is context sensitive. This means that alternative mappings using the same measures are not only possible but required for a more complete understanding of the observations and analyses of a particular context. This highlights the need for transparency in the mapping process, as each inference is, following Mislevy, Almond, and Lukas (2003), part of an evidentiary argument.

Table 4.1 Measures, targets and evidence types collected for analyses

Measures	Evidence targets	Evidence type
Measures of collaboration		
(a) Network density—degree to which team members worked together on tasks	C2, 3	Actions
(b) Time on tasks	C2, PS3, PL3	Actions
(c) Time series analysis of team participation	C1, C2	Actions
(d) Task and subtask organisation	C1, C2	Products
(e) Task and subtask sequencing and duration	C2	Actions
Measures of problem-solving		
(f) Time to respond to instructor feedback	PS4	Actions, communications
(g) Sequence and duration on tasks	PS1, C2	Actions, products
(h) Completion path analysis	PS3	Actions
(i) Time on task implication of blended learning: Do students working collaboratively spend time outside of class time and if so, when?	PS3	Actions
(j) Structural relationship of participation in outcomes-tagged activities with acquisition or demonstration of learning outcomes	PS1,2	Actions, Products
(k) Time series analysis of task structure	PS3, C3	Products
(l) Correlational or causal relationship of levels of participation in outcomes-tagged activities with acquisition or demonstration of learning outcomes	PS1, 2	Actions, products
Measures of personal learning		
(m) Access to learning outcomes via task structure	PL3	Products
(n) Team reflection and evaluation	PL1, PS4	Communications, Products
(o) Content analysis of communications related to task completion	PL1, 2, 3, 4 PS1, 4	Communications

The multi-to-multi mappable relationships of observations to evidence leading to interpretable and actionable information is a complex yet bounded scope for analyses. It is complex in the sense that reasonable people can diverge in their interpretations; more than one true statement may be possible for an inferential mapping. At the same time, the scope of interpretation is bounded by the domain model, making inappropriate interpretations less likely. For example, an observation and inference would not be expected to be reasonable evidence of a dimension outside of the domain (e.g. of an ability to draw or sing if an appropriate prompt and affordance for performance had not been provided). Analyses presented in this chapter and those outlined for future research utilise one or more of the evidence targets to make triangulated claims about student performance in terms of personal learning, collaboration and problem-solving.

4.3 Research Questions

The focus of exploratory data analysis in this research aimed to determine the challenges and potential of fine-grained time-sensitive analyses of collaborative problem-solving tasks to inform an understanding of the structural, correlational and causal relationships of students achieving learning outcomes. In particular, to what extent can network analyses and related measures assist in the characterisation and prediction of learning processes and learning outcomes (Ifenthaler, 2010b)? Guiding the research were five research questions concerning how network analysis can assist in characterising learning in a collaborative problem-solving context:

1. Task Participation—who does what to help the team accomplish its objectives, how do team members relate to and divide up the task, and which task activities and outcomes involved which team members?
2. Completion Paths—how do teams differ with respect to time to completion, what variability do they exhibit in starting and ending times, and sequence of tasks
3. Attention to Feedback—how do teams differ in responsiveness and the percentage of feedback used to improve, how do teams differ in the type of feedback requested and received
4. Use of Time—how do teams differ in their use of time during a long-term project with 24-7 access, which subtasks take the teams more time than others, how to the teams differ in overall time
5. Learning Outcomes—the extent of coverage of outcomes per team member, quality and amount of evidence of achievement of outcomes

4.4 Method

While this study is based on a small number of teams (N_t = 5) and participants (N = 18), the analysis is based on more than 1000 time-based data records that were automatically collected from a digital learning experience during a semester-long high school classroom project, 300 of which related to the highest and lowest performing teams. We selected the contrasting teams to test the feasibility and face validity of the data mapping system and analysis methods.

4.4.1 Participants and Context

Participants in the study were N = 18 students in their last year of high school enrolled in a semester-long Vocational Education and Training programme (VET) leading to a certificate in Business Practice with a focus on Health and Workplace Safety and Social Media in Communication. VET programmes provide students with learning experiences that are often tailored towards workplace experience, or niche subject content that is not covered in a traditional high school syllabus. Students self-formed into five teams of 3 or 4 members and chose an organisation that they wished to represent in a business scenario. The main task was to research the company and deliver a social media communications plan that effectively educated the company's employees on workplace health and safety legislation rights and responsibilities. The assignment was structured through a series of 17 primary tasks and 76 sub-tasks that created artefacts, including research, writing and design-based work.

The teacher created the project framework, tasks, sub-tasks and learning outcomes, using the Challenge platform, a web-based, mobile-ready application platform for active digital learning experiences and event-level data collection (Gibson & Jackl, 2015; Ifenthaler & Gibson, 2019) developed at Curtin University. Challenge integrates with Cisco Webex Teams (https://www.webex.com/downloads.html) to provide each team with telecommunications capability for working globally, including a whiteboard, file sharing and teleconference facilities automatically organised by the Challenge platform into the main deliverables in the curriculum design.

The students in this study used the platform to form teams, upload files, chat with team members and complete the assigned tasks for the project. The analyses presented here are based on data collected from student's interactions with the platform, in particular, the creation and submission of artefacts and other inputs required by the tasks, communications among the team members and with the teacher about how to organise the work, and the instructor-judged quality of the team's product as well as the team's self-evaluation of their project. Data for the research team's analyses were collected from log-files and evidence stored on the platform (e.g. uploaded files and the content of page interactions, chat discussions, and written responses to prompts). Analyses and findings of the research team were validated by inspection and protocol review by the instructor as well as by cross-validation of multiple measures presented below.

4.4.2 Data Handling and Analytics

The data used in this study for exploring group collaborative problem-solving was collected from Challenge platform and Webex Teams platforms, merged into one dataset, which was straightforward given the similarities in data structure. Raw transcripts (communications) and trace data (actions and artefacts) were capable of being downloaded at any time for any time frame. Each time that a team member (user) contributed towards an assignment artefact, an interaction transaction was captured by the platform. Data collected from the Challenge platform included (a) timestamp converted to local time, (b) the user responsible for the interaction, (c) the team of the user, (d) the task they were working on, (e) the artefact they were working on, (f) the content they provided to this artefact and (g) the status of the interaction noted as *visible* (current state), *archived* (saved previous edited version) or *published* (submitted as final state) content. In addition, the communication data among team members was also collected from the Webex Teams platform, including (a) timestamp converted to local time, (b) the user posting the message, (c) the team of the user, and (d) the message content posted to the group. A manually edited column for linking events to tasks and artefacts was manually added and applied to messages in the Webex Teams data where a student or teacher directly and unambiguously referred to a specific assignment task or artefact. This link allowed analysts to measure the effectiveness and response time to teacher feedback.

Networks modelling the interactions between individual students and artefacts were constructed for analysing individual participation and shared contribution in group collaboration. Two sets of nodes in the network include individual students and task artefacts. The links in the networks represent the interactions between agents and artefacts (Ifenthaler, 2010a). There is no link within the same set of nodes (e.g. students to students or artefacts to artefacts) in the network. Bipartite networks, a technique that has been widely used to present the affiliation relationship in social problems, such as personal recommendation, were constructed with this approach and used for measuring individual-level and group-level network structures of group collaboration. Since we are interested in studying the interactive relationships between students and sub-tasks in group collaboration, the number of samples of data per team is the product of the total number of sub-tasks and number of students (e.g. several hundred samples). Strength of connections between a student node and an artefact node were represented as a weighted line that summarises effort (e.g. time and number of interactions) and indicates the relative contribution of a student to an artefact. In addition, three levels of distributions were created to represent artefacts where one, two, or three people had interacted. Implications of the naturally occurring task distributions (e.g. for setting empirical probabilities for future studies) for assessment and social network analysis of collaborative problem solving are under preparation.

4.5 Results

Multiple measures were observed based on the data records as described and mapped to the conceptual framework measures, targets and evidence types (see Table 4.1). A mapping from the observations to research questions is presented along with a summary of the findings from the two case examples—a 'highest performing' team compared with a 'lowest performing' team from a cohort (see Table 4.2).

In the following sections, we briefly discuss the research questions, observations, measures and findings listed in Table 4.2.

Table 4.2 Mapping of research questions to measures and findings

Research questions	Measures	Findings: high versus low performing team	
RQ1 **Task participation**—who does what to help the team accomplish it objectives, how do team members relate to and divide up the task, and which task activities and outcomes involved which team members?	C1 (c, d)	Ample evidence, balanced task load	Some evidence, unbalanced task load
	C2 (a, b)	High density, high time on task	Low density, low time on task
	C3 (k)	All contributed	One person dominates team
RQ2 **Completion paths**—how do teams differ with respect to time to completion, what variability do they exhibit in starting and ending times, and sequence of tasks	PS3 (b, h)	Organised, compact use of time	Highly variable and gaps in use of time
	C2 (b)	Ample time on task	Lack of time on task
	C3 (k)	Tasks completed in order	Tasks completed out of order
	C1 (c)	Work starts early	Works starts and ends late
RQ3 **Attention to feedback**—how do teams differ in responsiveness and the percentage of feedback used to improve, how do teams differ in the type of feedback requested and received	PS4 (f)	1 day	8 days
	C2, C3 (f)	85% response	48% response
RQ4 **Use of time**—how do teams differ in their use of time during a long-term project with 24-7 access, which subtasks take the teams more time than others, how to the teams differ in overall time	PS1, PS2 (j)	All students equally engaged in all tasks	Fewer students engaged in fewer tasks
	PS3 (i)	Low but efficient time in both in and out of class	Least time and most inefficient both in and out of class
	PL2 (l, m, o)	Key subtasks had adequate time	Key subtasks missed
	PL4, PS4 (n, o)	Key reflection task engaged	Key reflection task missed
RQ5 **Learning outcomes**—the extent of coverage of outcomes per team member, quality and amount of evidence of achievement of outcomes	PL2 (j, l, m)	Individual work in early stages	No work in early stages
	PL3 (j)	All students	One student
	PS3 (k)	Task completion in 3 months	Task completion in 5 months

4.5.1 Task Participation

Some of the key questions about task participation in online collaborative learning are: ‘who does what’ to help the team accomplish it objectives, how do team members relate to and divide up the task, and which task activities and outcomes involved which team members? Group members ideally need to complete key assigned artefacts together in order to achieve the identified learning outcomes. For example, a team cannot acquire or demonstrate any state of collaboration (C1, C2, C3) if they work independently and do not share their work with each other. A visualisation of the bipartite networks for a high performing (HP) and low performing (LP) team shows 'who did what' (see Figs. 4.1 and 4.2).

The node sets of team members (e.g. person agents—red nodes) are presented in relationship to artefacts classified as 1-person (green nodes), 2-person (yellow nodes) and 3-person (blue nodes) artefacts. In the high-performance (HP) team, members worked on more 2-person artefacts compared to the low performance (LP) team. In addition, there was no 3-person artefact in the LP team, in spite of the fact that the relevant team evaluation artefact prompted all group members to participate.

Examining the extent to which team members worked together on tasks compared to work done on their own, the HP team showed a relatively balanced participation distribution of artefact creation by two members and fewer contributions by a third team member. Incidentally, this may be new objective evidence of interpersonal status hierarchies within social expectations states theory (Berger, Cohen, & Zelditch, 1966).

Importantly, there were ten instances where HP team members worked on 2-person artefacts and all team members took part in paired production activity. In addition, the HP team self-evaluation included participation by all members. The automatically documented evidence from the Challenge platform is thus strongly linked to the theoretical framework of personal learning in a collaborative problem-solving context.

In comparison, the lowest performing team exhibited a spread of individual workloads among team members but created only 4 artefacts in pairs (Fig. 4.2).

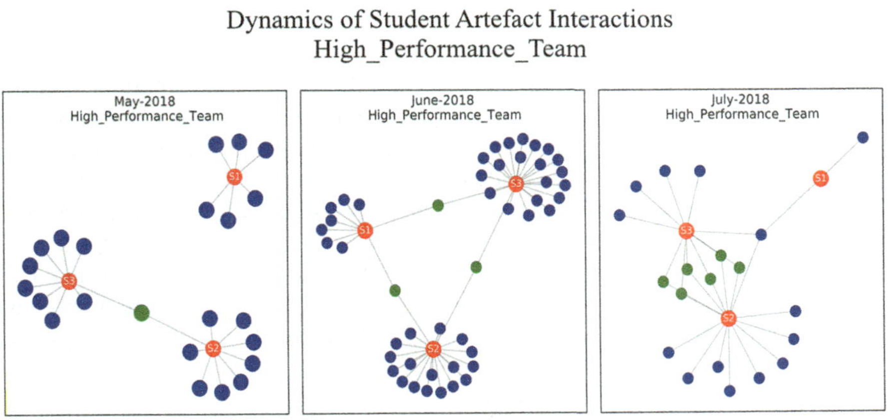

Fig. 4.1 Individual and team task participation in the HP team; semester summary

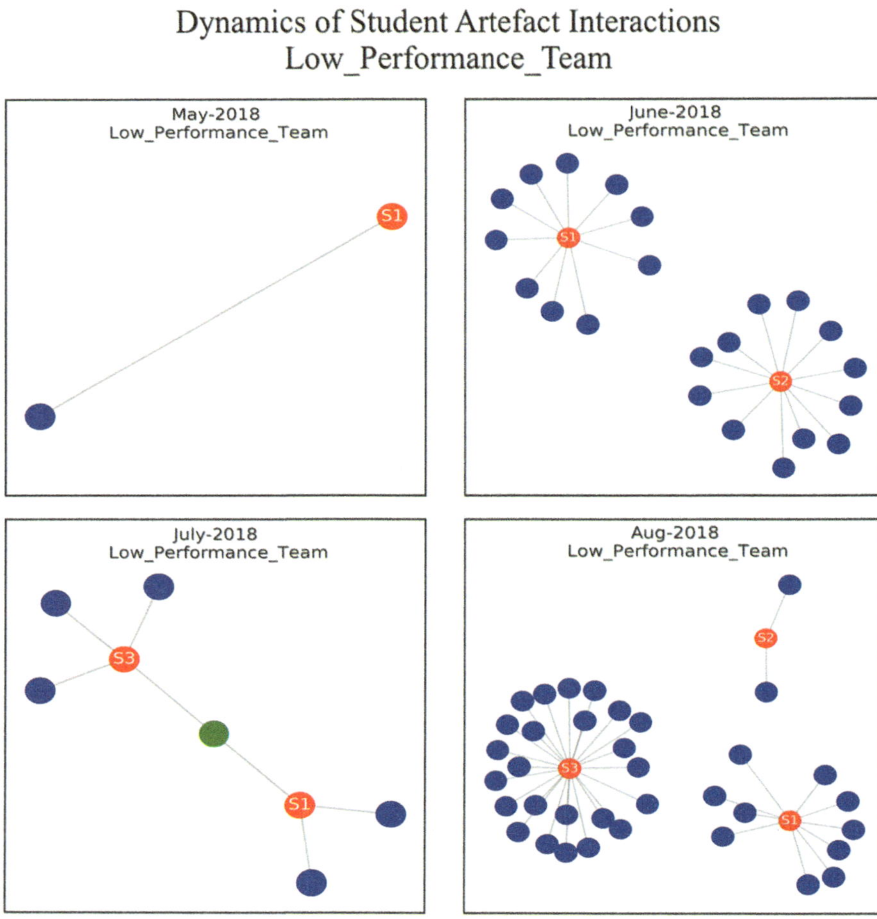

Fig. 4.2 Individual and team task participation in the LP team; semester summary

There was no instance of all team members working together on an artefact. Team self-evaluation, for example, was 'filled out' by only one team member. This suggests that while individual members took some appropriate actions to solve the problem (C2), there is a lack of evidence of an effort to establish a shared understanding (C1) and maintain team organisation (C3), and this is reflected in the team's overall low performance.

These network graphs of task participation and distribution (Figs. 4.1 and 4.2) are summary pictures of the semester-long project, so are missing important dynamic information, which we discuss below in the time series analyses of completion paths.

4.5.2 Completion Paths

The key research questions concerning completion paths are: how do teams differ with respect to time to completion and what variability do their completion paths exhibit in starting and ending times as well as the sequence of completion of tasks. As context, the tasks in Challenge were displayed in a listed order for teams to complete but students were free to start and finish tasks in whatever order they wanted. This provided an opportunity to analyse whether teams differed in their approach to taking appropriate action and planning and executing (C2 and PS3). The general trend in all groups was to start tasks in the expected order provided by the instructional framework that assumed reading from left to right and top to bottom, but the teams exhibited much more variability in the order of completed tasks (see Fig. 4.3).

Completion rank (order of completion) was computed based on the last time an artefact was touched by any team member (e.g. the ending rank in Fig. 4.3), thus capturing the order of any final check by the team. The research team did not consider the interval from first touch to last touch the actual time on task, because a student team could have conducted a last-minute final look at everything. Instead, each team's task interaction events resulted in a time ordered list of micro-durations of activity. Average duration and sequence order were calculated for each task and subtask, to identify where along the project completion path most of the team's work occurred. Sorting the tasks by their average sequence value produced an order in which teams started and completed work on the various subtasks.

In relation to the instructor's intended design (i.e. the expected rank in Fig. 4.3) both HP and LP teams exhibited starting ranks with positive correlations with the expected rank, which suggests that both teams started tasks following the sequence expected by the teacher's design. Using Kendall's Tau, a rank correlation coefficient, we compared the HP and LP teams with each other as well as the instructor's expected ranking. The starting rank of tasks in the HP team is highly and positively associated with the expected rank ($r = 0.838$, $p < 0.001$) while the LP team is moderately and positively associated ($r = 0.706$, $p < 0.001$). However, the end rank was significantly different. The HP team end rank was positively associated ($r = 0.691$, $p < 0.001$) with the expected rank, which is evidence that the high-performance team tended to complete tasks in the teacher's expected order. But in the LP team, *there was no statistical significance* between the end rank of tasks and the teacher's expected rank, which may have contributed to the team's lower quality of work.

The HP team evidenced one of the highest correlations to the benchmark ordering of task in both start and completion order. This suggests that the team methodically approached their work (C3) which may have assisted them in being a high performing team. The LP team on the other hand, exhibited more deviation from the suggested structure especially in the end ranks, evidenced also by their depth and timing of responses to instructor feedback (C2) and the timing of team member participation (C1). The LP team's ordering caused some tasks to be completed 'out of logical order.' For example, the LP team completed some of the research tasks (e.g. Social Media Sites Research) after some of the design-based tasks that were supposed to be

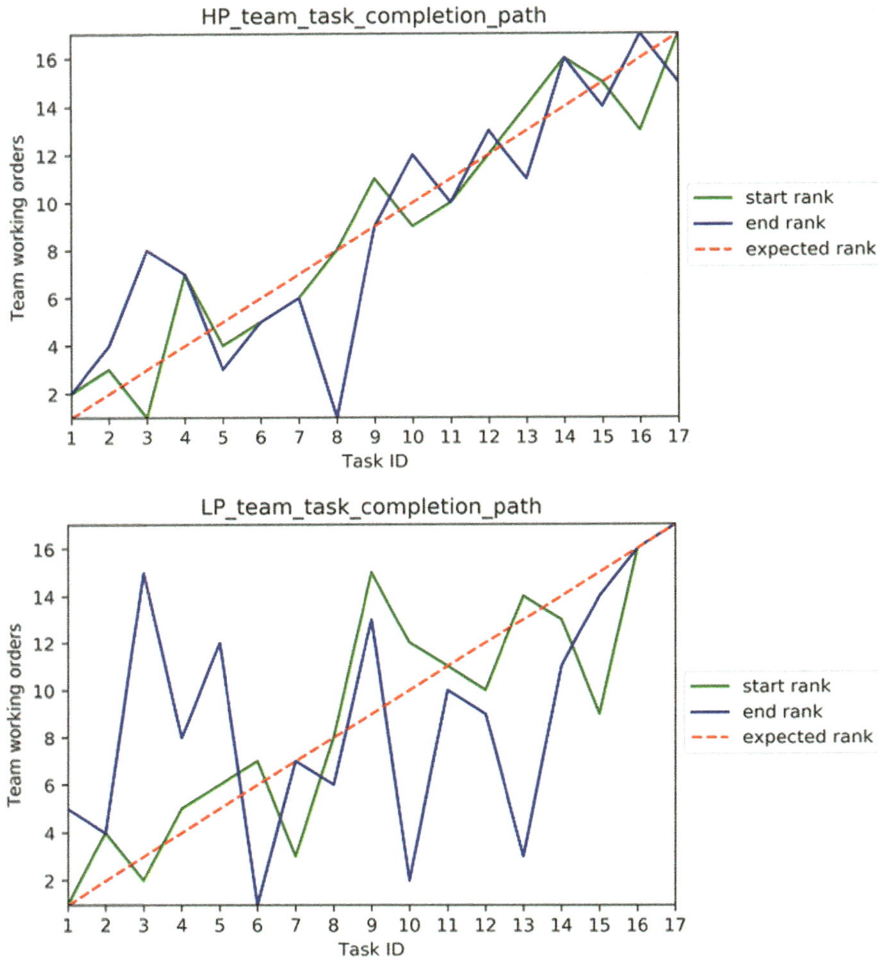

Fig. 4.3 Completion paths of the high performing (HP) versus low performing (LP) team. The HP team is positively correlated with the teacher's expected rank for starting and ending tasks, while the LP team is not significantly correlated

research-based (e.g. Social Media Summary) suggesting poor organisation and lack of cohesion among the group. This lack of structure (C1) and team cohesiveness (C3) helps explain why the LP team created a lower-quality final product.

4.5.3 Attention to Feedback

The research questions concerning using feedback to improve the team's work include: how do teams differ in terms of responsiveness and the percentage of feedback used to improve, and how do teams differ in the type of feedback requested and received.

There was a noticeable difference in the level of attention given to teacher feedback from both teams. 85% of teacher feedback addressed to the HP team had an identifiable user action response (i.e. a team member went back to the mentioned piece of work to make an edit). Comparatively the LP team only had an identifiable user action response to 48% of teacher feedback, meaning that over half of the teacher's feedback to the team was not acted upon. The LP team failed to take appropriate action to solve problems that arose (C2) (see Fig. 4.4).

The other component considered is the time taken to apply action to teacher feedback. As discussed in Sect. 4.5.4, the students and teacher had access to the platform 24-7 and teacher feedback was often provided outside of classroom hours. The HP Team had an average response time of 1 day between teacher feedback and identifiable user action response. The longest time delay for the HP team was 2 days, which indicates the group members were actively monitoring their progress (PS4) and were more engaged (C2).

The LP team on the other hand took on average 8 days to action teacher feedback, and the longest time delay was 15 days. It is evident that members within this team were either not monitoring teacher communication channels efficiently (PS4) or neglecting the need to revise and improve work. 63% of the feedback left unaddressed by the LP team related to research-based tasks (see Fig. 4.5).

Within this analysis it is also important to note that the HP team received no duplicate or follow-up reminders to action previously unaddressed feedback, and this is primarily due to the timely rate at which they responded to feedback (C2,

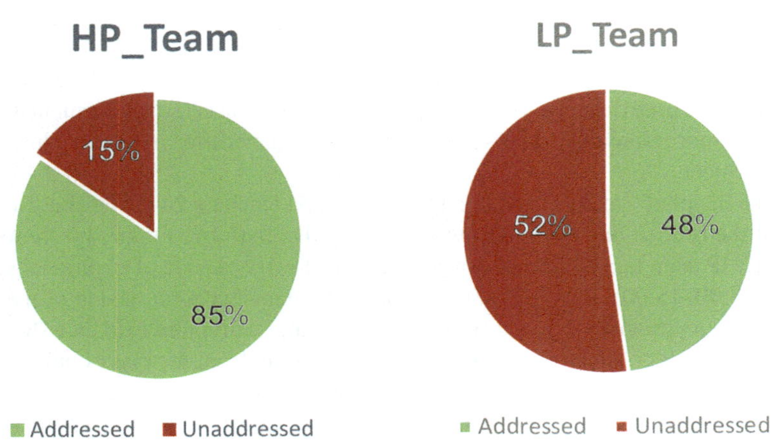

Fig. 4.4 Feedback response rates of HP team and LP team

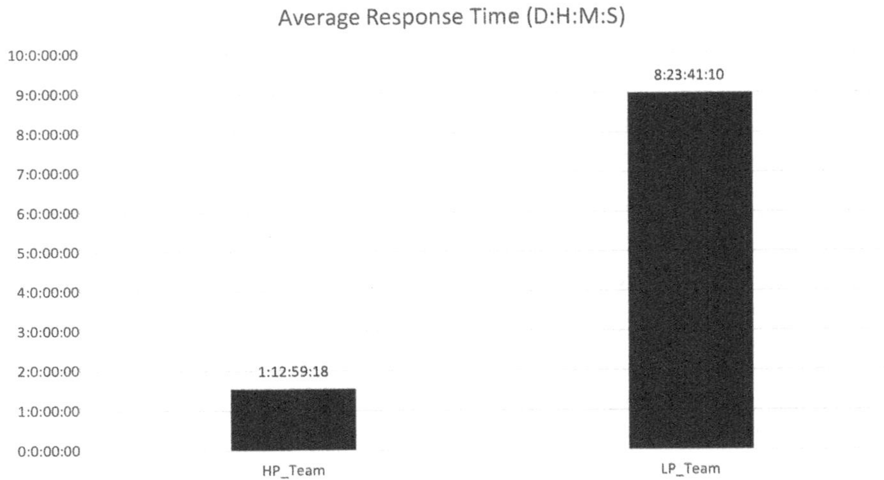

Fig. 4.5 Average time taken to respond to feedback for HP and LP teams

C3). Of the messages that remained unaddressed by the LP team, 41% of all artefacts referenced included multiple follow-up reminders from the teacher, referencing the previously unaddressed feedback (C2, C3). In one instance, the teacher followed up on unaddressed feedback left over a month prior that remained unaddressed.

The overall attention given to feedback clearly shows two different approaches to team problem solving (C2), team organisation (C3) and monitoring (PS4) between the HP and LP teams.

4.5.4 *Use of Time*

The primary research questions concerning use of time include: how do teams differ in their use of time during a long-term project with 24-7 access, which subtasks take the teams more time than others, and how to the teams differ in overall time spent on the project.

Since the platform was available to students and teachers 24-7, the analysis considered the time of day that students were using the platform. Figure 4.6 shows that only the HP team logged on and completed project work outside of traditional school hours (08:00–15:00). In addition, times that both teams worked outside of the class meeting hours provides evidence of blended learning, with more evidence shown by the HP team to coordinate and plan project effort away from the classroom (PS3).

Comparatively, all the efforts of the LP team occurred within traditional school hours. This suggests that the members of the LP team may have been unwilling, unorganised, or unmotivated (C3, PS3) to continue project momentum and effort away from the classroom.

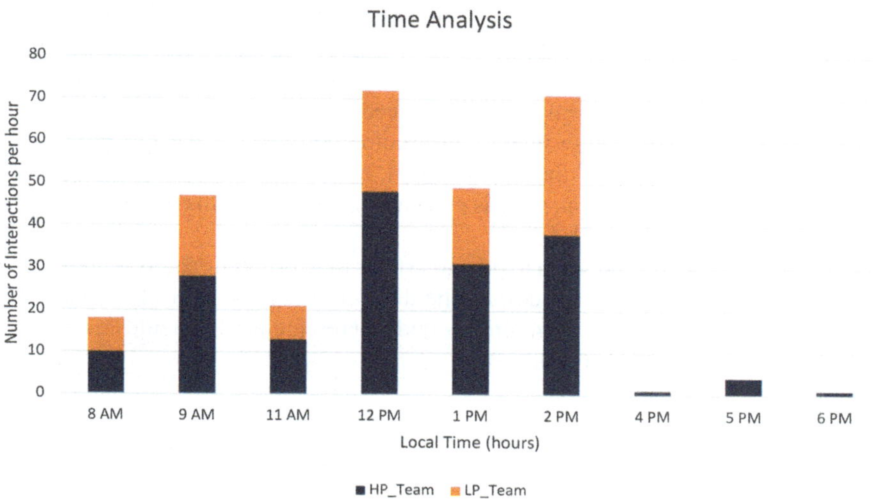

Fig. 4.6 Time of Day usage between HP and LP teams

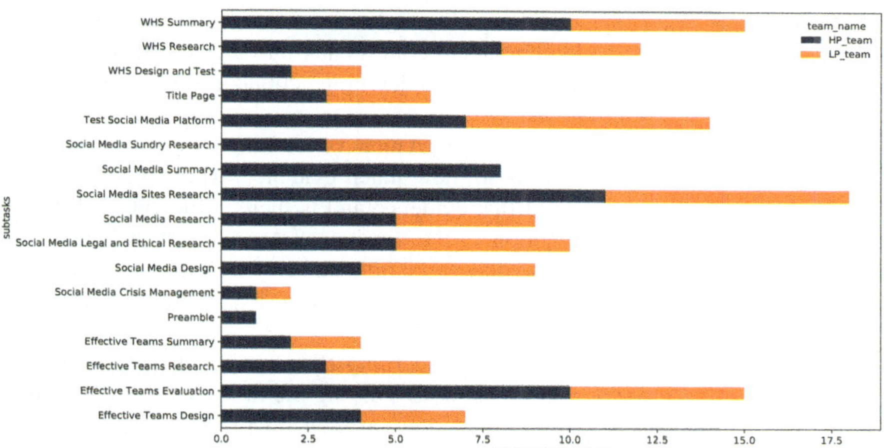

Fig. 4.7 Number of interactions per task for HP and LP teams

Another element considered was on which tasks and subtasks was the team effort placed, in order to identify the efficiency of time spent. The project 17 primary tasks varied in their expected level of detail and type of work required. Tasks were broadly categorised by the teacher as either research, design or written based. Figure 4.7 shows the number of interactions that both the HP and LP teams had with the 17 project tasks.

Overall, the HP team had 70% more interactions across all tasks than the LP team. This reinforces the inference and understanding that the HP team spent more time and effort working on the project than the LP team. In terms of interactions across specific tasks, the HP team recorded interactions against each of the 17 project tasks. The LP team had no interactions in both the 'Preamble' and 'Social Media Summary' tasks, two of the final tasks that needed to be completed. The LP team also responded to the team evaluation task before completing the full project, so combined with the network diagram in Fig. 4.2, showing only one team member worked on the team evaluation, there is evidence that the LP team did not effectively reflect on their work (PS4, PL4).

Both teams however did seem to prioritise their time proportionately on research-based tasks. This suggests that despite the differences in levels of effort and time between the two teams, both teams understood that research tasks should be assigned more time and effort (PS1, PS2).

4.5.5 Learning Outcomes

Ultimately, the goal of both planning and participating in learning experiences is the achievement of learning outcomes. Key questions are: the extent of coverage of outcomes per team member, and the quality and amount of evidence of achievement of the outcomes. Here, the domain model comes back into focus, because the outcomes of interest are growth in personal learning, collaboration and problem-solving skills and capabilities.

One component of measuring the achievement of learning outcomes is whether there is any evidence at all, and if so, how much evidence. At a deeper level the question is what is the quality of that evidence and is there sufficient evidence to infer anything about the acquisition of the outcomes? A record of evidence over time is thus needed to capture data during the full course of the experience, as some outcomes might be more engaged at different times of the experience and reiterated later in the experience (see Fig. 4.8).

The platform automatically collected learning outcomes evidence at both the individual and whole team level. In summary, during the first 4 weeks, individuals on the HP team began working immediately on their own in separate clusters of subtasks, evidence of (PL2) 'examining diverse concepts,' while the LP members team did nothing. All members of the HP were engaged in a relatively balanced number of subtasks throughout long term of the project, while the LP team in contrast had only one member highly engaged (arriving a month late to the team's work and doing a lot at the last minute), evidence of (PL3) 'articulating, applying and building an understanding' in one but not all team members. As a result, the HP team completed the project in 3 months while the LP team took 5 months to produce an inferior team result, evidence of differences in (PS3) 'planning and executing.'

The analysis also considered the level of contribution within a team in relation to outcomes associated with each task and subtask (Fig. 4.9). The measure of engagement with tasks entails at a minimum, a level of exposure to intended outcomes. While neither team had all members equally addressing all learning outcome

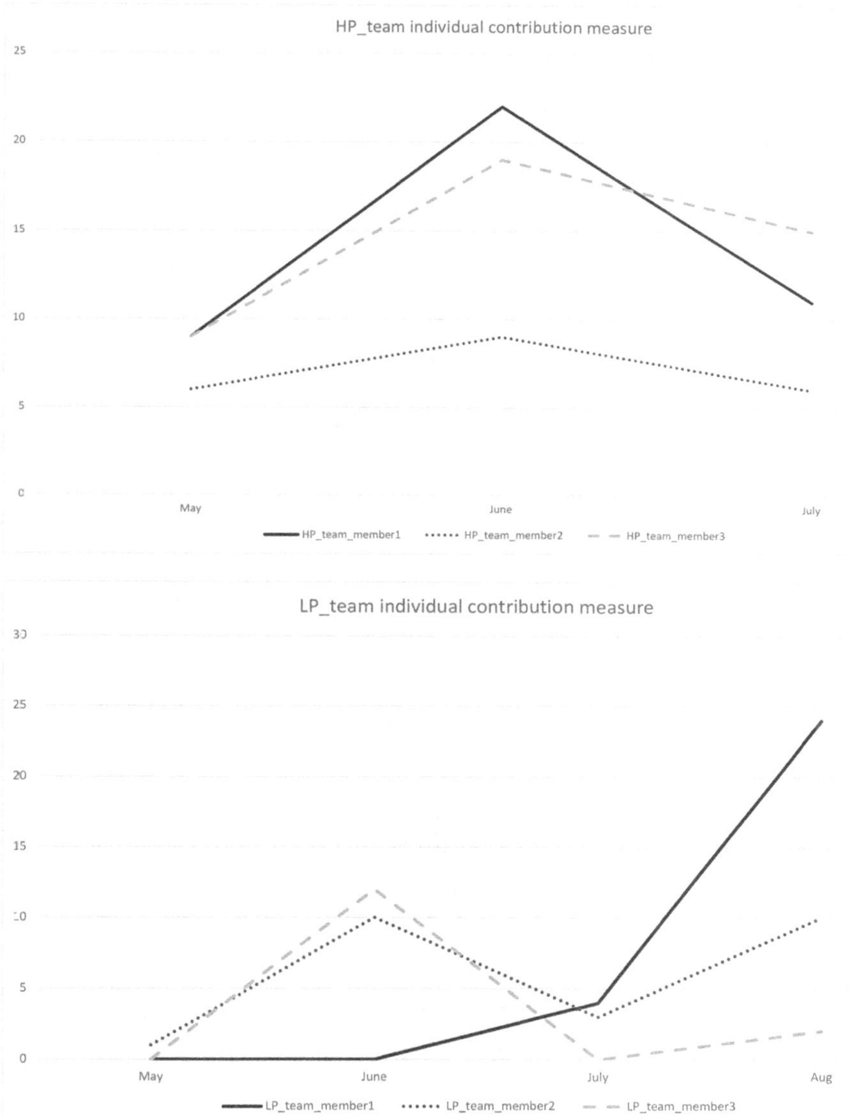

Fig. 4.8 Individual contributions per month for a low performing (LP) and high performing (HP) team. The LP team work peaked at the end and was 1 month late, while the HP work peaked in the middle and was completed on time

categories (which carries some implications for assessment that we have not addressed in this article), the HP team clearly occupied a larger surface area (i.e. greater exposure or *learning opportunity surface*) than the LP team. The exposure to learning outcome evidence meant that even the lower contributing members of

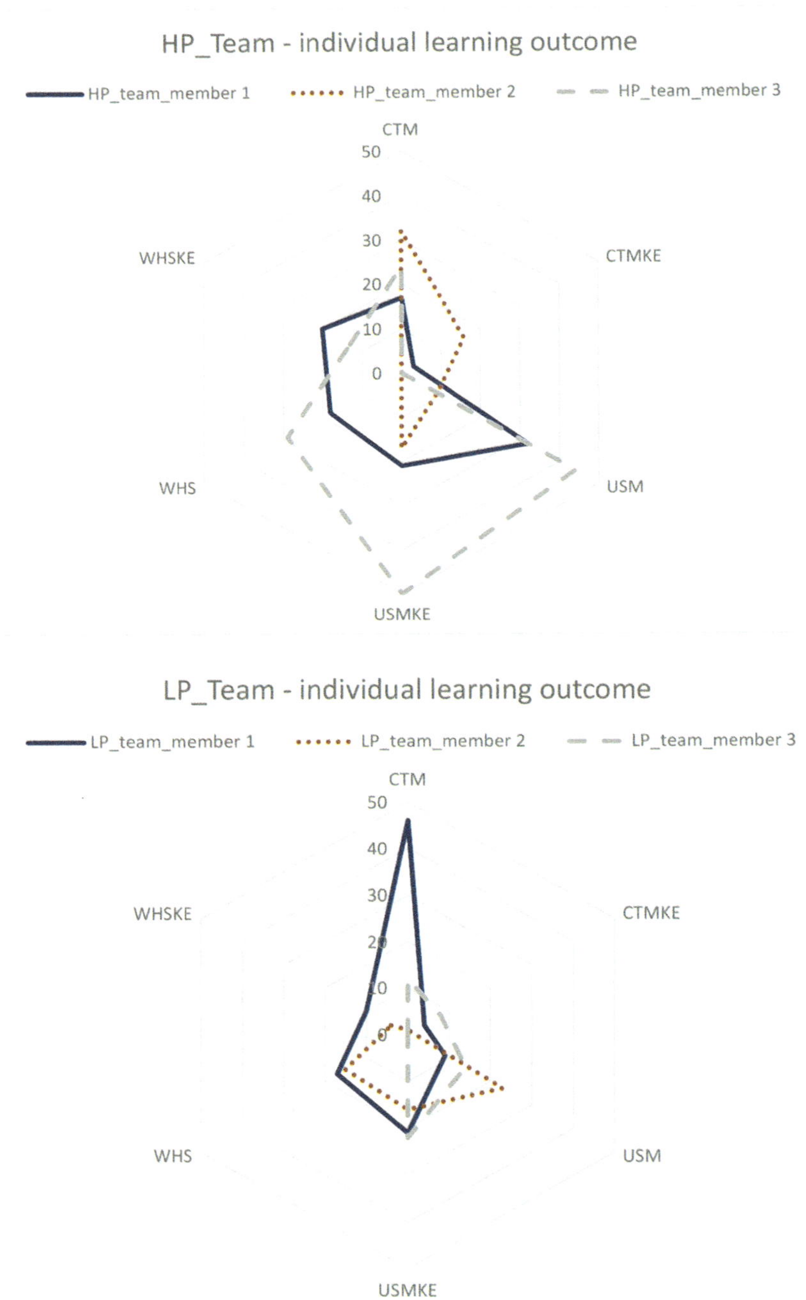

Fig. 4.9 Learning outcome coverage across both HP and LP teams

the HP team likely benefited in building more understanding (PL3) by being part of the stronger team.

Most of the 'Effective Teams' tasks that encouraged collaboration were tagged within the 'Contributing to Team Effectiveness' learning outcome category (CTM). Therefore, observers expect an even distribution of contributions linked to this learning outcome (e.g. all team members should contribute to reflection activities). But Fig. 4.9 makes clear that one team member was largely responsible for CTM activity in the LP team, which further enforces the belief that this team did not work as well collaboratively (C1, C2, C3).

4.6 Discussion

This research demonstrates a mapping of three forms of digital performance data—actions and use of resources, communications and constructed products—collected from a learning space designed for team learning. The data was analysed with network methods among others and was interpreted as evidence of individual and team behaviours linked to personal learning, collaboration and team problem solving. In this section we will reflect on the main findings concerning task participation, patterns and paths of completion, attention to feedback, use of time and learning outcomes and relate the findings to the theoretical foundation. To set the stage for the discussion, we first review network analytics and its contribution in this case to analyses of collaborative problem solving by supporting the creation of new measures of collaborative problem solving.

We have summarised our findings in Table 4.2 for quick inspection. Before we briefly discuss the findings, we want to summarise the gaps in the evidence system, also referred to as the Challenge data mapping system. The data for the evidence linkages has three components (see Table 4.1): *Outcome or Evidence Target* (e.g. PL1...PS4), *Measures* (e.g. a...o), and *Evidence Type* (e.g. actions, communications, products). Summary views of the linkage dataset point out different ways in which coverage may be lacking. For example, viewed from the standpoint of the three evidence types, actions and products have over 10 data linkages each, while communications has only 7. The reason for this is that automating the content of communications awaits further development of natural language processing analyses, which is only now getting underway. A weakness of our current analysis is thus that the content of communications is not considered here. Viewing the same dataset from the standpoint of the evidence targets or outcomes, four targets have only 1 or 2 data points each (i.e. C1, C3, PL2, PL4) while all other targets have from 3 to 7 data points. Increasing collection points to triangulate any related analyses will strengthen confidence and trust in the findings. Finally, from the viewpoint of the measures, the weakest are 'n and o', where content analysis of the content of communications could play an important role, so this limits the findings and discussion and cautions us to be conservative in making inferences that lack sufficient validation.

Task participation. A fundamental condition for collaboration is participation in shared work; for example, a team needs to build a common understanding of a problem or challenge, take actions to solve or address the problems, and maintain team cohesion (Roschelle & Teasley, 1995). There are both individual and team levels of analysis of these matters, since each individual team member plays a co-participating role that results in the wholistic group behaviour (Ifenthaler, 2014). Evidence of co-participation in a group task is one measure of an individual's contribution to the group and also indicates individual's opportunity to learn (Ifenthaler, 2014; Ifenthaler, Mistree, & Siddique, 2014). We have shown one way to automate the display of individual participation in tasks from a digital learning environment where the structure of a large assignment is broken into smaller tasks and the individual interactions of team with each task is then part of a wholistic picture of the evolution of the team's approach to the assignment via co-authoring efforts (e.g. creating or initiating new work, editing, commenting). The general picture comparing a high to a low performing team is that concerning the former, more people do more work, work together on more sub tasks, and establish deeper ties among all group members, while in the low performing case, team members work more independently, on fewer subtasks, and with little coordination. Other work patterns were also evident, including someone on a team who arrives to the work late and then does a lot at the last minute. The time-based fine grain data, when analysed as a bipartite (people and tasks) network at each slice of time, allows researchers to see the structure of relationships evolving over time, with several implications for task participation as one measure of collaboration.

Completion paths. How a team approaches a large assignment and its subtasks, particularly when the team plans and executes starting and finishing subtasks, is partial evidence of the team's problem-solving approach and capability (Eseryel et al., 2013; Mayer & Wittrock, 1996). Our data seems to indicate that a team that starts and completes subtasks in shorter periods of time and in a logical sequence outperforms a team that takes longer on each subtask, and that starts and completes tasks in a widely dispersed manner. We believe that some of the key subtasks being completed earlier would have created a foundation of knowledge that would have helped the low performing group to make better progress. In the near future, this analysis can also be automated and given to the teams to help them become aware of their 'team self-regulation' (Ifenthaler, 2012).

Attention to feedback. Evidence of a team responding in a timely way and specifically to feedback is linked to two parts of the domain model—examining diverse concepts, a dimension of personal development, (Friedrichs & Gibson, 2003) and monitoring and reflecting, a dimension of problem solving (Ifenthaler, 2014; Mayer & Wittrock, 1996). We found that a high performing team took a shorter amount of time to respond and addressed more of the topics of the feedback than a low performing team.

Use of time. Collaboratively taking appropriate action leaves trace evidence about the use of time by the team, as does the teams' planning and executing actions (Badke-Schaub et al., 2011). This study examined team approaches to the task from the perspective of the use of time. For example, the high and low performing teams differed in the levels of time spent on specific subtasks and what time of day their work occurred. The time analysis also presented a signal to learning designers about which parts of the overall task may have been missed or over-emphasised by the team.

Learning outcomes. If a participant on a team never engages with a particular subtask or other team member, the learning outcomes associated with the subtask might have been underrepresented (Eseryel, Law, Ifenthaler, Ge, & Miller, 2014; Ifenthaler, Gibson, & Zheng, 2020). If a participant works completely alone, for example, then there is no evidence in the digital record to support an inference of collaborative capabilities. As we are taking the stance of working from evidence and creating inferences from the best evidence we have, we expect that these findings will improve in detail and confidence as we add to the number of automated classifications and analyses supported by the system linking actions, products and communications to the intended learning outcomes of the content author.

4.7 Limitations and Conclusion

As outlined in the method, our study was based on a small number of teams. While this is not necessarily a limitation for this study, it is a factor that is driving future research. We have since grown the number of participants, teams and team learning contexts and are currently analysing data for a follow-up paper to test the consistency and extent of this chapters findings as the n sizes and contexts increase. There was also, in this early phase of research, limited external validity of the findings because the case was only about one classroom involving one teacher. We are currently addressing this limitation by expanding the number of teachers, student cohorts and individuals engaging as participants in team learning situations and expanding the ways we collect data via group and individual interviews, survey data and other mechanisms that enhance the insights found by computational methods.

A specific limitation in the data, which we are addressing for future experiments, is that all and any platform interactions by each student were given equal weight, which does not necessarily reflect the quality of effort (e.g. efficient or not) or effectiveness of the interaction (e.g. impactful or not) associated with someone's contribution. The research team sought external validation of the HP and LP teams from the classroom teacher in order to justify the assumption that the HP teams higher interaction count also demonstrated their higher level of performance over the LP team. In future studies, we plan to establish a broad multi-scaled set of measures associated with project tasks and subtasks complexity as well as contributions to allow analyses of the semantic import (e.g. topics covered in communications) and group influence (e.g. impact of an idea on subsequent communications and products) as well as the frequency of interaction and weight of contribution.

Finally, there was labour-intensive manual data processing and analysis during the early stages of the research. Several automated visualisations have since been created and implemented, which we hope will improve the speed, agility and size of the data sets.

Network analysis and graph theory have proven to be an appropriate analysis approach for educational applications. Pathfinder and combined techniques (Durso & Coggins, 1990; Ifenthaler, 2010b; Schvaneveldt, 1990) provide a reliable representation of knowledge structures and analysis of learning by using pairwise similarity ratings among concepts to create networks (Ifenthaler, 2010a). These networks are based on proximity data among entities and are determined by calculating the proximities that best fit within the network. Additionally, graph theory can be applied to almost every area of educational diagnostics (Ifenthaler, 2010b). Picard (1980) introduced a promising approach for the design and analysis of questionnaires using graph theory. Furthermore, graph theory has been successfully applied for instructional planning (Hsia, Shie, & Chen, 2008) and evaluation purposes (Xenos & Papadopoulos, 2007).

This study shows that network-based analyses provide an objective way to represent and evaluate many features of individual participation and contribution during collaborative problem-solving (Eseryel et al., 2013; Ifenthaler, 2014). Network analysis was also found useful for examining the intensity of team-level collaboration by utilising the density property of a bipartite network consisting of agents (team members) and artefacts (team tasks and work products). Furthermore, the analysis of dynamic team networks revealed the periodic changes of individual engagement and group coordination during each stage of a long-term team project, which provided information that in the future could be by instructors to deliver timely intervention and guidance (Ifenthaler, Gibson, & Dobozy, 2018).

Acknowledgements This research is supported by Curtin University's UNESCO Chair of Data Science in Higher Education Learning and Teaching (https://research.curtin.edu.au/unesco/). The research team wishes to thank Gabrielle Joliffe of Mount Lawley High School in Perth, for her invaluable insights as a classroom teacher and challenge content author.

References

Badke-Schaub, P., Neumann, A., & Lauche, K. (2011). An observation-based method for measuring the sharedness of mental models in teams. In M. Boos, M. O. Kolbe, P. M. Kappeler, & T. Ellwart (Eds.), *Coordination in human and primate groups* (pp. 177–197). Berlin: Springer.

Berger, J., Cohen, E., & Zelditch, M. (1966). Status characteristics and expectation states. In J. Berger & M. Zelditch Jr. (Eds.), *Sociological theories in progress* (Vol. 1, pp. 29–46). Boston: Houghton-Mifflin.

Cannon-Bowers, J. A., & Salas, E. (2001). Reflections on shared cognition. *Journal of Organizational Behavior, 22*(2), 195–202.

Clariana, R. B. (2010). Deriving individual and group knowledge structure from network diagrams and from essays. In D. Ifenthaler, P. Pirnay-Dummer, & N. M. Seel (Eds.), *Computer-based diagnostics and systematic analysis of knowledge* (pp. 117–130). New York: Springer.

Clarke-Midura, J., Code, J., Dede, C., Mayrath, M., & Zap, N. (2012). Thinking outside the bubble: Virtual performance assessments for measuring complex learning. In M. Mayrath, J. Clarke-Midura, D. Robinson, & G. Schraw (Eds.), *Technology-based assessments for 21st century skills* (pp. 125–148). Charlotte, NC: Information Age Publishers.

Diestel, R. (2000). *Graph theory*. New York, NY: Springer.

Durso, F. T., & Coggins, K. A. (1990). Graphs in social and psychological sciences: Empirical contributions to pathfinder. In R. W. Schvaneveldt (Ed.), *Pathfinder associative networks: Studies in knowledge organization* (pp. 31–51). Norwood, NJ: Ablex Publishing Corportion.

Eseryel, D., Ge, X., Ifenthaler, D., & Law, V. (2011). Dynamic modeling as cognitive regulation scaffold for complex problem solving skill acquisition in an educational massively multiplayer online game environment. *Journal of Educational Computing Research, 45*(3), 265–287.

Eseryel, D., Ifenthaler, D., & Ge, X. (2013). Validation study of a method for assessing complex ill-structured problem solving by using causal representations. *Educational Technology Research and Development, 61*(3), 443–463. https://doi.org/10.1007/s11423-013-9297-2

Eseryel, D., Law, V., Ifenthaler, D., Ge, X., & Miller, R. B. (2014). An investigation of the interrelationships between motivation, engagement, and complex problem solving in game-based learning. *Journal of Educational Technology & Society, 17*(1), 42–53.

Friedrichs, A., & Gibson, D. (2003). Personalization and secondary school renewal. In J. DiMartino, J. Clarke, & D. Wolf (Eds.), *Personalized learning: Preparing high school students to create their futures*. (p. 41–68.). Lanham, Maryland: Scarecrow Education.

Funke, J. (1991). Solving complex problems: Exploration and control of complex problems. In R. J. Sternberg & P. A. Frensch (Eds.), *Complex problem solving: Principles and mechanisms* (pp. 185–222). Hillsdale, NJ: Lawrence Erlbaum.

Funke, J. (2012). Complex problem solving. In N. M. Seel (Ed.), *The encyclopedia of the sciences of learning* (Vol. 3, pp. 682–685). New York, NY: Springer.

Gee, J. P. (2003). *What video games have to teach us about learning and literacy* (Vol. 1, p. 20). New York: Palgrave Macmillan.

Gibson, D. C., & Jackl, P. (2015). Theoretical considerations for game-based e-learning analytics. In T. Reiners & L. Wood (Eds.), *Gamification in education and business* (pp. 403–416). New York, NY: Springer.

Gibson, D., Irving, L., & Seifert, T. (2018). Assessing personal learning in online collaborative problem solving. In M. Shonfeld & D. Gibson (Eds.), *Collaborative learning in a global world* (p. 450). Charlotte, NC: Information Age Publishers.

Griffin, P., McGaw, B., & Care, E. (Eds.). (2012). *Assessment and teaching of 21st century skills*. Amsterdam: Springer.

Hsia, T. C., Shie, A. J., & Chen, L. C. (2008). Course planning of extension education to meet market demand by using data mining techniques—An example of Chinkuo technology university in Taiwan. *Expert Systems with Applications, 34*(1), 596–602.

Ifenthaler, D. (2010a). Relational, structural, and semantic analysis of graphical representations and concept maps. *Educational Technology Research and Development, 58*(1), 81–97. https://doi.org/10.1007/s11423-008-9087-4

Ifenthaler, D. (2010b). Scope of graphical indices in educational diagnostics. In D. Ifenthaler, P. Pirnay-Dummer, & N. M. Seel (Eds.), *Computer-based diagnostics and systematic analysis of knowledge* (pp. 213–234). New York, NY: Springer.

Ifenthaler, D. (2012). Determining the effectiveness of prompts for self-regulated learning in problem-solving scenarios. *Journal of Educational Technology & Society, 15*(1), 38–52.

Ifenthaler, D. (2014). Toward automated computer-based visualization and assessment of team-based performance. *Journal of Educational Psychology, 106*(3), 651–665. https://doi.org/10.1037/a0035505

Ifenthaler, D., & Gibson, D. C. (2019). Opportunities of analytics in challenge-based learning. In A. Tlili & M. Chang (Eds.), *Data analytics approaches in educational games and gamification systems* (pp. 55–68). Cham: Springer.

Ifenthaler, D., Mistree, F., & Siddique, Z. (2014). Learning how to learn in a team-based engineering eduction. *Interactive Technology and Smart Education, 11*(1), 63–82. https://doi.org/10.1108/ITSE-10-2013-0025.

Ifenthaler, D., Gibson, D. C., & Dobozy, E. (2018). Informing learning design through analytics: Applying network graph analysis. *Australasian Journal of Educational Technology, 34*(2), 117–132. https://doi.org/10.14742/ajet.3767

Ifenthaler, D., Gibson, D. C., & Zheng, L. (2020). Attributes of engagement in challenge-based digital learning environments. In P. Isaias, D. G. Sampson, & D. Ifenthaler (Eds.), *Online teaching and learning in higher education* (pp. 81–91). Cham: Springer.
Jonassen, D. H. (1997). Instructional design models for well-structured and ill-structured problem-solving learning outcomes. *Educational Technology Research and Development, 45*(1), 65–94.
Jonassen, D. H. (2011). *Learning to solve problems. A handbook for designing problem-solving learning environments*. New York: Routledge.
Mayer, R., & Wittrock, M. (1996). Problem-solving transfer. In D. Berliner & R. Calfee (Eds.), *Handbook of educational psychology* (pp. 47–62). New York, NY: Simon & Schuster Macmillan.
Mislevy, R. (2011). *Evidence-Centered Design for Simulation-Based Assessment*. Los Angeles, CA: The National Center for Research on Evaluation, Standards, and Student Testing.
Mislevy, R. J., Almond, R. G., & Lukas, J. F. (2003). A brief introduction to evidence-centered design. *ETS Report Series, 1*, 1–29.
Picard, C. F. (1980). *Graphs and questionnaires*. Amsterdam: North-Holland Publishing Company.
Roschelle, J., & Teasley, S. (1995). The construction of shared knowledge in collaborative problem-solving. In C. O'Malley (Ed.), *Computer-supported collaborative learning* (pp. 69–97). Berlin: Springer.
Schvaneveldt, R. W. (1990). *Pathfinder associative networks: Studies in knowledge organization*. Norwood, NJ: Ablex Publishing Corporation.
Seel, N. M., Ifenthaler, D., & Pirnay-Dummer, P. (2009). Mental models and problem solving: Technological solutions for measurement and assessment of the development of expertise. In P. Blumschein, W. Hung, D. H. Jonassen, & J. Strobel (Eds.), *Model-based approaches to learning: Using systems models and simulations to improve understanding and problem solving in complex domains* (pp. 17–40). Rotterdam: Sense Publishers.
Xenos, M., & Papadopoulos, T. (2007). Computer aided evaluation of higher education tutors' performance. *Studies in Educational Evaluation, 33*(2), 175–196.

Chapter 5
Experiences with Virtual Reality at Secondary Schools: Is There an Impact on Learning Success?

Thomas Keller and Elke Brucker-Kley

5.1 Introduction

Jaron Lanier who is considered a founding father of virtual reality (VR) defines VR among others as an "instrumentation to make your world change into a place where it is easier to learn" (Lanier, 2017). Thus, he describes the potential of immersive VR technologies, which offer 360-degree interactive three-dimensional stimulus environments to engage students in new learning experiences. But, today VR as a promising digital form of experience- and action-oriented learning is still in its infancy in secondary schools in Switzerland. VR learning units for fully immersive VR systems based on head-mounted displays and related to the current learning plans are not commercially available. Usability tests with three prototypically developed VR learning units at secondary schools in Switzerland have shown, however, that the use of VR is considered promising by teachers and that acceptance among students exists (Keller, Hebeisen, & Brucker-Kley, 2018a). However, investments in VR by educational institutions are questionable as long as there is no evidence that the use of immersive VR learning units has an effect on the achievement of learning goals. Beyond providing the basis for convincing business cases for VR-based learning, this research is motivated by the search for design criteria for effective VR learning units.

T. Keller (✉) · E. Brucker-Kley
ZHAW School of Management and Law, Winterthur, Switzerland
e-mail: kell@zhaw.ch; brck@zhaw.ch

D. Ifenthaler et al. (eds.), *Balancing the Tension between Digital Technologies and Learning Sciences*, Cognition and Exploratory Learning in the Digital Age,
https://doi.org/10.1007/978-3-030-65657-7_5

5.2 Research Objective and Approach

The aim of the described research effort is to assess the effect of prototypical VR learning units in mathematics on the learning success of secondary school students. In order to take into account both educational and usability-related criteria for the design and validation of the learning units a three-step research approach is applied and described in the following steps:

1. Prototyping: Design and implementation of a VR learning environment which makes use of the possibilities of immersive VR and implements the requirements of recognized mathematics didactics
2. Testing the usability of the VR learning environment in a controlled experiment with 20 secondary school students (Keller et al., 2018a)
3. Field experiment to evaluate the effect of the VR learning unit on the achievement of predefined learning goals with 67 secondary school students (Keller, Hagen, & Brucker-Kley, 2019)

5.3 State-of-the-Art

5.3.1 Immersive Learning

Learning in virtual worlds is closely linked to the concept of immersion. From a perceptual psychological point of view, immersion describes the feeling of presence in the virtual world, i.e., the illusion of actually being in a virtual reality and being able to interact with it (Cummings & Bailenson, 2016). Burdea & Coiffet speak (2003) of the so-called three I's—interaction, imagination, and immersion—which must be given in order for a virtual reality to feel real to the user.

From a technical point of view, the following properties must be fulfilled for successful immersion (Slater & Wilbur, 1997):

- If possible, the sensory impressions experienced by humans should be generated exclusively by the computer or one or more output devices
- As many senses as possible should be addressed
- The output devices should completely surround the user
- The output devices should enable a vivid depiction of reality

Currently, this degree of immersion is made possible in commercially available form by so-called Virtual Reality headsets. These VR glasses, which are worn as a head-mounted display (HMD), enable the wearer to move around in a 3D world with a 360-degree view and to experience virtual reality optically and acoustically isolated from the real outside world. This is what distinguishes fully immersive systems from simpler forms of virtual reality such as smartphone-based systems (e.g., Google Cardboard) or desktop-based virtual reality (e.g., Second Life).

Even though innovations in the immersive VR field are still driven by the gaming industry, serious fields of application are becoming increasingly apparent. For simulations in high-risk training areas (aerospace, medicine, emergency services, workspace safety) as well as for manufacturing and construction productive VR learning environments for vocational training exist (Höntzsch, Katzky, Bredl, Kappe, & Krause, 2013a). For primary, secondary, and higher education two meta-studies by (Mikropoulos & Natsis, 2011) and (Merchant, Goetz, Cifuentes, Keeney-Kennicutt, & Davis, 2014) conclude that the use of VR can be effective. The effectiveness of immersive 3D learning worlds depends to a large extent on their design being based on both educational and usability-related criteria. (Pirker, Gütl, Belcher, & Bailey, 2013) and (Fowler, 2015) discuss and evaluate models with regard to their applicability to immersive VR learning units. The focus of this valuable research is on the usability of VR learning from the perspective of university students, but not on the effect of VR on the achievement of learning goals by secondary school students. Another research gap arises from the rapidly advancing development of VR technologies and the resulting potential for more effective applications for education. The existing research on learning in virtual worlds and its effects mainly dates from 2006 to 2012 and refers to less immersive desktop-based 3D worlds like Second Life. These learning worlds are only partially comparable to learning units that are implemented for fully immersive VR.

5.3.2 Didactics of Mathematics for Digital Learning Environments

Krauthausen (2012, p. 3) criticizes the fact that computer-assisted learning programs in mathematics that are well-known and widespread on the market today pay too little attention to the current state of research and knowledge in mathematics didactics. The main criticism is that the focus is too much on technology instead of content and the programs thus contribute much to the media competence of learners but little to mathematics skills. He concludes that for the motivation of learners, mathematics itself should increasingly be presented in a way that makes the mathematical content exciting and captivating, rather than packaging the content in an exciting way (Krauthausen, 2012, p. 21). Burrill (2017, p. 316) mentions mathematical accuracy (fidelity) and user experience as central mathematics didactic principles. The mathematical accuracy means that the software should always be mathematically correct, the user experience should not prevent the user from working with the mathematical task and should promote mathematical thinking. Learners need to be able to make decisions to expand their thinking. This possibility is also closely related to the complexity of a task, which does not necessarily require complex mathematical requirements (Geiger, 2017, p. 289). According to Joubert (2017, p. 20), while working on a mathematical task, students apply means from so-called “Modes of Production.” These include acting (usually in the sense of indicating a solution), formulating (developing hypotheses, solution strategies, etc.) and validating (checking based on evidence, theorems, or explanations). Mathematical tasks are intended to encourage learners to do something mathematical and thus experience mathematics in the broadest sense (Joubert, 2017, p. 4). All tasks should always contain pragmatic and epistemological aspects. The epistemological aspects refer to the insights to be conveyed to learners while working on a task (Sinclair & Zazkis, 2017, p. 177), whereby the pragmatic value of a task is almost always equated with solving the task (Sinclair & Zazkis, 2017, p. 190). Laborde (2011, p. 82) supplements cognitive aspects (what kind of learning the task triggers in the learner at the current state of knowledge), didactic aspects (with what means the task is set) and instrumental aspects (which instruments the learner needs to solve the task).

With regard to pedagogy, Geiger (2017, p. 288) points out in the context of mathematics how eminently important it is to select, adapt, and implement the tasks in the learning environments. In this context, he stresses the importance of cooperation between teachers and researchers in order to anchor well-designed tasks with pedagogically correct approaches in the learning environments. Höntzsch et al. (2013b, p. 4) list the following measures to prevent learners from being overburdened in immersive learning environments:

- clear learning objectives, work orders and instructions,
- permanently available background information,
- hints and exercises that stimulate reflection (for example, setting a specific state of the simulation).

5.4 The VR Learning Units

Design and development of the VR learning environment was based on the ADDIE model according to (Olbrish, 2014). This model consists of five steps: analysis (problem definition), design, development, implementation (application in practice), and evaluation (demonstrating effect).

ADDIE follows game design principles in order to keep the tension of the learner in the so-called flow channel between under- and overchallenge. As recommended in the ADDIE model, the concept was set out in a so-called "Game Design Outline." This contains, among other things, the objective, storyline, teaching strategy, structure of the game, game components, results (including tracking) and content of the individual scenes (Olbrish, 2014, p. 51). The iterative implementation of the VR prototype was realized with the game engine Unity for a HTC Vive head-mounted display.

To evaluate the artifacts described above, the following design science evaluation methods according to (Hevner, March, Park, & Ram, 2004) were applied:

- Experimental (controlled experiment): usability testing with secondary school students
- Descriptive (substantiated discussion): The artifacts are compared with the state of the art, the requirements and the problem statement.
- In addition, both the iterations of the concept for the learning units and the iterations of the prototype were discussed with teachers in qualitative interviews

Based on the curriculum and the existing maths textbook (Affolter & Walt, 2017), the learning units described in the following section were selected for design and implementation in VR. The selection was driven on the one hand by the suitability of the learning content for immersive learning (imagination, interaction, immersion) and on the other hand by the recommendations and feedback of the involved teachers.

The learning environment is portrayed and can be downloaded as VR app for HTC Vive at www.neuelehrkonzepte.ch (Keller, 2017) for HTC Vive.

5.4.1 Unit 1: Introduction to VR

The first virtual learning unit is designed to introduce the students to the immersive experience and the interaction controls. The virtual reality is to be experienced for the first time and the basic interaction concepts for the other learning environments can be experienced and tested. This unit is provided with the intention to prepare the learners so they can fully concentrate on the task in the subsequent learning environments without being distracted by the controls. All learning units are experienced in a single player mode without interacting with other students or teachers. The student learns to move via teleportation, to grab and drop objects and to handle sliders. Duration is 5–10 min.

5.4.2 Unit 2 "Base Area * Height"

The aim of VR unit 2 is to illustrate the connection between the base area, height, and volume of a geometric body and to enable the students to experience their interrelation in three dimensions. Five different geometric bodies must first be placed on their base surface and then a given volume must be achieved by changing the base surface and height in an interactive manner with a slider. Duration is 10–15 min.

5.4.3 Unit 3 "So Big, So Small"

Unit 3 enables the student to interact with hollow masses and lengths in a playful manner. The user has to take different sizes, written as text signs (e.g., 1 L), from a shelf and assign them to a suitable everyday object (e.g., a carton of milk, an aquarium etc.). At the beginning of the unit all objects are displayed in the same size. Only when the solution is requested they take on their correct relative size and the differences become tangible. Duration is 5–15 min.

The next scene works similar, with the difference that linear measures are used (e.g., 1 m). These must be assigned to suitable lengths from everyday life (e.g., the

Fig. 5.1 Sample scene of VR learning unit 3 "So big, so small"

arm span of a person). Thus, length measurements from the tip of a pencil up to 2.5 laps in a sports stadium can be experienced by the student. Duration is 5–10 min (Fig. 5.1).

5.5 Usability Test

Since usability undoubtedly has an influence on the effectiveness of the VR learning units, the prototype was tested in a first phase at four different Swiss secondary schools in the canton of Berne with a total of 20 students. The aim of the usability tests was to identify factors which could possibly favor or hinder the use of the VR learning units with the target group of secondary students. Factors examined included the motivation of the students to learn in such a virtual environment, the learners' personal feelings about learning success, the personal wellbeing during the experience and the role of gamification elements (rewards). Furthermore, the tests helped to validate the didactic correctness and the quality of the implementation of the learning units.

5.5.1 Sample and Test Procedure

A relevant concern with the use of new technologies in education is that they may not only have potential but also disadvantages for students with special needs. The fact that the VR learning environment described in the previous section was

Table 5.1 Summary of participants of the usability tests

Grade	Age	Gender	Number
7	13–15	Female	1
		Male	7
8	14–16	Female	3
		Male	3
9	15	Female	1
		Male	5
	Total number of participating students		20

originally designed for pupils with special needs was a favorable factor for the usability tests. Discussions with the teachers of the participating four schools did show that the idea of inclusion had been implemented for about a year and that students with special needs were integrated into regular classes. In the subject of mathematics, diagnoses for isolated learning disorders (e.g., dyscalculia) were uncommon. Instead, the teachers spoke of partial weaknesses in mathematics. Students with partial weaknesses were integrated into regular mathematics lessons. As a result, the scope for the usability tests was extended. Students with partial weaknesses in mathematics were selected as participants regardless of whether they were diagnosed with a specific disorder. The decision as to who would participate in the evaluation was made by the teachers. Table 5.1 shows the age and number of participants by grade and gender.

The evaluation sessions at each of the four schools lasted between half a day and a full day (depending on the number of participants). If possible, a room was reserved for the entire duration of the evaluation. The test set-up consisted of a powerful PC with sufficient graphical performance, a monitor for observation by the supervisor and the "HTC Vive" system consisting of head-mounted display (HMD, "glasses"), the two controllers and the base stations for motion tracking. The students were taken out of class individually or in pairs and completed the various learning environments in a maximum of 45 min. Before entering the virtual world, the instructions including rules and rights were discussed and the most important operating elements explained. The coaching during the experience in the virtual learning environments was carried out by a researcher co-present as a supervisor outside of the virtual reality environment.

5.5.2 *Findings of the Usability Test*

Directly after completing the learning units all participants completed an anonymous paper-based questionnaire with 24 questions to evaluate their experience (see Tables 5.2 and 5.3). Questions 1 and 2 ask about the previous experience with virtual reality and video games. Questions 3 to 5 examine the operation and clarity of the tasks. The learning units 2 and 3 are evaluated more precisely with questions 6

Table 5.2 Usability testing—compilation of survey results questions 1–10

#	Age	Class	Sex	Maximalwert in cm (unterschiedlich auf Grund unterschiedlicher Skalierung verschiedener Drucker) Max.	Learning unit 1 VLU1	Learning unit 2 VLU2	Learning unit 3a VLU3	Learning unit 3b VLU3	How easy was the user interface to learn? 3.		How clear were the tasks? 4.		How much were you dependent on the support of the supervisor (coaching)? 5.		How well known were the tasks? Have you already solved similar tasks? 6.		How much did you like the learning environment? 7.		How difficult were the tasks for you? (inverse scale) 8.!		How did you solve your tasks? 9.		How much did you like the learning environment? 10.	
1	15	9	male	15.3	X		X	X	14.8	96.7%	15.3	100.0%	0.0	0.0%	14.6	95.4%							15.3	100.0%
2	14	8	male	15.3	X		X	X	7.6	49.7%	14.9	97.4%	7.5	49.0%	7.7	50.3%							14.9	97.4%
3	15	9	male	15.3	X	X	X	X	13.9	90.8%	14.9	97.4%	10.2	66.7%	14.7	96.1%	14.2	92.8%	1.7	11.1%	7.2	47.1%	13.9	90.8%
4	15	9	female	15.1	X	X	X	X	7.4	49.0%	14.3	94.7%	7.2	47.7%	7.5	49.7%	13.2	87.4%	10.0	66.2%	8.7	57.6%	11.9	78.8%
5	15	8	female	15.1	X	X	X	X	11.1	73.5%	12.9	85.4%	7.6	50.3%	14.2	94.0%	12.4	82.1%			5.5	36.4%	11.0	72.8%
6	14	8	male	15.1	X	X	X	X	12.7	84.1%	4.1	27.2%	7.5	49.7%	7.7	51.0%	9.2	60.9%	3.9	25.8%	5.3	35.1%	8.7	57.6%
7	15	9	male	15.1	X	X	X	X	12.6	83.4%	11.3	74.8%	6.5	43.0%	6.4	42.4%	11.4	75.5%	1.7	11.3%	11.8	78.1%	10.2	67.5%
8	15	9	male	15.1	X	X	X	X	7.5	49.7%	11.6	76.8%	9.7	64.2%	0.0	0.0%	11.3	74.8%	2.4	15.9%	15.0	99.3%	11.3	74.8%
9	15	9	male	15.1	X		X	X	13.8	91.4%	15.1	100.0%	11.2	74.2%	15.1	100.0%							12.1	80.1%
10	13	7	female	14.5	X	X	X	X	12.7	87.6%	7.2	49.7%	7.2	49.7%	7.2	49.7%	14.5	100.0%	6.3	43.4%	9.3	64.1%	14.5	100.0%
11	13	7	male	14.5	X	X	X	X	8.7	60.0%	14.5	100.0%	3.6	24.8%	6.6	45.5%	13.0	89.7%	5.1	35.2%	11.2	77.2%	13.1	90.3%
12	13	7	male	14.5	X	X	X	X	14.3	98.6%	14.4	99.3%	7.4	51.0%	10.3	71.0%	8.1	55.9%	6.1	42.1%	11.1	76.6%	14.3	98.6%
13	14	8	female	14.5	X	X	X	X	7.2	49.7%	10.5	72.4%	7.2	49.7%	9.7	66.9%	11.5	79.3%	6.2	42.8%	9.8	67.6%	8.8	60.7%
14	16	8	male	14.5	X			X	6.1	42.1%	7.0	48.3%	6.8	46.9%	7.7	53.1%							11.5	79.3%
15	15	8	female	14.5	X	X	X	X	10.0	69.0%	13.9	95.9%	7.4	51.0%	10.5	72.4%	8.8	60.7%	7.8	53.8%	0.0	0.0%	6.4	44.1%
16	14	7	male	14.5	X	X	X	X	11.2	77.2%	13.9	95.9%			0.7	4.8%	11.7	80.7%	3.5	24.1%			11.8	81.4%
17	13	7	male	14.5	X	X	X	X	8.4	57.9%	7.2	49.7%	7.3	50.3%	13.9	95.9%	11.5	79.3%	3.5	24.1%	14.3	98.6%	11.6	80.0%
18	15	7	male	14.5	X	X	X	X	11.3	77.9%	11.8	81.4%	8.3	57.2%	14.1	97.2%	13.7	94.5%	3.8	26.2%	8.9	61.4%	14.2	97.9%
19	13	7	male	14.5	X	X	X	X	13.8	95.2%	14.1	97.2%	9.0	62.1%	9.5	65.5%	13.0	89.7%	0.3	2.1%	10.5	72.4%	13.1	90.3%
20	13	7	male	14.5	X	X	X	X	14.5	100.0%	9.9	68.3%	9.5	65.5%	11.4	78.6%	14.3	98.6%	1.0	6.9%	7.2	49.7%	12.4	85.5%
								N		20		20		19		20		16		15		15		20
								Max		100.0%		100.0%		74.2%		100.0%		100.0%		66.2%		99.3%		100.0%
								Min		42.1%		27.2%		0.0%		0.0%		55.9%		2.1%		0.0%		44.1%
								Average		74.2%		80.6%		50.2%		64.0%		81.4%		28.7%		61.4%		81.4%
								Median		77.6%		90.1%		50.3%		66.2%		81.4%		25.8%		64.1%		80.8%
								Variance		3.8%		4.8%		2.6%		8.4%		1.8%		3.4%		6.6%		2.4%
								Standard deviation		19.4%		21.9%		16.2%		29.0%		13.4%		18.3%		25.6%		15.4%

Table 5.3 Usability testing—compilation of survey results questions 11–21

#	Age	Class	Sex	Maximalwert in cm (unterschiedlich auf Grund unterschiedlicher Skalierung verschiedener Drucker) Max.	Learning unit 1 VLU1	Learning unit 2 VLU2	Learning unit 3a VLU3	Learning unit 3b VLU3	How difficult were the tasks for you? (inverse scale) 11.!		Did you know the "Stellentafel" before? 12.		How easily did you recognize the objects? 13.		How much did it help you to see the items in full size to imagine the sizes? 14.		How was it for you in the virtual environment? (inverse scale) 15.!		How concentrated were you in the virtual environment compared to normal lessons? 16.		How well could you concentrate on the instructions of the supervisor? 17.		Wie sehr hat sich das Lernen nach Schule angefühlt? 18.		Did the awards in the main menu motivate you to solve further tasks or to repeat a task? 19.		How much do you think you have learned? 20.		Would you like to have classes in virtual reality again? 21.	
1	15	9	male	15.3	X		X	X	6.2	40.5%	12.3	80.4%	11.5	75.2%	7.5	49.0%	0.0	0.0%	10.3	67.3%	15.3	100.0%	5.5	35.9%	10.5	68.6%	11.5	75.2%	13.1	85.6%
2	14	8	male	15.3	X		X	X	6.7	43.8%	7.2	47.1%	12.6	82.4%	11.8	77.1%	0.2	1.3%	12.6	82.4%	14.7	96.1%	0.6	3.9%	7.3	47.7%	4.2	27.5%	14.9	97.4%
3	15	9	male	15.3	X	X	X	X	2.0	13.1%	12.4	81.0%	12.4	81.0%	15.3	100.0%	0.0	0.0%	15.2	99.3%	12.3	80.4%	3.0	19.6%	15.3	100.0%	11.4	74.5%	15.3	100.0%
4	15	9	female	15.1	X	X	X	X	7.1	47.0%	12.8	84.8%	11.0	72.8%	13.4	88.7%	4.1	27.2%	12.5	82.8%	12.5	82.8%	5.4	35.8%	13.8	91.4%	10.7	70.9%	11.7	77.5%
5	15	8	female	15.1	X	X	X	X	6.8	45.0%	1.2	7.9%	11.1	73.5%	14.0	92.7%	5.0	33.1%	11.5	76.2%	12.2	80.8%	12.9	85.4%	11.7	77.5%	10.4	68.9%	12.3	81.5%
6	14	8	male	15.1	X	X	X	X	3.8	25.2%	11.2	74.2%	5.6	37.1%	13.2	87.4%	1.3	8.6%	11.0	72.8%	12.0	79.5%	13.0	86.1%	10.6	70.2%	7.7	51.0%	14.7	97.4%
7	15	9	male	15.1	X	X	X	X	2.6	17.2%	0.6	4.0%	14.5	96.0%	7.1	47.0%	6.5	43.0%	7.1	47.0%	13.3	88.1%	11.3	74.8%	13.0	86.1%	10.1	66.9%	13.9	92.1%
8	15	9	male	15.1	X	X	X	X	2.4	15.9%	15.0	99.3%	11.3	74.8%	15.1	100.0%	10.2	67.5%	0.0	0.0%	14.9	98.7%	3.5	23.2%	13.6	90.1%	7.5	49.7%	11.5	76.2%
9	15	9	male	15.1	X		X	X	2.1	13.9%	13.8	91.4%	14.4	95.4%	14.8	98.0%	3.5	23.2%	4.1	27.2%	12.1	80.1%	8.2	54.3%	12.3	81.5%	3.0	19.9%	12.5	82.8%
10	13	7	female	14.5	X	X	X	X	6.3	43.4%	6.6	45.5%	7.1	49.0%	10.7	73.8%	0.0	0.0%	14.5	100.0%	11.2	77.2%	11.5	79.3%	14.5	100.0%	10.6	73.1%	14.4	99.3%
11	13	7	male	14.5	X	X	X	X	6.2	42.8%	0.1	0.7%	7.8	53.8%	10.2	70.3%	7.2	49.7%	14.4	99.3%	14.5	100.0%	5.6	38.6%	12.8	88.3%	6.8	46.9%	14.5	100.0%
12	13	7	male	14.5	X	X	X	X	3.9	26.9%	10.9	75.2%	14.3	98.6%	10.7	73.8%	0.4	2.8%	13.5	93.1%	14.3	98.6%	7.0	48.3%	14.4	99.3%	8.0	55.2%	14.5	100.0%
13	14	8	female	14.5	X	X	X	X	6.2	42.8%	6.8	46.9%	10.7	73.8%	9.5	65.5%	6.3	43.4%	12.7	87.6%	6.8	46.9%	12.5	86.2%	9.3	64.1%	9.4	64.8%	14.5	100.0%
14	16	8	male	14.5	X			X	6.2	42.8%	13.3	91.7%	7.4	51.0%	13.3	91.7%	0.4	2.8%	13.3	91.7%	13.1	90.3%	13.3	91.7%	13.6	93.8%	13.9	95.9%	13.1	90.3%
15	15	8	female	14.5	X	X	X	X	4.9	33.8%	12.5	86.2%	14.5	100.0%	11.4	78.6%	0.0	0.0%	12.2	84.1%	12.5	86.2%	2.5	17.2%	6.9	47.6%	8.4	57.9%	14.5	100.0%
16	14	7	male	14.5	X	X	X	X	3.9	26.9%			13.7	94.5%	13.8	95.2%	0.5	3.4%	13.5	93.1%	14.0	96.6%	13.8	95.2%	7.2	49.7%	14.1	97.2%	14.0	96.6%
17	13	7	male	14.5	X	X	X	X	3.5	24.1%	13.5	93.1%	13.6	93.8%	6.7	46.2%	0.2	1.4%	14.3	98.6%	10.8	74.5%	6.9	47.6%	7.0	48.3%	9.5	65.5%	14.3	98.6%
18	15	7	male	14.5	X	X	X	X	6.1	42.1%	8.4	57.9%	8.4	57.9%	6.7	46.2%	0.2	1.4%	13.0	89.7%	12.1	83.4%	3.0	20.7%	11.2	77.2%	10.2	70.3%	14.0	96.6%
19	13	7	male	14.5	X	X	X	X	1.3	9.0%	13.2	91.0%	13.7	94.5%	10.1	69.7%	0.3	2.1%	12.4	85.5%	8.5	58.6%	12.8	88.3%	0.6	4.1%	12.3	84.8%	13.4	92.4%
20	13	7	male	14.5	X	X	X	X	0.1	0.7%	6.6	45.5%	10.0	69.0%	10.2	70.3%	0.3	2.1%	13.8	95.2%	11.6	80.0%	13.4	92.4%	13.9	95.9%	10.8	74.5%	14.4	99.3%
								N		20		19		20		20		20		20		20		20		20		20		20
								Max		47.0%		99.3%		100.0%		100.0%		67.5%		100.0%		100.0%		95.2%		100.0%		97.2%		100.0%
								Min		0.7%		0.7%		37.1%		46.2%		0.0%		0.0%		46.9%		3.9%		4.1%		19.9%		76.2%
								Average		29.8%		63.4%		76.2%		76.1%		15.6%		78.6%		83.9%		56.2%		74.1%		64.5%		93.2%
								Median		30.3%		75.2%		75.0%		75.5%		2.8%		86.6%		83.1%		51.3%		79.5%		67.9%		97.0%
								Variance		2.1%		10.1%		3.5%		3.4%		4.4%		6.8%		1.9%		9.4%		6.1%		3.8%		0.7%
								Standard deviation		14.4%		31.7%		18.8%		18.4%		21.0%		26.0%		13.6%		30.7%		24.6%		19.4%		8.1%

to 14 (some questions appear duplicated because they refer to learning unit 2 and 3 separately). The remaining 10 questions deal with the general learning experience.

The answers to question 1 ($N = 20$) show that the majority of the participating students had little experience with VR before the experience (only one person had regular access to a system, eight students had already been in contact or had already seen a VR system).

The frequency with which the test persons play computer games (question 2, $N = 19$) varies greatly. However, no influence on the operation of the learning environments could be observed. Some of those who were very quickly familiar with the operation of the virtual learning environments state that they only very rarely deal with video games. Others, who state that they play a lot, had more trouble with the handling of the VR unit.

In question 5 ($N = 19$) all indicated that they were dependent on the support of the supervisor. Most of them, according to their own statements, needed only a few tips (12) or very little support (5). Two students state that they could not cope with the learning environment without support or at least were heavily dependent on the support of the supervisor.

All of them stated that they liked the two learning units up to perfect (median 81.4% and 80.8%, respectively; question 7 and 10, $N = 16$ and $N = 20$). Almost all of them indicate the level of difficulty (questions 8 and 11) between easy and medium. Only two describe the learning units as rather difficult ($N = 15$ and $N = 20$ respectively). The observations did show that few students were able to solve the tasks (especially learning unit 3) directly without having to think and correct their first solution. For learners who have succeeded in doing so, there is a small tendency for them to assess their concentration. (Question 16) and their learning success (Question 20) lower.

The place value chart (question 12, $N = 19$) implemented as a shelf in virtual learning unit 3, a didactic element of the known math textbook, was not recognized by 3 learners according to their statements. The others had already seen it, with a tendency to know it well. This didactic material is therefore also well recognized in its virtual form of presentation. Only 9 immediately recognized the everyday objects (question 13, $N = 20$). The others did not recognize all of them immediately. According to the observations, this particularly affects the 3D model of a child's arm span (1 m, often thought to be a doll not a person) and the syringe and ink cartridge of the hollow masses (even in original size the difference in size was often not recognized). Scales had been added as supporting aids during the implementation.

Seeing the objects in their original size has helped all participating students (mostly a lot) (question 14, $N = 20$). It therefore seems to have succeeded in making it possible to experience orders of magnitude that are difficult to comprehend on paper and to point out errors simply, comprehensibly, and impressively.

According to the learners (question 16, $N = 20$), learning in the virtual learning environment has a positive influence on concentration. Three learners state to have been more concentrated than in class, 14 students state that they have even been very concentrated. The 3 test participants (all from the ninth grade), who stated lower values, had the subjective impression that the tasks were rather too easy for them. This is also confirmed by their answers on the level of difficulty and subjectively perceived learning success.

What is surprising is the effect of the trophies which were implemented as a minimal form of gamification to increase the motivation of the learners. This was observed during the evaluation and is also clearly reflected in the questionnaire in the answers to question 19 ($N = 19$) (median 79.5%). Only one person states that the awards hardly motivated him. 10, on the other hand, have been very motivated, 5 also indicate a strong positive influence on motivation and the remaining 4 have at least been somewhat motivated.

With an overwhelmingly high value, all participants state that they would very much like to have lessons in virtual reality again (question 21, $N = 20$, median 97.0%, minimum 76.2%). The positive effect of the novel medium on motivation seems to be given. The fact that half of all learners say that learning in the virtual environment felt more than half (or even completely) like school (question 18, $N = 20$), and that everyone claims to have learned something (mostly much, median 67.9%) (question 20, $N = 20$) also points out that the interest in learning with VR is not just an escape from regular school.

5.6 Field Experiment

5.6.1 *Design of the Field Experiment*

In the field experiment a slightly modified version of the pre-test/post-test control group design according to (Campbell & Stanley, 1967) was applied. A written pre-test was conducted and graded at the beginning of the field experiment to assess the level of knowledge in relation to the learning goals covered by the learning unit (see Sect. 5.6.2). Then the independent variable, i.e. the exposure to the VR learning environment, was changed, and a post-test was performed. In addition to this second test, a third test was performed 1 month later to measure medium-term learning success. The external validity of the experiment was increased by not informing the students that they are part of a field experiment.

Four classes of a Swiss secondary school in the canton of Zurich served as comparison groups. This school was not part of the usability tests described in Sect. 5.5. In the public school system of the respective canton, secondary school classes are divided into three categories—Sek A, Sek B, Sek C—based on cognitive requirements, with A being the most demanding category. Two of the four classes participating in the field experiment belong to the higher performance level A and two classes to the lower performance level B. One of the two classes of each category was in the experimental group (VR unit) and the other one in the control group (regular unit). In total, the comparison groups included 87 students, 45 of whom belonged to a Sek A class and 42 to a Sek B class. Due to drop-outs in course of the 3 tests of the field experiment, 67 subjects could be included in the final analysis. Of the 67 valid subjects 34 were in the experimental group and experienced the VR unit, while the other 33 subjects were in the control group and attended a regular mathematics lesson targeted at the very same predefined learning goals (see Sect. 5.6.2).

The randomization in this field experiment was given by the class distribution of the school. Thus, the field experiment must be regarded as quasi-randomized.

A protocol was kept during the execution of the field experiment, which recorded exceptional events such as assistance or technical problems. Furthermore, the students of the experimental group assessed their subjective learning success and learning experience in a verbal interview based on a structured questionnaire directly after experiencing the VR learning units.

5.6.2 *Refinement of Learning Goals*

In order to ensure that the VR lesson and the conventional mathematics lesson have the same learning objectives and to check the level of knowledge in the pre-test and the two post-tests in a comparable way, verifiable learning goals had to be specified starting from the learning goals driving the design of the VR learning units. Ten cognitive learning goals on all six layers of the Bloom taxonomy of learning goals were defined in close collaboration with the involved teachers (Bloom, Krathwohl, & Masia, 1984):

1. Knowledge "involves the recall of specifics and universals, the recall of methods and processes, or the recall of a pattern, structure, or setting."
2. Comprehension "refers to a type of understanding or apprehension such that the individual knows what is being communicated and can make use of the material or idea being communicated without necessarily relating it to other material or seeing its fullest implications."
3. Application refers to the "use of abstractions in particular and concrete situations."
4. Analysis represents the "breakdown of a communication into its constituent elements or parts such that the relative hierarchy of ideas is made clear and/or the relations between ideas expressed are made explicit."
5. Synthesis involves the "putting together of elements and parts so as to form a whole."
6. Evaluation engenders "judgments about the value of material and methods for given purposes."

This resulted in the following learning goals matched to the six taxonomy levels according Table 5.4.

Table 5.4 Goal—taxonomy level mapping

Learning goal	Taxonomy level
The student knows the different units of hollow measures and can arrange them according to size	1 and 2
The student knows the different units of measures of length and can arrange them according to size	1 and 2
The student is able to determine/imagine the hollow measures of everyday objects without using any aids	3
The student is able to determine/imagine the measures of length of everyday objects without using any aids	3
The student is able to compare the hollow measures with each other and put them in relation to each other without using any aids	4
The student is able to compare and relate length measures to each other without using any aids	4
The student is able to assign hollow measures to everyday activities, for example, how much water she or he drinks per day	5
The student is able to assign length measures of lengths to everyday activities, for example, the length of the way to school	5
The student can check the correctness of a statement or task regarding hollow measures without using any aids	6
The student can check the correctness of a statement or task regarding measures of length without using any aids	6

5.6.3 Learning Unit Without VR

The conventional lesson without the use of VR for the control group was performed by the teachers themselves in order to maintain the authenticity of the field experiment. The content of the lesson and the paper-based exercise were based on the same math textbook chapters as the VR learning unit and were geared toward the same learning goals as listed above.

5.6.4 Evaluation of the Results of the Field Experiment

Both the immersive VR learning experience and the conventional teaching lesson have resulted in short and medium-term learning success for the students in both comparison groups.

The subjects in the experimental group (with VR) between the pre-test (test 1) and the first post-test (test 2) achieved an average learning success of 1.38 points (19.14%). In the control group, the average learning success between the same tests was 1.30 points (18.49%). The medium-term learning success, which is measured by the difference between test 1 and the second post-test (test 3), averages 0.64 points (8.88%) for the subjects in the experimental group. The subjects in the control group recorded an average learning success of 0.45 points (6.40%) between the same tests (Fig. 5.2).

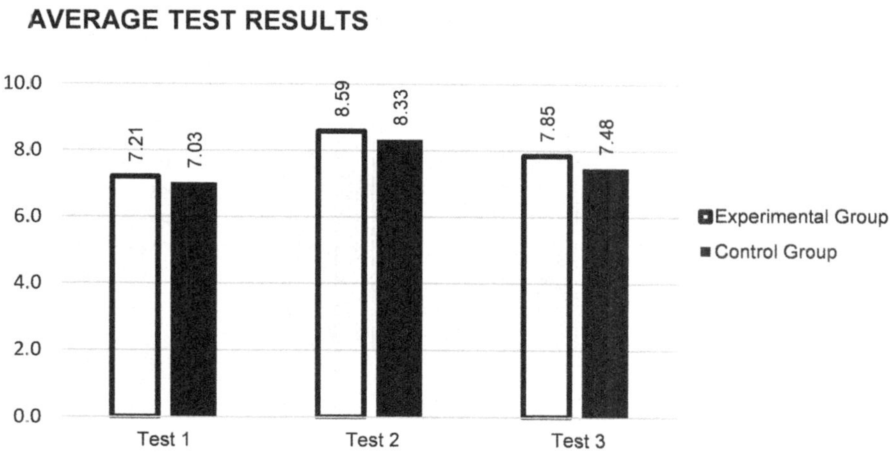

Fig. 5.2 Average values of test results of experimental and control group

Table 5.5 Comparison between experimental with control group

Change in learning success	Experimental group	Control group	Difference	Statistical significance
From test 1 to test 2	1.38	1.30	0.08	58.24%
From test 1 to test 3	0.64	0.45	0.19	64.71%

If the average test results of the subjects in the experimental and control groups are compared, the difference between test 1 and test 2 is 0.08 points. The statistical significance is 58.24%, which means that the effects of the two teaching methods do not differ. Similar observations can be made for the measured average learning outcomes between test 1 and test 3. There it is a difference of 0.19 points with a statistical significance of 64.71% (Table 5.5).

In summary, this means that both immersive VR and the conventional teaching lesson have resulted in statistically verifiable learning success for the test persons. If the learning outcomes of the different groups are compared, however, there are no differences. Both ways of teaching are to be regarded as equivalent in terms of both short-term and medium-term learning success in this setting. However, there are tendencies, if the questions of the random sample tests are considered individually, that the tasks with reference to the measures of length were better solved by the test persons of the experimental group and the tasks with reference to the hollow masses were better solved by the test persons of the control group.

In order to be able to make more precise statements about the learning success of the test persons, the test results of the comparison groups were analyzed by the performance level of students given by their affiliation to the categories of Sek A (high) and Sek B (lower). Their average test results show that there are no striking differences in the achievement of measurable learning goals of the test subjects between test 1 and test 2 (Fig. 5.3). However, if one looks at test 3, it is noticeable

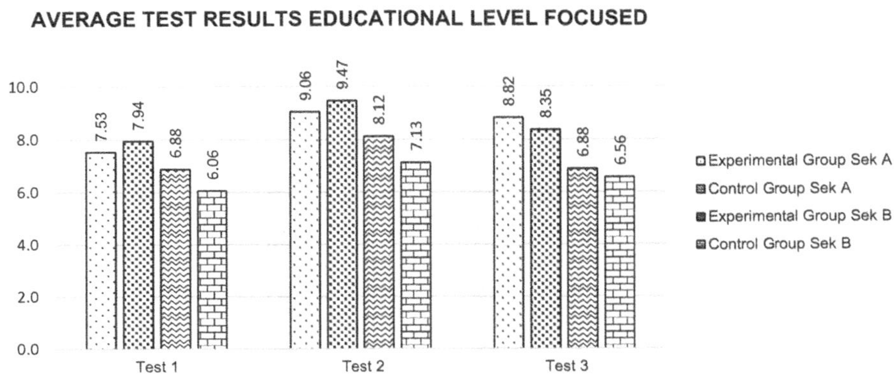

Fig. 5.3 Average test results by performance levels

Table 5.6 Learning success between test 1 and test 2

Comparison of subjects of Sek A	Experimental group	Control group	Difference	Statistical significance
Learning success test 1 to test 3	1.24	0.41	0.83	88.50%

Table 5.7 Comparison of learning success between levels

Comparison of experimental group	Subjects of Sek A	Subjects of Sek B	Difference	Statistical significance
Learning success test 1 to test 3	1.24	0.00	1.24	99.01%

that the test subjects in Test Group Sec A achieved the highest average score, which was not the case in the previous two tests.

The average learning success between test 1 and test 3 was 1.53 points (16.33%) for the subjects in the Sek A test group. The average learning success achieved by the test subjects in the Sek A control group was 0.41 points (5.16%), which is considerably lower.

If the average test results between test 1 and test 3 of the subjects in the Sek A category of both comparison groups are compared, the difference between the average learning successes is 0.83 points (Table 5.6). The statistical significance is 88.50%. This means that the difference is not statistically relevant, but there is a tendency that should be further investigated.

If the average learning success of the subjects in Sek A is compared with that of the subjects in Sek B of the experimental group between test 1 and test 3, this shows a difference of 1.24 points. In addition to the comparatively high learning success of the subjects in Sek A, this is mainly due to the fact that the subjects in Sek B did not achieve any learning success in these tests. With a statistical significance of 99.01%, this difference is considered statistically relevant (Table 5.7).

In summary, this means that immersive teaching methods led to verifiable short and medium-term learning success for subjects in Sek A, in contrast to the students

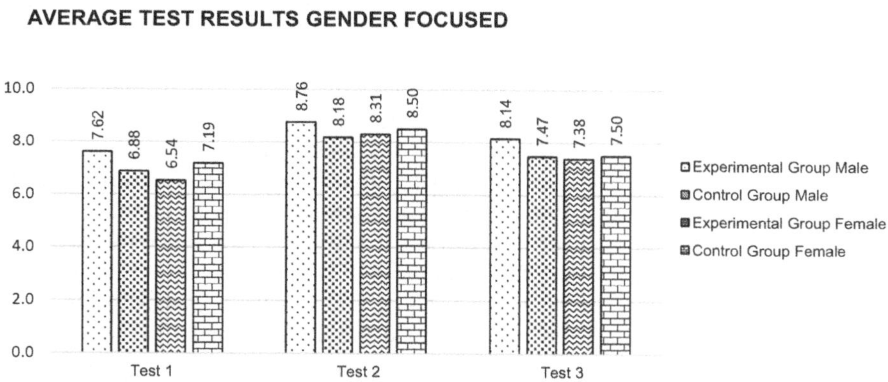

Fig. 5.4 Average test results by gender

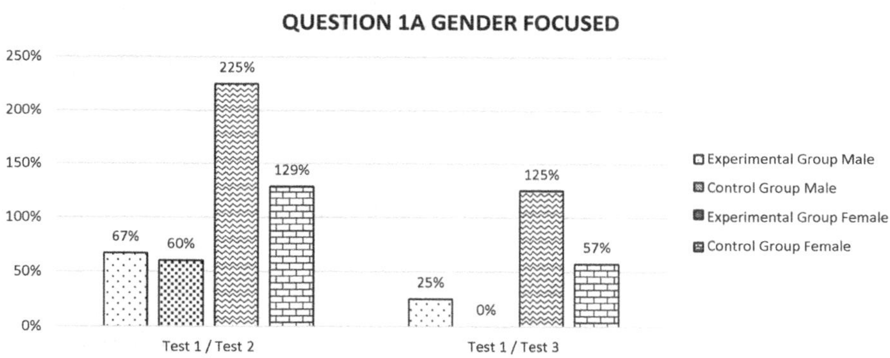

Fig. 5.5 Question 1A visualized by gender

in Sek B, who only achieved short-term learning success. Consequently, there is a difference between the two performance categories levels with regard to medium-term learning success with VR learning units.

In a next step the test results of both control and experimental groups were analyzed by gender. As can be seen in Fig. 5.4, the average test results are similar. The female students in the experimental group who achieved the highest learning success between test 1 and test 2 with 1.77 points (27.06%) are notable. This finding is particularly evident when the focus is placed on the individual questions of the random sample tests. For example, the female respondents in the experimental group in question 1a, which related to the measures of length in the learning unit, have by far the highest average learning success with 225.00% between test 1 and test 2.

As shown in Fig. 5.5, the female respondents in the control group are far below this with an average learning success of 128.57%. The male subjects in the experimental group were also lower with an average learning success of 66.67%. Similar observations can be made for the learning success between test 1 and test 3. With 125.00% learning success, the female subjects in the experimental group are far ahead of the female subjects in the control group with 57.14% and the male subjects in the experimental group with 66.67%.

If, on the other hand, the average learning success between test 1 and test 2 of the female subjects in the test and control groups are compared, the difference is 0.46 points, with the test group scoring higher. The statistical significance is 78.39%. A similar observation can be made between test 1 and test 3, where the difference is also 0.46 points. Here, the statistical significance is somewhat lower at 74.82%. This means that both learning units are of equal value for the female subjects.

If the average learning successes of the male and female subjects in the comparison group are compared, there is a difference of 0.63 points, with the female subjects being higher.

In summary, this means that there are no differences between the subjects in the experimental and control groups when analyzed by gender. The effects of the VR and the conventional learning units are to be considered as equivalent. In addition, there are no differences between male and female subjects who used the VR unit. However, there is a tendency that the female subjects benefited more from the VR units than the male subjects.

5.6.5 *Observations and Feedback of Students*

The positive effect on motivation and possible new experiences mentioned in the literature was confirmed by the observations and the verbal survey among the students experiencing the VR learning units. Almost all students have worked in a very concentrated manner and state this in the questionnaire. The potential for addiction quickly became apparent in course of the experiment. Questions were asked about the possible use of VR technology for video games and many stated in the answers to the questionnaire that they already spend a lot of time with video games every day. A constructive discussion with the learners about addictive media behavior and content in private use appears to be sensible and, alongside clear rules of conduct seems relevant to counteract problematic media behavior.

The need for coaching, help, and feedback in the VR learning units also became evident. Most learners say that they were dependent on the support of the physically present supervisor and were able to concentrate on them. These feedbacks indicate, that most of the subjects could be kept in the flow channel (Olbrish, 2014, see Sect. 5.4) during the evaluation, as the observations of the supervisor during the experiences also confirm. The successful implementation of the ADDIE model could also be concluded from the minor correlation that has been recognized in the answers on task difficulty, concentration, and learning success. The positive effect on self-assessed learning success and motivation mentioned in the literature can be confirmed. With a few exceptions, all learners indicated a positive learning success in the responses to the questionnaire. Many learners were able to observe how they could expand their personal experience space (eureka moment). However, empirical proof is still lacking.

5.7 Conclusions and Outlook

The analysis of the field experiment showed that both learning units resulted in short and medium-term learning success for the test persons, but that there were no differences between the effects of the two teaching approaches. A difference can be observed between the educational levels. In contrast to the students in Sek B, the students in Sek A had a better learning success in the medium term.

Statistically, there are no significant differences by gender. However, there are indications that female subjects benefit more than male subjects. This tendency could not be statistically proven in the context of the present study. It concerns both short and medium-term learning success. This is a finding that could be further investigated within the framework of future research.

The effectiveness of the tested VR learning units differs by content. Measures of length performed better than hollow measures. It remains to be examined whether this really has to do with the specific content or with the specific design of the respective learning units.

Furthermore, an important limitation must be mentioned. There is no doubt that the novelty of VR has an influence on the test persons. This can have a positive effect (motivation), but also a negative one (distraction). In future field experiments this can only be addressed if the students work with VR learning units or other VR applications for a longer time in advance that have no relation to the content of tested learning units.

Finally, one must conclude once more that the use of VR learning units does not per se lead to better learning success (Keller et al., 2019; Keller, Hebeisen, & Brucker-Kley, 2018b). As with conventional teaching methods, the quality and design of such VR learning units play a decisive role. Exploratory experiments that explore the influence of known and new design criteria on different learning outcomes beyond measurable test results are considered very relevant. In particular, the design and the effect of a help and feedback system for VR learning environments appeared to be a promising research topic in both the usability test and the field experiment.

References

Affolter, W., & Walt, M. (2017). mathbuch IF: Begleitband für die integrative Förderung Klassen 7–9 (1. Auflage). Klett und Balmer Verlag.

Bloom, B. S., Krathwohl, D. R., & Masia, B. S. (1984). *Taxonomy of educational objectives. The classification of educational goals: Cognitive domain handbook 1*. New York: Longman.

Burdea, G., & Coiffet, P. (2003). *Virtual reality technology* (2nd ed.). Hoboken, NJ: J. Wiley-Interscience.

Burrill, G. (2017). Designing interactive dynamic technology activities to support the development of conceptual understanding. In A. Leung & A. Baccaglini-Frank (Eds.), *Digital technologies in designing mathematics education tasks: Potential and pitfalls* (Vol. 8, pp. 303–328). Basel: Springer.

Campbell, D. T., & Stanley, J. C. (1967). *Experimental and quasi-experimental designs for research (2. Print)*. Boston: Houghton Mifflin Comp.

Cummings, J. J., & Bailenson, J. N. (2016). How immersive is enough? A meta-analysis of the effect of immersive technology on user presence. *Media Psychology, 19*(2), 272–309. https://doi.org/10.1080/15213269.2015.1015740
Fowler, C. (2015). Virtual reality and learning: Where is the pedagogy? *British Journal of Educational Technology, 46*(2), 412–422. https://doi.org/10.1111/bjet.12135
Geiger, V. (2017). Designing for mathematical applications and modelling tasks in technology rich environments. In A. Leung & A. Baccaglini-Frank (Eds.), *Digital technologies in designing mathematics education tasks: Potential and pitfalls* (Vol. 8, pp. 285–302). Basel: Springer.
Hevner, A. R., March, S. T., Park, J., & Ram, S. (2004). Design science in information systems research. *MIS Quarterly, 28*, 75–110.
Höntzsch, S., Katzky, U., Bredl, K., Kappe, F., & Krause, D. (2013a). Simulationen und simulierte Welten. In Lehrbuch für Lernen und Lehren mit Technologien: 2. Auflage, p. 327.
Höntzsch, S., Katzky, U., Bredl, K., Kappe, F., & Krause, D. (2013b, January 1). Simulationen und simulierte Welten: Lernen in immersiven Lernumgebungen [InternetDocument]. Retrieved from http://l3t.tugraz.at/index.php/LehrbuchEbner10/article/download/102/108
Joubert, M. (2017). Revisiting theory for the design of tasks: Special considerations for digital environments. In A. Leung & A. Baccaglini-Frank (Eds.), *Digital technologies in designing mathematics education tasks: Potential and pitfalls* (Vol. 8, pp. 17–40). Basel: Springer.
Keller, T. (2017). Neue Lehrkonzepte an Schulen – Einsatz von Virtual Reality. Retrieved February 25, 2020, from http://neuelehrkonzepte.ch/
Keller, T., Hagen, F., & Brucker-Kley, E. (2019). A field study about the impact of a VR learning unit. In *CELDA 2019 16th International Conference Cognition and Exploratory Learning in Digital Age, Cagliari, Italy, 7–9 November 2019* (pp. 307–314). IADIS Press.
Keller, T., Hebeisen, A., & Brucker-Kley, E. (2018a). Integration of children with special needs in mathematics through virtual reality, pp. 30–37. Retrieved from http://www.iadisportal.org/digital-library/integration-of-children-with-special-needs-in-mathematics-through-virtual-reality
Keller, T., Hebeisen, A., & Brucker-Kley, E. (2018b). Integration of children with special needs in mathematics through virtual reality, pp. 30–37. Retrieved from https://digitalcollection.zhaw.ch/handle/11475/16019
Krauthausen, G. (2012). Digitale Medien im Mathematikunterricht der Grundschule. https://doi.org/10.1007/978-3-8274-2277-4
Laborde, C. (2011). Designing substantial tasks to utilize ICT in mathematics lessons. In A. Oldknow & C. Knights (Eds.), *Mathematics education with digital technology* (pp. 75–83). New York: Bloomsbury. https://doi.org/10.5040/9781472553119
Lanier, J. (2017). Dawn of the new everything: Encounters with reality and virtual reality. Retrieved from https://us.macmillan.com/dawnofthenewe verything/jaronlanier/9781627794091/
Merchant, Z., Goetz, E. T., Cifuentes, L., Keeney-Kennicutt, W., & Davis, T. J. (2014). Effectiveness of virtual reality-based instruction on students' learning outcomes in K-12 and higher education: A meta-analysis. *Computers & Education, 70*, 29–40. https://doi.org/10.1016/j.compedu.2013.07.033
Mikropoulos, T. A., & Natsis, A. (2011). Educational virtual environments: A ten-year review of empirical research (1999–2009). *Computers & Education, 56*(3), 769–780. https://doi.org/10.1016/j.compedu.2010.10.020
Olbrish, K. (2014). *Immersive learning: Designing for authentic practice*. Alexandria, VA: ASTD Press.
Pirker, J., Gütl, C., Belcher, J. W., & Bailey, P. H. (2013). Design and evaluation of a learner-centric immersive virtual learning environment for physics education. In A. Holzinger, M. Ziefle, M. Hitz, & M. Debevc (Eds.), *Human factors in computing and informatics* (pp. 551–561). Berlin: Springer.
Sinclair, N., & Zazkis, R. (2017). Everybody counts: Designing tasks for TouchCounts. In A. Leung & A. Baccaglini-Frank (Eds.), *Digital technologies in designing mathematics education tasks: Potential and pitfalls* (Vol. 8, pp. 175–192). Basel: Springer.
Slater, M., & Wilbur, S. (1997). A framework for immersive virtual environments five: Speculations on the role of presence in virtual environments. *Presence: Teleoper. Virtual Environment, 6*(6), 603–616. https://doi.org/10.1162/pres.1997.6.6.603

Chapter 6
Pre-service Teachers' Adoption of a Makerspace

Junko Yamamoto

6.1 Introduction

Teachers are tasked with preparing their students to be successful in the workforce for a lifetime. Since society and necessary skills change rapidly, lifelong learning, creativity, innovation, and adaptation of new ideas needs to be embedded in instructional strategies within teacher education. Learners need to develop an aptitude for dealing with uncertainty, and to make a mental effort to acquire new skills. In order to facilitate classroom activities that stimulate creativity and innovation, teachers need to be prepared and willing to teach themselves as new methods, ideas, and tools are presented to them. Consequently, the International Society for Technology in Education (ISTE) published the 2017 Standards for Educators that expect teachers to be able to facilitate activities that nurture creativity and innovation (International Society for Technology in Education, 2017).

The idea that the education should prepare a future workforce is not new. For example, Trilling and Fadel published a book about twenty-first century skills in 2009. In this book, they stated that the nature of work has changed from routine manual or routine cognitive work to being more reliant upon expert thinking. According to Trilling and Fadel (2009), schools need to teach critical thinking, creativity, collaboration, communication, information literacy, media literacy, technology literacy, and flexibility. Routine jobs are more likely to be replaced by automated machinery or artificial intelligence than jobs that require expert thinking and creativity. The implication being that a future workforce needs to be innovative. Since the book was published in 2009, society has indeed witnessed routine work, such as cash register checkout and highway toll payment, being replaced by machines.

J. Yamamoto (✉)
Slippery Rock University of Pennsylvania, Slippery Rock, PA, USA
e-mail: junko.yamamoto@sru.edu

D. Ifenthaler et al. (eds.), *Balancing the Tension between Digital Technologies and Learning Sciences*, Cognition and Exploratory Learning in the Digital Age,
https://doi.org/10.1007/978-3-030-65657-7_6

Hence, schools need to educate students to be inventive, and also to quickly adopt innovation.

Innovation refers to creating something new, which can be a new idea, designing new ways of completing a task, or making something that no one has made before. Innovation does not have to involve technology (Krueger, 2019). Designing and making something new, using digital or non-digital tools and materials promotes a leaner's innovative ability. Likewise, using a tool in a way that no one has used it before, even if the tool has been available for a long time, is an innovation.

Working with hands to create something new stimulates the brain (McQuinn, 2018) and promotes creativity and innovation. Because of this, the maker space movement has gained momentum in the United States. Although 3D printers, cutting machines, and robotics kits are found in a makerspace, digital technology is not required for a maker space. Conventional tools such as a sewing machine (Mann, 2018) and woodworking tools can also be a part of a maker space.

The purpose of this paper is to show how future teachers can use a makerspace to convert an abstract concept into a concreate visual product. Pre-service teachers in an instructional technology class created digital or physical materials that represent someone that they find inspiring. They did so in an environment in which they had access to high tech tools and low-tech materials. To simulate real choices that in-service teachers make, pre-service teachers were not required to use a specific tool, but instead were granted access to familiar materials as well as novel devices. Since in-service teachers choose to learn to use new technology tools on their own, pre-service teachers in the class also had the option to use tools and materials of their choice. The study collected the participant's comments to capture their insights into their attitudes about adopting something new, and how the perceived usefulness or perceived ease of use of materials and tools influenced their decision to adopt a new technology or a new procedure.

6.2 Literature Review

6.2.1 ISTE Standard

Teacher education is often driven by standards. Standards for teacher education function as accountability measures, ensuring that educators are preparing teacher candidates to have the ability to successfully prepare future generations for their jobs (Chung & Kim, 2010). Aligning instruction to standards in teacher preparation also demonstrates a high quality teacher education (Murray, 2001). ISTE Standards for Educators includes categories such as "Designer: Educators design authentic, learner-driven activities and environments that recognize and accommodate leaner variability," and "Facilitator: Educators facilitate learning with technology to support student achievement of the ISTE Standards for Students." These standards call for creating personalized learning experiences, designing authentic learning activities, encouraging independent learning, challenging learners to use the design process, and modeling creativity (ISTE, 2017).

6.2.2 Makerspace

A makerspace can facilitate personalized learning experience, authentic activities, independent learning, and the design process (Taheri, Robbins, & Maalej, 2020). A makerspace is a set up that allows students to create digital or physical objects to express their understanding. Students share materials and collaborate in the space (Trust, Maloy, & Edwards, 2018). Some argue that making does not need to involve high technology: students can make their mental representation models or prototype of a product using common materials, such as cardboard and duct tape (Maughan, 2018), index cards, craft sticks, pipe cleaners and modeling clay (McGlynn & Kelly, 2019). On the other hand, Valente and Blikstein (2019) recommend combining digital and physical materials. They also state that it is important to carefully balance advanced technology and traditional material. The concept of makerspace is more about the act of making, rather than a physical space (Trust et al., 2018).

Technology integration in classrooms has shifted from teacher-centered models to student-centered models (Muilenburg & Berge, 2015; Passehl-Stoddart, Velte, Henrich, & Gaines, 2018), and makerspace is in line with a student-centered approach. The makerspace process in which learners invent and test their prototypes, promotes creativity and innovation (Woods & Baroutsis, 2020). Moreover, reflection, comprehension, and theorization about why and how something works are important both during and after making (Valente & Blikstein, 2019).

Designing a makerspace does not start with which tools to use, but with what is desirable for learners to accomplish. For example, rather than thinking, "we must use a 3D printer," and forcing technology onto an educational setting, the designer should think, "I would like my learners to create a 3D model that represents XXX," or, "How can my students identify a problem and solve it?" or "How will my students improve this product?" When a teacher sets up the context, and students determine which tools to use, the technology user realizes "resource fluency" (Muilenburg & Berge, 2015).

When multiple tools are available, learners should be able to choose what they want to use. Learners' choices and innovative use is captured in the Technology Integration Matrix (TIM) (https://fcit.usf.edu/matrix/matrix/). The matrix shows 5 levels of technology integration in classrooms: entry, adoption, adaptation, infusion, and transformation. When students choose the tools they want to use to meet their goals, the manner of use is considered to be in the infusion level. When there is a classroom culture in which learner invents a new way of using a tool, the teacher is reaching to the transformation level of TIM. For instance, when a teacher assigns students to create a 3D model that moves, students should have a choice to use a variety of tools, such as cardboard with Hummingbird, LEGO EV3, or wood combined with metal and rubber. Students can use the materials that teachers provide, ones they bring from home, or both.

6.2.3 Theoretical Frameworks: Diffusion of Innovation and Technology Acceptance Model

Teachers' willingness to try something new, or using a conventional tool in an innovative manner, is important for creating a learner-driven environment. Unfortunately, the availability of new ideas, resources, or tools does not automatically warrant their use by the majority of the population. Many innovations require a lengthy period from when it becomes available to when they are adopted. In other words, not all people adopt innovations when they become available. Some choose not to use tools at all. In general, 2.5% of a population are innovators, and only about 10–25% of a population are early adopters of an innovation (Rogers, 1983). Rogers (1983) explains that 13.5% of a population adopt an innovation at early stage. Then next 34% are the early majority. It is followed by the late majority, which make up for 34%. Finally, 16% of the population is the laggards, or the ones that fall behind others (p. 246–247). Although Rogers published the book decades ago, human nature has not changed. There are some individuals that are interested in learning something new, and there are those who resist change.

Diffusion is the process in which an innovation is communicated and adopted. Individuals go through an innovation-decision process to decide if they want to adopt or reject new ideas, resources, or technology tools. The innovation-decision process commands mental work. There are individual differences in the commitment to learning, especially in problem-solving when encountering a glitch. Adopters need to cope with uncertainty and need to have the motivation to seek out innovation. One's level of education and social status may affect their attitude toward adopting innovation, but age is not an influential factor. In addition, earlier adopters have a greater ability to deal with abstraction, and have a greater rationality than later adopters (Rogers, 1983).

The Technology Acceptance Model (TAM) explains the diffusion of innovation through the lens of who is willing to adopt the technology. As stated earlier, the availability of technology does not automatically result in its adoption. There are two factors of innovation that must be considered for the technology to be adopted: perceived usefulness and perceived ease of use. Self-efficacy, or belief that one can succeed on a task, influences perceived ease of use (Davis, 1989). Self-efficacy involves motivation, cognition, and self-regulation. Those with high self-efficacy tend to be more willing to take risks, and keep trying when they experience difficulties. On the other hand, those who doubt their ability to cope with a task shy away from it (Bandura, 1994). In fact, self-efficacy plays a key role in an educator's decision to adopt technology (Joo, Park, & Lim, 2018).

Perceived usefulness and perceived ease of use increase an educator's willingness to adopt a new technology tool (Flavell, Harris, Price, Logan, & Peterson, 2019; Kukul et al., 2018; Shittu, Kareem, Obielodan, & Fakomogbon, 2017). Furthermore, those who enjoy challenges may be more willing to try something new. Teachers' psychological factors, ranging from a willingness to take risks, fear of failure, a lack of confidence, and general technology anxiety influence perceived usefulness and perceived ease of use (Flavell et al., 2019). TAM is a widely accepted model to examine a population's intention to adopt technology (Albarghouthi, Qi, Wang, & Abbad, 2020; Lazar, Panisoara, & Panisoara, 2020; Nadlifatin, Ardiansyahmiraja, Persada, Redi, & Lin, 2020). Figure 6.1 shows the Technology Acceptance Model.

6.2.4 Research Questions

Research questions include:

1. Do education majors increase their knowledge about makerspace after the instruction?
2. What type of tools do research participants choose to use?
3. Will there be research participants who use tools that they have never used without the instructor's assistance?
4. Will there be research participants who use tools that they have never tried with the instructor's assistance?
5. Will there be research participants who does not use tools that they have never used even when assistance is available?
6. Do perceived usefulness and perceived ease of use influence the decision to use new tools?

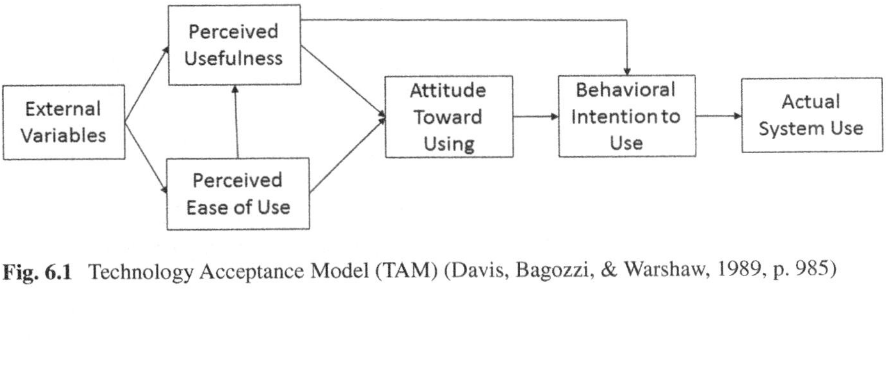

Fig. 6.1 Technology Acceptance Model (TAM) (Davis, Bagozzi, & Warshaw, 1989, p. 985)

6.3 Method

6.3.1 Research Participants

Research participants were pre-service teachers who were enrolled in an undergraduate level instructional technology class in a teacher preparation program in the United States of America during the fall 2018 semester. There were 26 who participated in this study. The participants included 4 Middle Level (Grade 4–8) Math Education majors, 3, Middle Level (Grade 4–8) Science Education majors, 2 Middle Level (Grade 4–8) English Education majors, 6 Secondary (Grade 7–12) English Education Majors, 6 Secondary (Grade 7–12) Social Studies Education majors, 2 French Education Majors, 1 double major of Secondary English Education and French Education, 1 Art Education major, and 1 Pre-Math Education major.

There were 10 freshmen, 11 sophomores, 2 juniors, 1 senior, and 2 post baccalaureate students. In this teacher preparation program, students take the instructional technology class before they take a teaching methodology class. Also, a typical semester schedule of a freshman or a sophomore is populated by general education courses, or courses that any majors would take. Hence, they have taken a limited number of content area courses and/or pedagogy courses prior to the study.

6.3.2 Instruments

A Likert-type survey was used with the following statements:

- I know what a makerspace is.
- I know how to design makerspaces.
- I know how to use a makerspace.

Value assigned to the self-evaluation were:

Strongly Agree = 5; Agree = 4; Neither Agree or Disagree = 3; Disagree = 2; Strongly Disagree = 1.

A t-test was used to calculate the mean difference between the pre-instructional survey and post-instructional survey. This instrument was used for research question 1.

Moreover, participants wrote reflections with these prompts:

- "Which tools did you choose?" This prompt was used for research question 2.
- "Why did you use the tools?" This question was used to determine participants' beliefs about the usefulness and ease of use.
- "Did you know how to use the tools?" This question was asked to find out if participants used a new tool.
- "Did you decide to use the tools because you saw it being used during the project explanation?" This question was asked to find out if participants perceived the

usefulness of tools when they saw the instructor model the use before the participants decided which tools to use.
- "How do you rate yourself about exploring new tools and learning how to use them? Elaborate your answer." This question examines participants' attitudes about willingness to learn new tools.

6.3.3 Instruction

The participants were introduced to the concept of makerspace and used a makerspace for 275-min class sessions. After they used the makerspace, they presented their products to peers and explained to them what they represented. Idealistically, the future teachers would create a teaching material in their instructional technology class. However, with 21 out of 26 research participants being freshmen and sophomores, the instruction needed to adjust according to the prior content and pedagogical knowledge of the participants. In the past, the instructor made projects more open-ended by asking pre-service teachers to choose something to teach for their certificate area. However, this open-ended assignment overwhelmed freshmen who had very limited content knowledge, especially when they were taking general education classes, instead of their content area classes. Hence, the instructor decided to make the project semi close-ended where students were to select a hero from their subject areas.

The task was to create a presentational material that represents a hero, or someone who made a significant contribution to their area of certification. The instructor chose the task because the research participants would be likely to have some prior knowledge about the person that they would like to present about, and if they chose someone that they have positive emotions about, they would likely be motivated to do further research. The students were asked to determine their heroes of their teaching certificate area, research about their heroes, and make something to present their heroes to the class. Any subject area has someone who made a significant contribution. When each student identifies someone he or she looks up to, he/she is likely to have positive emotion on the topic. Internal motivation facilitates "active engagement, deeper understanding, and a desire to learn more" (Trilling & Fadel, 2009, p. 33).

Since the project required the participants to look up new methods and tutorials, a training to ensure that they have this skill was necessary. Hence, they completed Google Fundamentals Training Level 1 (https://teachercenter.withgoogle.com/fundamentals/course) prior to the project. They participated in Unit #1 (Get Ready to Use Technology in Classroom) and Unit #2 (Expand Your Access to Help and Learning). When an open-ended task is assigned, it is important to show sample products in the beginning of the project. The instructor presented heroes in different subject areas using a variety of technology tools. At this stage, the concept of the makerspace was introduced to the students. The instructor showed two examples:

The first example was made for world language education. The instructor visited a native speaker of Spanish with her iPhone, and recorded a brief biography of Roberto Clemente. Clemente was a Puerto Rican baseball player who played for the Pittsburgh Pirates. He not only overcame racism to become a popular baseball player in the United States, but also made an impact on Puerto Rican Society. The instructor uploaded the biography recording to Dropbox, and set the audio file to be publicly accessible. She then used Cricut, a cutting machine, to create 3-D model of Roberto Clemente, representing a baseball player bearing number 21, a Puerto Rican flag, and an iron bridge. The instructor then created a QR code to access the Clemente's biography recording and the Spanish language biography from Smithsonian Institution Traveling Exhibition Service (http://www.robertoclemente.si.edu/spanish/virtual_legacy.htm). She pasted the QR codes to the 3-D model.

The second example was made for social studies education majors. The process of creating the example started with a story about Marcus Tullius Cicero attempting to defend the ideal of the Republic when the climate of Roman politics was moving toward dictatorship. The story ended with Cicero's assassination by Mark Anthony. The instructor made the story into a digital movie using Windows Movie Maker, and uploaded it to YouTube.

As the instructor showed two examples, she explained that research on the content was the driving force for making both. She advised the students to spend time on increasing content knowledge before selecting which tools to use. She also explained that since the objective is to explore innovation and creativity, everyone would be free to choose materials that they want to use. Students were also invited to seek out help in adding an audio recording, a video, and a QR code to their work. The instructor also informed the students that they will have access to a Cricut cutting machine, cardstock, vinyl, scissors, glue, color pencils, a paper cutter, and a laminator. Students reported to two class sessions where the instructor set up the makerspace. They then presented their heroes to the class.

6.4 Results

Research question 1: Do education majors increase their knowledge about makerspace after the instruction?

Paired sample *t*-tests were calculated in order to compare values between pre-instructional time and post-instructional time. Table 6.1 shows the result of the *t*-tests.

The result of the t-tests indicate that the concept of makerspace was a new idea to the research participants. All names that are used in this manuscript are pseudonyms. Mary stated,

"Before this assignment, I was unaware of makerspace and how it works. This tool essentially brings ideas to life, encourages collaborations, and promotes creativity. I am now more familiar with this kind of environment and it is something I

Table 6.1 Mean comparisons between pre-instructional and post-instructional self-assessments

	N	Mean pre (SD)	Mean post (SD)	Mean post- mean pre	*T*	Sig. (2 tailed)
I know what a makerspace is	26	2.4231 (1.30)	4.6923 (0.68)	2.2692	8.092	0.000
I know how to design a makerspace	26	2.1538 (1.19)	4.5769 (0.70)	2.4231	9.726	0.000
I know how to use a makerspace	26	2.3462 (1.16)	4.6538 (0.69)	2.3076	9.129	0.000

Table 6.2 Tools and materials used

Tools used	Frequency
Internet search for content	12
Construction paper	5
Glue	4
Google Docs	3
Scissors	3
Microsoft word	2
iMovie	2
Cricut design space	2
Laminator	2
Popsicle sticks	1
Paper trimmer	1
Paint	1
Styrofoam balls	1
Drawing supplies	1
PowerPoint (to create a digital movie)	1
QR code maker	1

would implement into my own classroom. Reading and exploring the makerspace website to find ideas helped to enhance my understanding."

Research question 2: What type of tools do research participants choose to use?

In this study, the participants freely chose to use tools that they wanted to use. Table 6.2 shows tools and materials that the participants chose to use to create their hero presentation.

The data indicates that 12 out of 26 students said that they used the Internet to search for content. The total number exceeds 26, the number of research participants, because the participants used multiple items to complete their projects.

Research question 3: Will there be research participants who use tools that they have never used without the instructor's assistance?

This group represent early adopters. Reflective writing was used to answer this question. Students' statements in this section represent the research participants who tried new tools or methods. Linsey made a movie using PowerPoint. Although she knew how to use PowerPoint, she did not know how to create a movie. She decided that she wanted to create a video about Sir Isaac Newton when she saw the instructor's sample movie about Marcus Tullius Cicero. However, while the instructor told the class that she made her movie with Windows Movie Maker, it was Linsey's initiative to turn PowerPoint slides into a movie. She wrote, "I knew how to upload videos to YouTube prior to this project. However, I did not know how to turn a PowerPoint into a slide show and create voice narrations. I had to turn to the Internet to find instructions on how to do so. … I am always open to new and exciting technologies or tools."

Janet also taught herself how to make a movie, but she chose iMovie. She "was not very familiar with it. This project gave me the opportunity to really learn how to use iMovie effectively and I now know how to use it better than I did." Jackson also used iMovie for the first time. He stated, "Using iMovie was a totally new experience. I had never uploaded a video to YouTube. Lastly, I had never had the need to record audio on my laptop, so I was unfamiliar with how to use the laptop microphone. … While learning how to use iMovie was a slight struggle at times, I was able to learn enough to create a decent finished product."

Beth and Amber stated that they used Google search, YouTube tutorial videos, and Pinterest to look for ideas prior to deciding which tools they wanted to use. Amber said, "I had to research and look up tutorials on how to create the project that I was working on."

Melinda, a French Education Major, took the idea of combining paper cut with Cricut machine and a QR code maker from the instructor's Roberto Clemente presentation. She created a graphic representation that was structured similar to the one that the instructor presented, but chose to present about Charles Aznavour, a French Armenian singer. She created 2 QR codes that allow the audience to access Charles Aznavour's 2 famous songs. She said she regularly used YouTube and Pinterest when she needs to learn something new, and successfully completed the project without the instructor's assistance.

Elaine, a French and English Education Major, decided that she wanted to create something about Antoine de Saint-Exupéry, a French writer. She wanted to create a planet mobile to represent scenes from *Le Petit Prince*, Saint-Exupéry's famous book. She stated, "Although I did not use any tool out of the ordinary, I did have to research and look up tutorials on how to create the project that I was working on. I never made a planet mobile: therefore, this project created a new experience for me."

Research question 4: Will there be research participants who use tools that they have never tried with the instructor's assistance?

This group belong ether in early majority, according to Rogers' Diffusion of Innovation framework. Mary stated, "Because I became familiar with makerspace in class and motivated to take advantage of this workspace since I knew what I was getting into. If I was unaware of how this tool worked, I many have been hesitant to utilize it and may have missed out on a great, creative opportunity."

Jillian, a math education major, became interested in using Cricut Design Space when she saw the instructor's sample during the project explanation. She is a math education major, and decided to create a display on Pythagoras. She stated,

"I did not know how to use the Cricut Design Space or the laminator, but through this project I was able to learn how to use two different tools that I would have never had the chance to learn. I really liked Dr. XXX's Roberto Clemente example presentation. She also explained how she used a Cricut for the images. Knowing what a Cricut was, but not how to use it I knew would be a challenge. I wanted to challenge myself and use something that I have never used before. I also wanted to use the laminator because I saw how nice Dr. XXX's project looked, and I knew in the future I would be using a laminator in my class. …The Cricut took a few moments of trial and error, but I believe I learned a lot from that experience. I was able to use my problem-solving skills to discover how to use the program, and how to troubleshoot."

David, who is "open to learning about new tools" decided to use Cricut after the instructor suggested it. He made a visual presentation of Joshua Chamberlain, capturing his leadership during the Battle of Gettysburg in the American Civil War. The instructor recommended that he add an eagle shaped cut out of gold vinyl and helped him use Cricut. He stated,

"I used this tool because it helped me to create a nice eye-catching piece for my project. The precise cuts and glossy sheen of the vinyl material make it stand out among the other parts of the project, and crated a wonderful centerpiece for the project that other methods may not have achieved. I honestly say that it helped immensely."

Research question 5: Will there be research participants who do not use tools that they have never used even when assistance is available?

Statements in this group showed the characteristics of late majority or laggards. Late majority can be someone who did not try to do something new during the project, but wished they had tried after they saw peers' successes. Heather wrote,

"I wanted to spend time on the small details of the project instead of trying to learn something brand new and get caught up in the process. With that being said, after walking around and seeing everyone else's project I have learned so much and cannot wait to try out some of their ideas… after seeing all of the other projects I am inspired in so many ways."

Kim said,

"I feel like I could have done better by trying to use a tool like Cricut. Cricut is something I have never used before and wish I would have tried it." Megan said, "I could have been a little more creative and gone out of my comfort zone to create this project."

On the other hand, laggards are the ones who refuse to try new tools or methods even after they see peers' successes. Geoff explained he choose to use the material because "they were familiar to me … and are very convenient." Mike said, "I tend to stay away from exploring new tools because of trying to learn how to use it is sometimes difficult and I would rather use that I know and works best for me. Makes it easier on me and being able to get things done efficiently without the headache of trying to learn how to use something new."

Research Question 6: Do perceived usefulness and perceived ease of use influence the decision to use new tools?

Comments about perceived usefulness included (the author italicized relevant text):

- "I used the laminator for the finishing touches and *to enhance the quality* of my work."
- "*I thought that QR code will be useful*. If I had not seen the model of Roberto Clemente that was presented to the class I would have had no idea that I could have done that and I would have not used it in my project."
- Comments by Jillian and David in Research Question 4 indicate that they decided to use Cricut because they saw the use of the cutting machine improve the quality of a finished product.

Comments about perceived ease of use included (the author italicized relevant text):

- "I was able to work out how to use the QR Code Maker *with ease*."
- "I tend to stay away from exploring new tools because trying to learn how to use it is sometimes *difficult*."
- "Cricut *seems complicated* [hence, I did not use it]; I would like to become more familiar with it. A fellow student taught me how to use the laminating machine. *I liked this tool because it is simple*, yet it makes a project appear professional."

However, some strived to keep trying when they encountered difficulties. For example, Janet's quote in Research Question 3 shows, "...While learning how to use iMovie was a slight struggle at times, I was able to learn enough to create a decent finished product."

6.5 Conclusion

6.5.1 Implications and Discussion

This research used Diffusion of Innovation theory and Technology Acceptance Model as frameworks. According to Rogers (1983), the decision making process requires mental work. The framework also implies that adopters tend to have some tolerance for uncertainty. Although Rogers suggested the framework three decades ago, his work is still applicable because human nature has not changed. There are those who enjoy the challenge of learning something new and those who avoid the mental work. Likewise, some have a higher tolerance to uncertainty than others. In this study, two participants saw that the instructor used Movie Maker to create Marcus Tullius Cicero's biography movie, and decided to make a digital movie. One used iMovie, and the other turned PowerPoint into a movie. Both had never created a digital movie before, but they searched for tutorials from YouTube and Pinterest and taught themselves how to create a digital movie. One participant used Cricut and QR code creator for the first time, and completed the project without the instructor's assistance. They represent the research participants who were willing to seek out new ideas and teach themselves.

Two participants reflected that they decided to used Cricut after they saw the Roberto Clemente 3D model. They had access to a laptop with Cricut Design Space, machine, and a variety of cardstock paper and vinyl to cut with the machine. They completed their project with instructor's help to use the machine. They represent the population that is willing to adopt innovation when assistance is available.

The third group decided not to try anything new, but wished they had tried after they saw peers' successful adoption. Statements from this group indicate the characteristics of late adopters. They did not want to spend the time to learn Cricut while they used a makerspace. However, after they viewed their peer's finished products, they wished they had used the tool.

Finally, the fourth group self-reflected that they prefer to stay in their comfort zone, and would like to avoid, the "headache of trying to learn how to use something new." This group has the characteristics of laggards. The research participants' comments support the conclusion that perceived usefulness and perceived ease of use influence the decision to adopt a new technology tool. However, attitudes about the perceived ease of use differ among those who are willing to learn on their own, those who learned to use a tool with assistance, and the who avoided using new tools.

In this study, comments from the participants implied that some find dealing with the unknown and troubleshooting to be rewarding, while others shy away from the process of figuring out something new. In order to promote innovation, the former attitude needs to be encouraged in an educational setting. When learners venture into an unknown territory, they are likely to experience some failures. Learners need the guidance to form the attitude to focus on the process, instead of an end-product with a deadline. Unfortunately, children and youth may come from an environment in which failures are stigmatized (Androutsos & Brinia, 2019).

6.5.2 Limitations

This research used the Diffusion of Innovation theory as a framework. According to Rogers (1983), the decision making process requires mental work. The framework also imply that adopters tend to have some tolerance for uncertainty, and the motivation to seek innovation. This research showed that not all are willing to try new tools or methods when they become available. However, the research did not measure personal attributes, such as the willingness to make a cognitive effort, tolerance to uncertainty, and the motivation to seek out new tools or methods. It is suggested, therefore, that these personal attributes should be measured in future studies.

This research also used the Technology Acceptance Model, which suggests that the perceived usefulness and perceived ease of use of technology influence the decision to adopt a new technology tool. While some studies, such as Çakiroglu, Gökoglu, and Öztürk (2017) and Joo et al. (2018) use surveys to measure perceived usefulness and perceived ease of use, the present study did not because the participants were encouraged to freely choose materials and tools. Collecting data on perceived usefulness and perceived ease of use of specific maker space tools, and examining how self-efficacy influences perceived ease of use, is a meaningful area of a future study.

6.5.3 Suggestion for a Future Research

Muñoz-Carril, González-Sanmamed, and Fuentes-Abeledo (2020) used the Technology Acceptance Model in the form of Likert-type survey and statistical analysis to show that self-efficacy influences perceived usefulness and perceived ease of use when future teachers were asked to use blogs. The current study did not require anyone to use a specific technology tool, but asked the participants to choose their tools to express their ideas. A study with a task, not centered around a tool, with a statistical analysis to show the relationship between self-efficacy, the decision to use a new tool, perceived usefulness, and perceived ease of use, would be insightful. Such a study would be meaningful in respect to the Technology Integration Matrix, which encourages educators to give their learners the autonomy to choose their technology tools. Choosing tools, however, can be an overwhelming process. Moreover, the tool that they choose may not work the way they wanted. How the attitudes about taking risks, failure, and uncertainty influence the decision to learn to use new tools, or use it in an innovative manner, should be a part of a future study.

References

Albarghouthi, M., Qi, B., Wang, T. C., & Abbad, M. (2020). ERP adoption and acceptance in Saudi Arabia higher education: A conceptual model development. *International Journal of Emerging Technologies in Learning, 15*, 110–120. https://doi.org/10.3991/ijet.v15i15.12039

Androutsos, A., & Brinia, V. (2019). Developing and piloting a pedagogy for teaching innovation, collaboration, and co-creation in secondary education based on design thinking, digital transformation, and entrepreneurship. *Education Sciences, 9*(2), 113. Retrieved from https://doi.org/10.3390/educsci9020113

Bandura, A. (1994). Self-efficacy. In V. S. Ramachaudran (Ed.), *Encyclopedia of human behavior* (Vol. 4, pp. 71–81). New York: Academic Press.

Çakiroglu, Ü., Gökoglu, S., & Öztürk, M. (2017). Pre-service computer teachers' tendencies towards the use of mobile technologies: A technology acceptance model perspective. *European Journal of Open, Distance and E-Learning, 20*(1), 175–190.

Chung, H., & Kim, H. (2010). Implementing professional standards in teacher preparation programs in the United States: Preservice teachers' understanding of teaching standards. *KEDI Journal of Educational Policy, 7*(2), 355–377.

Davis, F. D. (1989). Perceived usefulness, perceived ease of use, and user acceptance of information technology. *MIS Quarterly, 13*(3), 319–339.

Davis, F. D., Bagozzi, R. P., & Warshaw, P. R. (1989). User acceptance of computer technology: A comparison of two theoretical models. *Management Science, 35*(8), 982–1003.

Flavell, H., Harris, C., Price, C., Logan, E., & Peterson, S. (2019). Empowering academics to be adaptive with eLearning technologies: An exploratory case study. *Australasian Journal of Educational Technology, 35*(1), 1–15. https://doi.org/10.14742/ajet.2990

International Society for Technology in Education. (2017). ISTE standards for educators. Retrieved from https://www.iste.org/standards/for-educators

Joo, Y. J., Park, S., & Lim, E. (2018). Factors influencing preservice teachers' intention to use technology: TPACK, teacher self-efficacy, and technology acceptance model. *Educational Technology & Society, 21*(3), 48–59.

Krueger, N. (2019, April). Innovate, iterate, educate: Sometimes teachers' small ideas make the biggest difference. *Empowered Learner, 2*(4), 17–21.

Kukul, V., Ünal, M., Karataş, S., Kılıç Çakmak, E., Yılmaz, A., & Ömeroğlu, E. (2018). Analysis of teachers' opinions in the context of acceptance of technology. *Journal of Kırşehir Education Faculty, 19*(3), 2055–2067. https://doi.org/10.29299/kefad.2018.19.03.007

Lazar, I. M., Panisoara, G., & Panisoara, I. O. (2020). Digital technology adoption scale in the blended learning context in higher education: Development, validation and testing of a specific tool. *PLoS One, 15*(7), e0235957. https://doi.org/10.1371/journal.pone.0235957

Mann, L. (2018). Making a place for makerspaces in information literacy. *Reference & User Services Quarterly, 58*(2), 82–86.

Maughan, S. (2018, December 14). School & library spotlight fall 2018: Q & A with Nicholas Provenzano. Publishers Weekly. Retrieved from https://www.publishersweekly.com/pw/by-topic/childrens/childrens-authors/article/78847-school-library-spotlight-fall-2018-q-a-with-nicholas-provenzano.html

McGlynn, K., & Kelly, J. (2019, September). Making it work: Incorporating design thinking into all areas of instruction to fit the needs of unique learners. *Science Scope, 43*(2), 20–25. https://doi.org/10.2505/4/ss19_043_02_20

McQuinn. C. (2018). The brain science of making. School Library Journal. Retrieved from https://www.slj.com/?detailStory=brain-science-of-making

Muilenburg, L. Y., & Berge, Z. L. (2015). Revisiting teacher preparation: Responding to technology transience in the educational setting. *The Quarterly Review of Distance Education, 16*(2), 93–105.

Muñoz-Carril, P. C., González-Sanmamed, M., & Fuentes-Abeledo, E. J. (2020). Use of blogs for prospective early childhood teachers. *Educación XX1, 23*(1), 247–273. https://doi.org/10.5944/educXX1.23768

Murray, F. B. (2001). From consensus standards to evidence of claims: Assessment and accreditation in the case of teacher education. *New Directions for Higher Education, 113*, 49–66. https://doi.org/10.1002/he.4.abs

Nadlifatin, R., Ardiansyahmiraja, B., Persada, S. F., Redi, A. A. N. P., & Lin, S. (2020). The measurement of university students' intention to use blended learning system through technology acceptance model (TAM) and theory of planned behavior (TPB) at developed and developing regions: Lessons learned from Taiwan and Indonesia. *International Journal of Emerging Technologies in Learning, 15*(9), 219–230. https://doi.org/10.3991/ijet.v15i09.11517

Passehl-Stoddart, E., Velte, A., Henrich, K. J., & Gaines, A. M. (2018). History in the making: Outreach and collaboration between special collections and makerspaces. *Collaborative Librarianship, 10*(2), 133–149.

Rogers, E. M. (1983). *Diffusion of innovations* (3rd ed.). New York: The Free Press.

Shittu, A. T., Kareem, B. W., Obielodan, O. O., & Fakomogbon, M. A. (2017). Investigating predictors of pre-service science teachers' behavioral intention toward e-resources for teaching. *Contemporary Educational Technology, 8*(2), 142–157.

Taheri, P., Robbins, P., & Maalej, S. (2020). Makerspaces in first-year engineering education. *Education Sciences, 10*(1), 8. https://doi.org/10.3390/educsci10010008

Trilling, B., & Fadel, C. (2009). *21st century skills. Learning for life in our times*. San Francisco: Jossey-Bass.

Trust, T., Maloy, R. W., & Edwards, S. (2018). Learning through making: Emerging and expanding design for college classes. *TechTrends, 62*, 19–28. https://doi.org/10.1007/s11528-017-0214-0

Valente, J. A., & Blikstein, P. (2019). Maker education: Where is the knowledge construction? *Constructivist Foundations, 14*(3), 252–262.

Woods, A., & Baroutsis, A. (2020). What's all the full about makerspaces and making? *Practical Literacy: The Early & Primary Years, 25*(2), 39–41.

Chapter 7
Relationship Between Learning Time and Dimensions of a Learning Organization

Vaclav Zubr

7.1 Introduction

According to Senge, a learning organization can be defined as: ' ... *an organization where people constantly improve their skills and achieve the results that they really desire, where they find support, new and dynamic models of thinking, where collective thinking and inspiration are welcome and where people still learn how to learn together*' (Senge, 2016). A learning organization is made up of basic components: organization, people, knowledge and technology. Individual components support each other in learning, which is the essence of a learning organization (Serrat, 2017). Several factors contribute to the good functioning of the learning organization concept, such as management, learning communities, internal compliance, empowerment of individuals, organizational culture, self-development, teamwork, information sharing, knowledge creation, facilitating leadership, building reliable dimensions of learning or innovation (Zubr, Mohelska, & Sokolova, 2017).

In the book Powerful learning, Brandt (1998, p. 51) states some characteristics of learning organizations, e.g.: '*incentive structure that encourages adaptive behavior, challenging but achievable shared goals, exchange information frequently with relevant external sources, get feedback on product and services, continuously refine their basic processes, have a supportive organizational culture, are "open systems" sensitive to the external environment*'.

Learning becomes an integral part of the whole work process, work and learning are interconnected in the process of continuous improvement. A key aspect of organizational learning is interaction between individuals (Yadav & Agarwal, 2016). In organizations, learning is implemented at the individual, group and organizational

V. Zubr (✉)
University of Hradec Kralove, Hradec Králové, Czechia
e-mail: vaclav.zubr@uhk.cz

D. Ifenthaler et al. (eds.), *Balancing the Tension between Digital Technologies and Learning Sciences*, Cognition and Exploratory Learning in the Digital Age,
https://doi.org/10.1007/978-3-030-65657-7_7

levels. Learning takes place with individuals, teams, subdivisions of the organization and at the level of the entire organization and cooperating organizations. Organizations focus on organizational learning at the systems level (Birdthistle, 2009). Organizational learning is the central activity of a learning organization and it is a process where the meaning of information is created, information is collected, interpreted and distributed. Considering the learning support level of the learning organization, several learning organization action requirements can be identified, such as: providing strategic guidance for learning, supporting employees in sharing visions, building systems to capture and share learning, fostering collaboration and team learning and to support dialogue and create opportunities for continuous learning. The result of meeting these requirements is continuous learning and transformation of the organization (Jamali, Sidani, & Zouein, 2009). It should be noted that organizational learning is not only a complicated and unplanned process, but also an interactive, dynamic, continuous, evolving, alerting and effective that is influenced by the knowledge base or cultural resources (Saadat & Saadat, 2016).

Organizational learning has a positive impact on gaining a long-term competitive advantage, improving performance, strengthening human resources, creativity and innovation, and accelerating the process of change to a learning organization (Saadat & Saadat, 2016). Vega Martinez, del Carmen Martinez Serna, and Parga Montoya (2019) states that commitment to learning has a positive influence on small and medium enterprises organizational performance and competitiveness (Vega Martinez et al., 2019).

According to Tichá (2005), companies that want to take learning as part of a strategy should ensure the proper functioning of learning-related characteristics such as: business strategy enables learning, all or most members of the organization can contribute to strategy creation, use of other organizations' experience to own learning, atmosphere supporting learning, opportunity for personal development for everyone (Tichá, 2005).

After 2010, the concept of a learning organization has become the subject of many studies around the world (Zubr, 2019). In all these studies, the Dimensions of a Learning Organization Questionnaire (DLOQ) is used to evaluate the learning organization created by Watkins and Marsick (Watkins & Marsick, 1993). Most often, studies are focused on the education sector and the banking sector (which is close to IT) (Zubr, 2019). Nevertheless, because in the Czech Republic is possible that there will be only a small number of respondents from the banking sector, we decided to focus this study on the IT sector and the education sector in the Czech Republic. The study focused on education sector was specified only on secondary schools due to the expectation that respondents, due to their age composition, will pay more attention to education than to primary schools (Louws, Meirink, van Veen, & van Driel, 2017).

7.1.1 Dimensions of a Learning Organization Questionnaire and Schools as a Learning Organization Scale

According to Marsick and Watkins (2003), there are seven dimensions of learning organizations that represent organizations' efforts to create learning opportunities for all employees, to create a platform that supports dialogues, responses and experiments among members, as well as team learning, vision sharing or strategic leadership (Norashikin, Safiah, Fauziah, & Noormala, 2016). These seven dimensions include: Create Continuous Learning Opportunities, Promote Inquiry and Dialogue, Encourage Collaboration and Team Learning, Create Systems to Capture and Share Learning, Empower People towards a Collective Vision, Connect the Organization to its Environment and Provide Strategic Leadership for Learning. Based on some studies, it was discovered that only two dimensions of a learning organization led to higher organizational performance—namely, Promote Inquiry and Dialogue and System Connection. The remaining dimensions have no effect on organizational performance (Norashikin et al., 2016). Moreover, studies previously conducted prove correlations between dimensions of learning organization and perceived financial, mission and adaptive performance (Watkins & Kim, 2018). According to Kim, Watkins, and Lu (2017), knowledge performance is positively affected by learning organization, financial performance is positively affected by knowledge performance and the relationship between financial performance and learning organization is fully mediated by knowledge performance (Kim et al., 2017).

The individual dimensions can be viewed from the human (Create Continuous Learning Opportunities, Promote Inquiry and Dialogue, Encourage Collaboration and Team Learning, Empower People towards a Collective Vision) and the structural level (Create Systems to Capture and Share Learning, Connect the Organization to its Environment, Provide Strategic Leadership for Learning) (Birdthistle, 2009).

DLOQ is developed on the basis of a theoretical framework combining four articles: organizational learning, workplace learning, learning climate and learning structure perspective (Kim, Egan, & Tolson, 2015). For example, this questionnaire can be used to evaluate an organization as a learning organization.

In the basic version, the questionnaire contains 42 questions, according to the authors' recommendation, it can be shortened to 21 questions to maintain the validity of the data obtained. To maintain the questionnaire's validity, a reverse translation, expert review and Cronbach's alpha coefficient should be performed to ensure that dimensional reliability is not significantly lower than the actual work validation reliability. The answers to the questions are given on a six-point Likert scale, where 1 represents 'strongly disagree' and 6 'strongly agree' (Watkins & O'Neil, 2013).

In 2018, the OECD published a relatively extensive publication focusing on 'Developing Schools and Learning Organizations in Wales'. M. Kools participated in the OECD study and in 2020 Kools et al. published an article, where he states the scale used in the Wales study. The 'Schools as a Learning Organization Scale' takes into account seven action-oriented dimensions that, according to Kools and Stoll (2016), a school as a learning organization should meet. The published scale has 69 items, the five-point Likert scale is used for the answering ('strongly disagree'—'disagree'—'neutral'—'agree'—'strongly agree') (Kools & Stoll, 2016) (Kools et al., 2020).

7.1.2 *Small and Medium Enterprises*

Small and medium-sized enterprises are defined by the number of employees, up to 250 people. Looking closer, they can be divided into tiny enterprises (1–9 employees), small enterprises (10–49 employees) and medium-sized enterprises (50–249 employees) (Czech Statistical Office, 2005) (Czech Statistical Office, 2013).

These enterprises are of relatively high importance within the Czech economy. According to a report by the Association of Small and Medium-Sized Enterprises and Self-Employed Persons of the Czech Republic from May 2019, small and medium-sized enterprises represented 99% of all companies in the Czech Republic (tiny enterprises 95.5%, small enterprises 3.7%, medium-sized enterprises 0.8%) and they employed two million people (61% of all employees) (Economic Diary, 2019). In 2017, the share of small and medium-sized enterprises was approximately 40% of gross domestic product and accounted for more than half of Czech exports (Finance.cz, 2017). In 2017, the share of employees in small and medium-sized enterprises in the total number of employees in the business sphere was 58.0%. Small and medium-sized enterprises in the Czech Republic represent more than one million economic entities as a whole, they are significant employers', a driving force for the business sector, growth, innovation and competitiveness. These enterprises are actively supported by the state, e.g. in the form of projects and various programs (Ministry of Industry and Trade, 2017). Small and medium-sized enterprises also have access to financial programs at European Union level. In the event of a crisis, small and medium enterprises slow down the economic downturn and help maintain stability. Their main advantage is their flexibility (Industry Reports, 2019). Total IT accounts for 3.6% of the business sector, in IT the tiny enterprises (40,232) and small enterprises (1463) are the most represented (Czech Statistical Office, 2017).

If the principles of learning organization are applied in small and medium-sized enterprises, we can see a positive effect, for example, on employee creativity (Herawati, Lupikawati, & Purwati, 2018) or sales and employment growth (Michna, 2009). According to Michna (2009), the growth of sales and employment was mainly associated with dimensions '*dialogue and empowerment of the employees, collaboration, team learning, leaders 'attitudes*'. The same survey shows that organizations that achieve a higher level of organizational learning are likely to achieve higher performance (Michna, 2009). The long-term growth and survival of small and medium enterprises is supported by inter-organizational learning and if the enterprises process internally the external knowledge, they can improve, e.g. the customer relationship, human resource, finance or organizational development (Alashwal, Low, & Kamis, 2019).

7.1.3 Schools as Learning Organizations

Education is an important sector in all countries, producing professionals in various fields. However, the quality of schools across countries varies as well as the gender representation of teachers or the statutory teachers' salaries. When comparing primary and lower secondary education teachers, the teachers in lower secondary education have 10% less teaching time than primary teachers (in the Czech Republic at least 30% less than primary teachers) and teachers at lower secondary schools are older than teachers at primary schools. '*Around one-third of primary and lower secondary teachers are 50 years old and over on average across OECD countries*'. (OECD, 2018). We can expect older teachers to be more experienced and if we are talking about self-study of teachers, based on the previous studies the beginning teachers prefer observing colleagues, university courses or interaction with experienced colleagues. On the opposite experienced teachers prefer '*sharing and collaborative initiatives, experimenting, and reading professional literature*' (Louws et al., 2017). According to OECD Indicators: Education at a Glance 2019 publication are teachers at secondary schools in Czech Republic 30 years old and older, so it can be expected to be closer to learning as experienced teachers (OECD, 2019).

The culture of a learning organization in a secondary school environment is not defined anywhere, but the definition of a learning organization culture at universities can be applied. Schools characterized by a learning organization culture support continuous learning for the sustainable improvement of teaching and learning. The knowledge gained leads to the education and support of individual development, teamwork and leadership in order to fulfil the institution's mission (Ponnuswamy & Manohar, 2016). Schools then have the ability to adapt to new environments and, through learning, find their way to carry out their visions.

Brandt (1998) states characteristics of schools being learning organizations. These characteristics are (Brandt, 1998):

- *'They have an incentive structure that encourages adaptive behavior.*
- *They have challenging but achievable shared goals.*
- *They have members who can accurately identify the organization's stages of development.*
- *They gather, process, and act upon information in ways best suited to their purposes.*
- *They have an institutional knowledge base and processes for creating new ideas.*
- *They exchange information frequently with relevant external sources.*
- *They get feedback on products and services.*
- *They continuously refine their basic processes.*
- *They have a supportive organizational culture.*
- *They are "open systems" sensitive to the external environment, including social, political, and economic conditions'.*

Kools and Stoll (2016) proposed a model of the school as a learning organization and characterized in the model the characteristics of the school as a learning organization. This model consists of seven action-oriented dimensions, namely (Kools & Stoll, 2016) (Kools et al., 2020):

- *'developing a shared vision centered on the learning of all students,*
- *creating and supporting continuous learning opportunities for all staff,*
- *promoting team learning and collaboration among all staff,*
- *establishing a culture of inquiry, innovation and exploration,*
- *embedding systems for collecting and exchanging knowledge and learning,*
- *learning with and from the external environment and larger learning system,*
- *modelling and growing learning leadership*'.

Based on the theoretical background, the main research question of this survey was to determine: Does a relationship between learning time and the dimensions of the learning organization score exist?

7.2 Method

7.2.1 Participants

In 2018, a cross-sectional questionnaire was conducted focusing on small and medium-sized enterprises in the IT sector in the Czech Republic. A total of 2884 respondents from small and medium-sized enterprises from the Czech Republic were addressed with a focus on IT activities.

In 2019, a cross-sectional questionnaire was conducted focusing on secondary schools across the Czech Republic. Altogether 1304 representatives of secondary schools were addressed.

Both studies included respondents of both sexes and older than 21 years old.

7.2.2 Materials

A shortened 21 questionnaire version of the DLOQ was used, including 7 dimensions in Czech language. To maintain the questionnaire's validity, the questionnaire was translated by two independent translators from English to Czech and then back to English. At the same time, the preservation of the questionnaire's meaning was assessed. Total reliability was 0.933 in 2018, in 2019 total reliability was 0.941. Respondents evaluated the individual dimensions on a 6-point Likert scale. The final version of the questionnaire was created using 'docs.google.com'.

7.2.3 *Design and Procedure*

In total, two cross-sectional questionnaires were conducted in the Czech Republic in 2018 (December 2017–February 2018) and 2019 (January 2019). Contacts to small and medium-sized enterprises were obtained from the Albertina database for trade and marketing (Albertina for Trade and Marketing, 2019), contacts to secondary schools were obtained from the secondary school databases in the Czech Republic at www.stredniskoly.cz (List of Schools, 2019). The reference to the online questionnaire was sent to respondents' e-mail addresses. In 2018, approximately 250 e-mails sent were returned as undeliverable after sending due to the absence of the e-mail address, 25 respondents responded to the e-mail with the response that they no longer operate the business. In 2019, 91 e-mails were returned due to the absence of the given e-mail address, 1 respondent directly rejected the survey. Respondents who did not complete the questionnaire were reminded every 14 days.

The data obtained was analysed using Microsoft Excel 2016 and IBM SPSS Statistics Version 24 using descriptive statistics, parametric and non-parametric assays at confidence levels $\alpha = 0.01$ and $\alpha = 0.05$. The Cronbach reliability coefficient was calculated for each dimension using IBM SPSS Statistics version 24.

7.3 Results

In 2018, 201 respondents from small and medium-sized enterprises in the IT sector in the Czech Republic participated in the study (questionnaire returns were 6.97%). The respondents consisted of 137 men and 64 women. The respondents most frequently reported employment in the organization within 5 years (32.8%), followed by 11–15 years (21.4%).

In 2019, 121 respondents from secondary schools in the Czech Republic participated in the study (9.28% return). The respondents consisted of 45 men and 76 women. The most frequently reported employment periods were less than 10 years (32.2%), followed by 11–20 years (27.27%). The respondents' demographic profile is showed in Table 7.1.

When comparing the demographic profile of respondents, it is clear that respondents from organizations employing up to 50 employees were dominantly represented in the survey in both years, similarly to the representation of common employees and executives. From this viewpoint, both surveys can be compared, although they were conducted in different sectors. The respondents' profile in terms of learning time per month is also similar. While the IT sector has the highest number of employees dedicated to learning 1–10 hours per month, teachers spend usually 11–20 hours a month self-educating. Self-education is mostly applied by e-learning focused on current topics related to a particular job (Table 7.1).

Table 7.1 The respondents' demographic profile (own processing)

	Number of respondents	
	2018 ($n = 201$) n (%)	2019 ($n = 121$) n (%)
Age		
21–30 years	30 (14.9)	4 (3.3)
31–40 years	76 (37.8)	17 (14.0)
41–50 years	57 (28.4)	29 (24.0)
51–60 years	32 (15.9)	57 (47.1)
Over 61 years	6 (3.0)	14 (11.6)
Organization size		
Up to 10 employees	65 (32.3)	0
Up to 50 employees	91 (45.3)	73 (60.3)
Up to 250 employees	45 (22.4)	47 (38.9)
More than 250 employees	0	1 (0.8)
Position in employment		
Staff member	72 (35.8)	34 (28.1)
Executive member	129 (64.2)	87 (71.9)
Learning time		
0 hours per month	9 (4.5)	1 (0.8)
1–10 hours per month	109 (54.2)	40 (33.1)
11–20 hours per month	48 (23.8)	54 (44.6)
21–35 hours per month	15 (7.5)	20 (16.5)
More than 36 hours per month	20 (10.0)	6 (5.0)

Table 7.2 Comparison of respondents' responses with different type of school (own processing)

Hours a month	Grammar school	Secondary vocational school	Secondary vocational school focused on crafts
0	0	0	1
1–10	16	21	3
11–20	16	34	4
21–35	10	10	0
36 or more	2	2	2

It should be noted that schools are traditionally divided according to their type and not according to the number of employees. Therefore, a closer look at the learning of employees in education is interesting, if the respondent is traditionally divided according to the type of school (grammar school, secondary vocational school, secondary vocational school focused on the crafts). The results show that most learning is provided by employees of secondary vocational schools, followed by employees of grammar schools. In secondary vocational schools focused on the crafts, employees rarely attend to learning (Table 7.2).

As shown in Tables 7.3 and 7.4, when comparing the average values of the two dimensions, it is clear that the results are very similar and there is no statistically significant difference in the t-test at the significance level $\alpha = 0.05$ ($p = 0.06$–0.96).

Table 7.3 Comparison of respondents' responses with different education intensity D1–D4 (own processing)

	Average of D1		Average of D2		Average of D3		Average of D4	
Hours a month	2018	2019	2018	2019	2018	2019	2018	2019
0	2.74	2.67	3.46	1.67	3.78	3.00	3.15	3.33
1–10	4.48	4.46	4.35	4.11	4.13	4.18	3.56	3.95
11–20	4.39	4.50	4.44	4.32	4.35	4.40	3.49	4.16
21–35	4.60	4.18	4.58	3.67	4.44	4.09	3.62	3.72
36 or more	5.10	4.83	4.73	4.50	4.17	4.50	3.55	4.28

Table 7.4 Comparison of respondents' responses with different education intensity D5–D7 (own processing)

	Average of D5		Average of D6		Average of D7	
Hours a month	2018	2019	2018	2019	2018	2019
0	3.11	4.33	3.37	3.33	3.29	3.00
1–10	4.34	4.63	3.89	4.19	4.28	4.50
11–20	4.40	4.79	3.95	4.60	4.46	4.83
21–35	4.62	4.60	4.08	4.35	4.38	4.22
36 or more	3.55	4.50	4.40	4.72	4.67	4.44

The lowest $p = 0.06$ can be seen for dimension 4 Create Systems to Capture and Share Learning, while the highest $p = 0.96$ can be seen for dimension 7 Provide Strategic Leadership for Learning. Generally speaking, with education up to 20 hours a month, the assessment of dimensions in both sectors is increasing in both years.

When comparing the average values of individual dimensions in terms of human and structural level, we can observe that dimensions from human level: 1 Create Continuous Learning Opportunities, 2 Promote Inquiry and Dialogue, 3 Encourage Collaboration and Team Learning, and 5 Empower People towards a Collective Vision receive higher ratings, while dimensions, 4 Create Systems to Capture and Share Learning, and 6 Connect the Organization to its Environment from the structural level have lower ratings.

When comparing the time spent on learning for IT sector staff and executive members, it is clear that common staff in the IT sector is dedicated to learning for 1–20 hours per month compared to teachers as staff members. However, teachers self-educate for at least an hour a month. For education of executives in the education sector we can see more respondents who spend 11–20 and 21–35 hours per month, while IT sector executives most often devote 1–10 hours per month to learning (Fig. 7.1).

7.4 Discussion

In the Czech Republic, only a few studies of a learning organization have been performed using DLOQ. (Zubr, 2019) The IT sector and the education sector were selected for this study. The IT sector in the Czech Republic represents an important

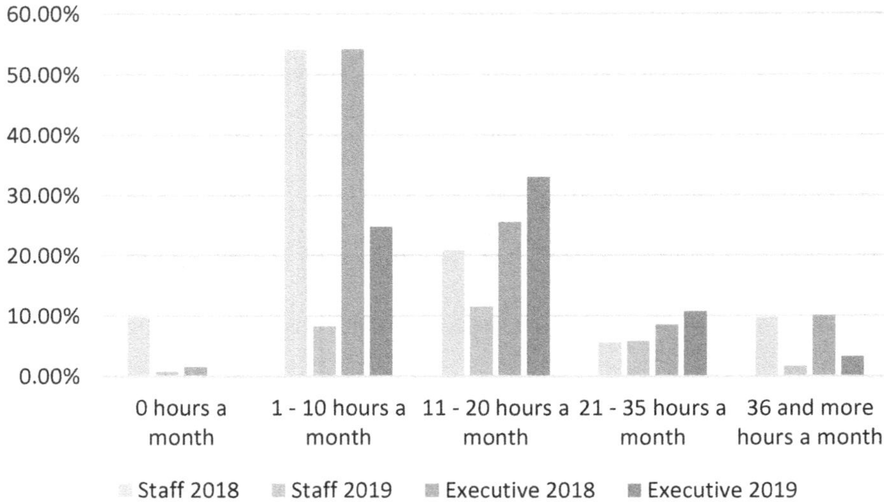

Fig. 7.1 Learning time (per month) by worker's position (own processing)

employer within small and middle-sized enterprises (Ministry of Industry and Trade, 2017), the education sector keeps its significance in all countries thanks to its function of producing quality workers.

A total of 201 respondents participated in the survey in 2018, 121 respondents in 2019. Due to the percentage of the questionnaire return (2018: 6.97% vs. 2019: 9.28%), these surveys can be compared for the number of respondents. Looking at the composition of respondents, the individual years are quite different. In 2018, 137 men and 64 women were included in the survey, while in 2019, 45 men and 76 women participated in the survey. Therefore, the gender ratio is reversed in the 2 years. This is determined by the sectors in which the survey was conducted. While in 2018, there were approximately 3 women for every 10 men in the Czech Republic's IT sector, in the area of education this ratio is reversed, i.e. for every 10 women there is approximately 2.7 men (Czech Statistical Office, 2018).

In both surveys, organizations with less than 50 employees are the most represented, with 45.3% of these enterprises being represented in 2018 and 60.3% of enterprises in 2019. Small and medium-sized enterprises in 2018 are the second most represented group of small firms with less than 10 employees. No school with less than 10 employees participated in the survey and only 1 school said it had more than 250 employees. Given the nature of the schools addressed, it can be assumed that the response from 'more than 250 employees' was wrong. Based on available data, a total of 44,993 teachers worked in Czech secondary schools in the 2017/2018 school year and there are currently around 1400 secondary schools in the Czech Republic (Czech News Agency, 2019) (Ministry of Education, Youth and Sport, 2014) (List of Schools, 2019). Therefore, there is an average of around 32 teachers per school, which is in line with the study's outcome (the dominant representation of schools with less than 50 employees).

Both surveys show that approximately 78% of employees in both sectors spend 1–20 hours a month on self-education. In the education sector, only one respondent stated that they did not spend 1 hour per month with education. This result is expected in the education sector—in order to ensure the quality of teaching students, it is necessary that educators continue to educate themselves and monitor current changes in the subjects taught. Therefore, it is not surprising that 16.5% of teachers say they spend 21–35 hours a month on self-education. When comparing the time of self-education among ordinary and senior executives, it is clear that ordinary workers in the IT sector educate more often than ordinary staff in the education sector. By contrast, education managers educate more than IT sector workers. This result is surprising in view of the above-mentioned assertion of maintaining a certain quality of student education, which is being looked after by ordinary staff in education.

If we compare the different dimensions of the learning organization in relation to the time devoted to learning, it is clear that the results are very similar for both years. The biggest difference between the average dimension rating is for Dimension 4: Create Systems to Capture and Share Learning, where p is approaching 0.05 ($p = 0.06$). Dimension 4 is generally assessed by the higher education staff as a higher average score than in the IT sector. This result is expected due to the need for teachers to be constantly educated, a number of training events for teachers and hence systems that make learning easier to share. It should also be noted that non-teaching topics and seminars can also be included in teacher education for secondary school teachers, e.g. in healthcare, a relatively sophisticated network of learning sharing systems can be recorded. It can be argued that the average rating of individual dimensions increases with learning times up to 11–20 hours per month, while in the IT sector, for most dimensions, the average rating for 21–35 hours per month may increase further. The smallest difference between learning time and dimensional assessment for each sector was noted for Dimension 7: Provide Strategic Leadership for Learning. This result indicates that there is an effort to establish a learning organization in both sectors.

If we compare the average values of individual dimensions in terms of human and structural level in relation to the time devoted to learning, the dimensions from human level have higher ratings than two dimensions from the structural level. This result is probably due to the fact that in the human level there are mainly dimensions focused on learning.

The results show that most learning is provided by employees of secondary vocational schools and in the opposite in the secondary vocational schools focused on the crafts employees rarely attend to learning. This distribution of results is probably due to the focus of the school. While in the secondary vocational schools it is necessary to supplement not only new trends in practice, but also in theory (e.g. medical schools), in secondary vocational schools focused on crafts the learning is focused primarily on students' practical skills and self-education of teachers can probably represent more insights into news in practice than theoretical education.

When comparing the average rating of individual dimensions of this study with similar studies abroad from the banking sector, the average rating in IT sector in the Czech Republic in 2018 is higher for employees who are engaged in self-education at least an hour a month (Berberoğlu & Emine, 2011) (Soahib, Ihsaan, Yousaf, & Majeed, 2014). Compared to other studies abroad, the same result was also achieved in the education sector (Abo Al Ola, 2017). According to Voolaid data, the average score of dimensions of the study from Czech Republic is higher than international average in 2013 (Voolaid, 2013).

7.4.1 Limitations

The main limitation of this study is a relatively low number of respondents in both sectors. The low response rate is attributed to a number of other questionnaire surveys that are addressed to respondents by various students and statistical firms. Due to reluctance to respond to another questionnaire survey, respondents in several cases directly refused to participate in the survey and did not complete the questionnaire even after repeatedly addressed. Unfortunately, it is not possible to address each respondent personally due to the inclusion of respondents from the whole Czech Republic. In future studies it would be appropriate to get a larger sample of respondents to gain a more objective view of the issue.

The setting of the study in the Czech Republic can be perceived as a slight limitation and it would be interesting to carry out the same study in other country of the European Union.

The limitation of the study in 2019 can also be considered the focus only on secondary schools and not on all levels of education. It is likely that the results would be slightly different when more than one level of education was included. On the other hand, it is not possible to include in the study all schools in the Czech Republic due to the lack of clarity of the data obtained.

It is a matter of discussion whether the use of DLOQ in the education sector is appropriate or whether it is more appropriate in the future to use the 'Schools as a Learning Organization Scale' applied by Kools et al. in schools in Wales (Kools et al., 2020).

To obtain a more comprehensive view of the issue, it would be appropriate to obtain data from employees from different levels of individual companies (both IT and education sector). However, as the questionnaires are anonymous, it would not be possible to match employees from one company, therefore, possibly another deeper study would be conducted in the future.

7.5 Conclusion

Based on a comparison of the results from individual studies conducted in 2018 and 2019, it can be argued that the introduction of the concept of a learning organization in both sectors in the Czech Republic is on the right track. According to the average evaluation of individual dimensions, it is clear that a lot of practices and processes of learning organization are applied in the examined sectors.

The assessment of individual dimensions implies that the time devoted to self-education is directly related to the assessment of individual dimensions, and there is leadership-supporting learning in organizations. Based on the results of this study, we can say that the relationship between learning time and the dimensions of the learning organization score exist. It could be said that if companies effectively support the training of employees and the sharing of their visions, it will lead to an overall improvement in the quality of the company in many respects.

Comparing two different sectors brings the possibility to learn from mistakes and weaknesses in one sector or to use the positive functional approach of the other sector. Sectors could also join together and work together on common weaknesses (e.g. Dimension 4). A change in the approach to evaluating employee education (e.g. setting up systems to measure the difference between actual and expected performance, publishing the evaluation to all employees or balancing the time and resources spent on education) would contribute to improving Dimension 4 evaluation in both sectors.

Improving the evaluation of individual dimensions in both sectors would bring some benefits of a learning organization concept as improving the competitive advantage by engaging with the external community and recognizing employee initiative, improving job satisfaction of employees and thus increasing their performance.

Although these studies have been carried out in two sectors, it is necessary to map other employment sectors in the Czech Republic in the future, or to compare them to those abroad. In the future, it would also be appropriate to conduct deeper studies on specific organizations, where the results would be evaluated both from the perspective of employees and from the perspective of the organization's management.

Acknowledgements The paper was written with the support of the specific project 2020 grant 'Determinants of Cognitive Processes Impacting the Work Performance' granted by the University of Hradec Kralove, Czech Republic and thanks to help of students František Hašek and Anna Borkovcová.

References

Abo Al Ola, L. M. (2017). The availability of dimensions of learning organizations questionnaire (DLOQ) in educational college at Taif University from the perspective of employees. *Journal of Educational and Psychological Sciences, 18*(1), 447–487. https://doi.org/10.12785/jeps/180114

Alashwal, A. M., Low, W. W., & Kamis, N. A. M. (2019). Effect of inter-organizational learning on construction SMEs performance. *IOP Nocf Series: Materials Science and Engineering, 495*(2019), 012043. https://doi.org/10.1088/1757-899X/495/1/012043

Albertina for Trade and Marketing. (2019). Retrieved from http://www.albertina.cz/?gclid=Cj0KCQiAzrTUBRCnARIsAL0mqcz2seJjjxmISfi7QlvHHAwN-ZnZrWpd4Botnujmtq08Zio1vx0BfBMaAu-bEALw_wcB

Berberoğlu, A. & Emine, Ü. (2011). Relation between organizational learning and organizational commitment: Case study of a private bank in North Cyprus. In *Proceedings of the European Conference on Intellectual Capital—Nicosia,* Cyprus, University of Nicosia, Cyprus (pp. 87–93).

Birdthistle, N. (2009). *Family businesses and the learning organisation: A guide to transforming the family business into a learning organisation.* Saarbrücken: VDM Verlag Dr. Müller Aktiengesellschaft & Co. KG.

Brandt, R. (1998). *Powerful learning.* Alexandria: Association for Supervision and Curriculum Development.

Czech News Agency—Czech newspaper (2019). Retrieved from https://www.ceskenoviny.cz/zpravy/ve-skolach-pribyva-ucitelu-podle-reditelu-je-jich-ale-stale-malo/1737874

Czech Statistical Office. (2005). Retrieved from https://www.czso.cz/csu/czso/cri/male-a-stredni-podniky-jejich-misto-a-role-v-ceske-ekonomice-2005-rhybfgzbj0

Czech Statistical Office. (2013). Retrieved from https://www.czso.cz/documents/10180/20534676/116111a.pdf/9c378e0f-d77a-4f21-bf3e-e4ed35cb1122?version=1.0

Czech Statistical Office. (2017). Retrieved from https://www.czso.cz/csu/czso/odvetvi-informacni-ekonomiky.

Czech Statistical Office. (2018). Retrieved from https://www.czso.cz/documents/10180/91606479/2501321924.pdf/a84d8e9a-6f38-455d-a524-e55ed62d9fde?version=1.1

Economic Diary. (2019). Retrieved from https://ekonomickydenik.cz/wp-content/uploads/2019/05/03_zdenek-tomicek_amsp.pdf

Finance.cz. (2017). https://www.finance.cz/501303-male-a-stredni-podniky/

Herawati, Y., Lupikawati, M., & Purwati, E. (2018). The influence of learning organization and organizational stress on performance of creative industry SMEs with the creativity of its owners/actors as mediator. *RJOAS, 73*(1), 65–78. https://doi.org/10.18551/rjoas.2018-01.09

Industry Reports. (2019). Retrieved from https://www.reportazezprumyslu.cz/cs/novinky/1173-male-a-stredni-podniky-citliva-pater-ekonomiky

Jamali, D., Sidani, Y., & Zouein, C. H. (2009). The learning organization: Tracking progress in a developing country. A comparative analysis using the DLOQ. *The Learning Organization, 16*(2), 103–121. https://doi.org/10.1108/09696470910939198

Kim, J., Egan, T., & Tolson, H. (2015). Examining the dimensions of the learning organization questionnaire: A review and critique of research utilizing the DLOQ. *Human Resource Development Review, 14*(1), 91–112. https://doi.org/10.1177/1534484314555402

Kim, K., Watkins, K. E., & Lu, Z. (2017). The impact of a learning organization on performance: Focusing on knowledge performance and financial performance. *European Journal of Training and Development, 41*(2), 177–193. https://doi.org/10.1108/EJTD-01-2016-0003

Kools, M., & Stoll, L. (2016). *What makes a school a learning organisation? A guide for policy makers, school leaders and teachers.* Paris: OECD Publishing.

Kools, M., Stoll, L., George, B., Steijn, B., Bekkers, V., & Gouëdard, P. (2020). The school as a learning organisation: The concept and its measurement. *European Journal of Education, 55*, 24–42. https://doi.org/10.1111/ejed.12383

List of Schools. (2019). http://www.seznamskol.eu/typ/stredni-skola/

Louws, M. L., Meirink, J. A., van Veen, K., & van Driel, J. H. (2017). Teachers' self-directed learning and teaching experience: What, how, and why teachers want to learn. *Teaching and Teacher Education, 66*, 171–183. https://doi.org/10.1016/j.tate.2017.04.004

Marsick, V., & Watkins, K. (2003). Demonstrating the value of an organization's learning culture: The dimensions of the learning organization questionnaire. *Advances in Developing Human Resources, 5*(2), 132–151. https://doi.org/10.1177/1523422303005002002

Michna, A. (2009). The relationship between organizational learning and SME performance in Poland. *Journal of European Industrial Training, 33*(4), 356–370. https://doi.org/10.1108/03090590910959308

Ministry of Education, Youth and Sport. (2014). Retrieved from http://www.msmt.cz/vzdelavani/skolstvi-v-cr/statistika-skolstvi/vykonova-data-o-skolach-a-skolskych-zarizenich-2003-04-2013

Ministry of Industry and Trade. (2017). Retrieved from https://www.mpo.cz/assets/cz/podnikani/male-a-stredni-podnikani/studie-a-strategicke-dokumenty/2018/10/Zprava_MSP_2017.pdf

Norashikin, H., Safiah, O., Fauziah, N., & Noormala, A. I. (2016). Learning organization culture, organizational performance and organizational innovativeness in a public institution of higher education in Malaysia: A preliminary study. *Procedia Economics and Finance, 37*, 512–519. https://doi.org/10.1016/S2212-5671(16)30159-9

OECD. (2018). How do primary and secondary teachers compare? *Education indicators in focus, No. 58*, OECD Publishing. Retrieved from https://doi.org/10.1787/535e7f54-en

OECD. (2019). *Education at a glance 2019: OECD indicators*. Paris: OECD Publishing. https://doi.org/10.1787/f8d7880d-en

Ponnuswamy, I., & Manohar, H. L. (2016). Impact of learning organization culture on performance in higher education institutions. *Studies in Higher Education, 41*(1), 21–36. https://doi.org/10.1080/03075079.2014.914920

Saadat, V., & Saadat, Z. (2016). Organizational learning as a key role of organizational success. *Procedia—Social and Behavioral Sciences, 230*, 219–225. https://doi.org/10.1016/j.sbspro.2016.09.028

Senge, P. (2016). *Fifth discipline: The art and practice of the learning organization*. Praha: Management Press.

Serrat, O. (2017). *Knowledge solutions*. Singapore: Springer.

Soahib, M., Ihsaan, M., Yousaf, J., & Majeed, A. (2014). Factors affecting the organizational learning: A study of banking sector of Pakistan. *International Journal of Management and Organizational Studies, 2*(2), 16–22.

Tichá, I. (2005). *Learning organisation*. Praha: Alfa Publishing.

Vega Martinez, J. E., del Carmen Martinez Serna, M., & Parga Montoya, N. (2019). Dimensions of learning orientation and its impact on organizational performance and competitiveness in SMES. *Journal of Business Economics and Management, 21*(2), 395–420. https://doi.org/10.3846/jbem.2020.11801

Voolaid, K. (2013). *Measurement of organizational learning of business schools*. [Thesis on Economics H33, TUT Press]. Tallinn.

Watkins, K., & Kim, K. (2018). Current status and promising directions for research on the learning organization. *Human Resource Development Quarterly, 29*, 15–29. https://doi.org/10.1002/hrdq.21293

Watkins, K. E., & Marsick, V. J. (1993). *Sculpting the learning organization: Lessons in the art and science of systematic change*. San Francisco, CA: Jossey-Bass.

Watkins, K. E., & O'Neil, J. (2013). The dimensions of the learning organization questionnaire (the DLOQ): A nontechnical manual. *Advances in Developing Human Resources, 15*(2), 133–147. https://doi.org/10.1177/1523422313475854

Yadav, S., & Agarwal, V. (2016). Benefits and barriers of learning organization and its five discipline. *IOSR Journal of Business and Management, 18*(12), 18–24. https://doi.org/10.9790/487X-1812011824

Zubr, V. (2019). Studies with dimensions of learning organization questionnaire—Research study. Hradec economic days (pp. 592–600). Hradec Kralove.

Zubr, V., Mohelska, H., & Sokolova, M. (2017). Factors with positive and negative impact on learning organisation. In *Double-blind peer-reviewed proceedings of the international scientific conference Hradec economic days* (pp. 980–985). Hradec Kralove: University of Hradec Kralove.

Part III
Cases from Higher Education and Further Education

Chapter 8
Learning Analytics Dashboard Supporting Metacognition

Li Chen, Min Lu, Yoshiko Goda, Atsushi Shimada, and Masanori Yamada

8.1 Introduction

Mastering self-regulated learning (SRL) can have a profound and positive effect on academic achievement by improving test grades and academic skills in the short term and helping students develop life-long learning skills, which is the function of education (Zimmerman, 2002). Academic self-regulation helps learners transform their mental abilities into academic skills, which has led researchers to attribute individual differences in learning to students' inability to self-regulate (Zimmerman, 1998, 2002). Within the framework of SRL, students recognize their own strengths and weaknesses in learning, set and revise their learning goals accordingly, and then work to achieve those goals (e.g., Zimmerman, 2002). SRL requires heightened metacognition and the motivation to orient thoughts, emotions, and behaviors to attaining goals (Tobias & Everson, 2009; Zimmerman, 1998, 2002).

L. Chen (✉)
Graduate School of Human-Environment Studies, Kyushu University, Fukuoka, Japan
e-mail: chenli@mark-lab.net

M. Lu · M. Yamada
Faculty of Arts and Science, Kyushu University, Fukuoka, Japan
e-mail: lu@artsci.kyushu-u.ac.jp; mark@mark-lab.net

Y. Goda
Graduate School of Instructional Systems, Kumamoto University, Kumamoto, Japan
e-mail: ygoda@kumamoto-u.ac.jp

A. Shimada
Faculty of Information Science and Electrical Engineering, Kyushu Universityy, Fukuoka, Japan
e-mail: atsushi@ait.kyushu-u.ac.jp

D. Ifenthaler et al. (eds.), *Balancing the Tension between Digital Technologies and Learning Sciences*, Cognition and Exploratory Learning in the Digital Age,
https://doi.org/10.1007/978-3-030-65657-7_8

Self-regulation is a multidimensional process comprising personal (emotional, cognitive), behavioral, and contextual components (Zimmerman, 1986, 1989). In SRL, students apply their cognitive learning strategies to specific tasks in academic contexts. When students are able to monitor and reflect on their own detailed learning processes, their self-confidence, satisfaction, and motivation improve (Schunk, 1982; Zimmerman, 2002). This makes metacognition—self-awareness of the thinking process—the key to enabling learners to make sense of their own learning processes.

Advances in information and communication technology (ICT) have expanded our ability to collect and analyze a widened variety of educational data. For example, the learning analytics (LA) approach measures, collects, analyzes, and reports on learners and learning environments, seeking to optimize learning and develop educational models (Baker & Siemens, 2014). In SRL, the LA approach can be useful for recording and monitoring students' learning processes and identifying patterns that can inform and empower instructors and learners.

In this chapter, we provide the theoretical background of metacognition in SRL, visualization, and learning analytics dashboards (LADs). We discuss some of the important factors and principles in the design and use of LADs and describe the design and development of a LAD to measure students' metacognition. Therefore, based on the purpose of this chapter, we put forward the following research question: what are important factors in designing a LAD supporting metacognition?

8.2 Theoretical Background

8.2.1 Metacognition in Self-Regulated Learning

SRL strategies are "actions and processes directed at acquiring information or skill that involve agency, purpose, and instrumentality perceptions by learners" (Zimmerman, 1989, p. 329). In SRL, which includes both metacognition and motivation dimensions, students monitor learning behaviors regarding their learning goals and reflect on learning effectiveness, which will improve their motivation to continue their strategies and approach (Zimmerman, 2002).

There are two dimensions to metacognition: *knowledge about cognition* and the *self-regulation of cognition* (Flavell, 1979; Hofer, Yu, & Pintrich, 1998; Jacobs & Paris, 1987).

Regarding the former, there are three general types of knowledge about cognition: declarative, procedural, and conditional knowledge (Jacobs & Paris, 1987). *Declarative knowledge* means awareness and understanding of the basic strategies; *procedural knowledge* means knowing how to use these strategies (e.g., skimming, underlining, summarizing, using context, or finding the main idea while reading); and *conditional knowledge* means knowing when and why to use particular strategies in specific conditions that influence learning goals, tasks, or contexts (Hofer et al., 1998; Jacobs & Paris, 1987; Weinstein, Acee, & Jung, 2011). Regarding the latter, many models of the self-regulation of cognition describe three general strategy phases: planning, monitoring, and regulating (e.g., Hofer et al., 1998; Pintrich, 1999; Pintrich, Smith, Garcia, & McKeachie, 1993; Tobias & Everson, 2009).

During the planning phase, first, learners set learning goals related to their extrinsic motivation such as the score and certifications, or intrinsic benefits of learning, including the improvement of skills and knowledge, or specific course contents (Lee, Watson, & Watson, 2019). When setting plan according to the specific learning goal, learners divide the overall goal into several manageable sub-goals, arranging these sub-goals according to priority and a timetable, overviewing the texts and asking themselves questions before intensively reading the texts to improve their understanding of the texts, and conducting task analyses of problems (Hofer et al., 1998; Zimmerman, 1989). The aims of planning activities are to activate learners' prior knowledge and learning experience, organize the learning materials, and prepare for a better understanding of the contents (Hofer et al., 1998).

During the monitoring phase, learners use many ways to track their progress: assess their attention during individual lectures and learning activities, distinguish between what they knew or learned with what they don't know or haven't learned yet, evaluate their understanding using questions or self-tests (Tobias & Everson, 2009), schedule and manage their learning time (Lan, 1998), observe or record learning behaviors according to the learning goals or criteria set during the planning phase (Belfiore & Hornyak, 1998; Zimmerman, 1989), and adjust their test-taking strategies in situ to accommodate the time allocated (Hofer et al., 1998). Through self-monitoring, learners can better recognize their strengths or limitations in learning, enabling them to regulate their cognitive strategies or learning behaviors during the next phase (Hofer et al., 1998; Winne & Stockley, 1998). Although monitoring is primarily the transitional phase between planning and regulation, it must also take place to a lesser extent during both planning and regulation to allow adaptation based on self-critique (Belfiore & Hornyak, 1998).

During the regulation phase, learners read or re-read the contents they are not familiar with, adjust their learning pace based on their learning abilities, review additional materials, or use such test strategies as skipping the questions they don't know (Hofer et al., 1998; Thiede, Griffin, Wiley, & Redford, 2009). Regulation activities help learners determine appropriate learning or test strategies based on the self-evaluation and causal attribution results, which can improve their learning behaviors (Hofer et al., 1998; Zimmerman, 2002).

These self-regulation processes are teachable, so it is important provide learners with instructions and guidance based on their strengths and limitations. For example, McDowell (2019) provided students with interventions including a daily query about students' daily plans and the use of some cognitive strategies and near real-time feedback concerning their learning strategy use and personalized learning questions. As their results indicated, the interventions promoted some cognitive and metacognitive strategies, which were beneficial for learning success.

To support learners' metacognition, this paper discusses the possibilities and effectiveness of using a LAD to display learners' information on their learning activities, focusing on the dimensions of planning, monitoring, and regulation.

8.2.2 *Metacognition and Learning Analytics*

It is not easy to measure learners' metacognition, which includes the use of several learning strategies and processes according to the tasks (Zimmerman, 2002). Self-reported inventories are commonly used (Dinsmore, Alexander, & Loughlin, 2008). However, this has limitations since learners cannot observe or remember all the behaviors or events in their learning processes. Moreover, what learners perceive about their learning processes will be influenced by their previous experiences and memories, their teachers' instructional designs, the various learning environments, and learning biases (Tobias & Everson, 2009; Winne & Nesbit, 2009). Therefore, using self-reporting as a method of metacognition measurement can raise questions relating to validity, candor, and accuracy.

The LA approach has shown significant potential in collecting educational log data during the learning processes while minimizing the limitations of self-reported inventories. For example, system operations can be collected as learning logs that show interactions between users and systems or between multiple users; this can show how the learners use or change their cognitive strategies (Tobias & Everson, 2009).

Many previous studies have used the LA approach to collect and analyze learning logs for learning behaviors and found them to be important indicators of learning performance and learning awareness in SRL. For example, Chen, Goda, Shimada, and Yamada (2020) found that students' learning behaviors, which were identified by learning logs, were the indicators to predict learning performance and SRL awareness. Moreover, students' learning performance and SRL awareness were affected by different learning behaviors related to different cognitive strategies during in-class and out-of-class activities. Similarly, Yamada et al. (2017) examined the relationships between learning behaviors related to reading e-learning materials, learning performance, and SRL factors. The results of their study indicated different relationships between SRL factors such as intrinsic value and specific learning behaviors with positive correlations among high performers and negative relationships among low performers. Rodrigues, Ramos, Silva, Dourado, and Gomes (2019) also used learning behaviors as indicators to predict learning performance. They developed a model of predicting learning performance based on indicators of behavior related to SRL and presented the results of the prediction in a dashboard to instructors to help monitor students' learning process.

In addition to learning logs representing learning behaviors, in some studies, qualitative data was also collected and analyzed. For example, Chiu and Fujita (2014) analyzed data from online forums regarding learners' messages and discussions to show how people influence one another through their interactions. As the results indicated, both individual characteristics and the message attributes regarding cognition and social metacognition had effects on subsequent new information and theoretical explanations.

LA has many advantages, including the continuous and automatic collection of data in ubiquitous learning environments without the interference of learners' engagement or external instructors (Yamada et al., 2016; Yin et al., 2015). Teachers can understand learner's learning behaviors out of class and informal learning settings (Shimada et al., 2015). Van Leeuwen (2015) found the LA approach to be effective in supporting teachers in diagnosing learners' learning progress and status by collecting information to an understandable and manageable level. Moreover, Fincham et al. (2019) focused on learners' learning strategies patterns regarding SRL via trace data collected from online learning environments. They identified patterns in student learning behaviors and confirmed relationships between learning strategies and learning outcomes, as well as the effects of the intervention learning strategies. The results would facilitate further studies concerning how learners adjust their learning strategies and a more accurate assessment of how interventions affect learning behaviors.

8.2.3 Learning Analytics Dashboards Supporting Metacognition

A LAD is a visualization technique that displays learning traces based on LA data to help learners monitor and reflect on their own learning processes and progress toward learning goals (Klerkx, Verbert, & Duval, 2017; Verbert et al., 2014). By integrating LA approach with the concept of a learning dashboard, it becomes possible to provide a visual display of the data, revealing patterns in learners' progress and behaviors that might otherwise go undetected (Teasley, 2017). LADs can help learners recognize what they have been doing and what they should do by bringing subtle information to the foreground.

Many previous studies have focused on the design and development of LADs and their use in educational settings, examining their objectives, the target users, the data LADs use, and how that data is visualized. Most of the research on LADs has focused on monitoring learning data for instructors and learners as the main target users, aiming to improve their teaching and learning (Schwendimann et al., 2016). LADs are generally used to support online or blended learning settings (Schumacher & Ifenthaler, 2018; Verbert et al., 2014). The data sources incorporated into LADs include the logs used to track computer-mediated user activity (e.g., frequency of interacting with the system), artifacts used or produced by learners (e.g., analyses of learners' contents), test or self-assessment results, and some recommended information based on the results (Bodily, Ikahihifo, Mackley, & Graham, 2018; Kim, Jo, & Park, 2016;Schwendimann et al., 2016 ; Verbert et al., 2014). For these data sources, the most common method for tracking data for LADs is capturing learners' relevant actions through application logs (Schwendimann et al., 2016; Verbert et al., 2014). The results of learning data are represented using various visualization types such as bar charts, line graphs, tables, pie charts, and network graphs (Roberts, Howell, & Seaman, 2017; Schwendimann et al., 2016).

For example, Kim et al. (2016) examined the effects of a student-facing LAD, focusing on dashboard usage frequency, dashboard satisfaction, and learning achievement in a higher education online course. The LAD used in their experiment visualized students' online activities during the course, such as log-in times and frequencies, number of log-ins, and the times spent using the dashboard. Their results supported the effectiveness of using dashboards in learning achievement, although dashboard usage frequency did not appear to have a significant impact on the students' learning achievement. They found a slight positive correlation between satisfaction with the LAD and learning achievement. Based on their findings, Kim et al. (2016) suggested the necessity of finding ways to motivate learners to use the LAD more consistently, taking into consideration the learners' different academic achievement levels.

Similarly, Bodily et al. (2018) designed and developed a student-facing LAD to support students' online learning in a higher education course. Their study used two separate dashboards: a content recommender dashboard to support students' content knowledge and a skills recommender to improve students' metacognitive strategies. Their study confirmed that while the students had positive perceptions of the dashboards, their inconsistency in usage remained a challenge.

Therefore, when determining what should be integrated into a LAD, it is necessary not only to consider how to improve the learning effectiveness but also how to motivate learners to use the LAD consistently. Teasley (2017) pointed out that rather than "one size fits all" displays that provide learners with the same format for all feedback, personalized displays might be more effective at increasing the frequency of use. Additionally, the LADs should provide easily understood information using simple comparisons, summaries of the data, feedback, advice on how to improve, and concise discussions of how to interpret the results.

Roberts et al. (2017) emphasized the importance of involving learners in functional design when developing LADs. They reported that many factors affect learners' perceptions of LADs. For example, students want easy access to the resources they need (e.g., numerical or graphical statistics) to compare their results with those of other learners; privacy protections, such as anonymized comparisons; and automated alerts. The main users of LADs are teachers and students, with less frequent use by administrators and researchers (Schwendimann et al., 2016). Ahn, Campos, Hays, and DiGiacomo (2019) designed LADs to support middle school mathematics teachers' pedagogical practices, integrating multiple layers of educational settings and suggestions from researchers, designers, developers, programmers, partner teachers, instructional coaches, and district leaders.

Drawing on the insights gleaned from these previous studies, we designed and developed a LAD to support metacognition. The next section introduces the design, development, and evaluation of the LAD.

8.3 An Example of a Learning Analytics Dashboard Supporting Metacognition

8.3.1 Design of the Metaboard

In order to support learners' metacognition, it is important to help learners monitor their learning behaviors and learning processes, and assess their own learning achievement by comparing with the criteria or with others' achievements, and regulate their learning based on the results of monitoring (Belfiore & Hornyak, 1998; Zimmerman, 2002). Therefore, it is considered effective to provide learners with specific information regarding their own activities and others' activities, to help them make sense of the gap between themselves and others. It is excepted that learners can regulate their learning behaviors when they understand their strengths and limitations, especially on the occasion of self-learning, which lacks instructors' feedback and assessment. In this paper, we introduce a LAD of our own creation that we named the "Metaboard," designed to display the details of digital learning activities to help learners monitor and compare their activities and regulate their learning effectively (Chen, Lu, Goda, & Yamada, 2019; Lu, Chen, Goda, Shimada, & Yamada, 2020a, 2020b). Learners are expected to regulate their reading behaviors after using the Metaboard that visualizes ordinal learning behaviors such as reading learning materials.

As previously stated, in the metacognition related to SRL, learners monitor their learning activities in relation to learning goals or criteria set in the planning phase, then regulate their learning by revising their strategies or methods or making other adjustments. When monitoring learning, rather than focusing directly on learning outcomes, it is essential to make sense of the processes of learning activities (Zimmerman, 1998). Supporting learners' self-monitoring activities means helping them with their self-observing or self-recording processes, such as observing or recording whether specific behaviors occurred or whether specific learning achievements met the criteria or learning goals (Belfiore & Hornyak, 1998).

To support learners' metacognition, we designed the Metaboard to provide information about how learners read educational content using BookRoll (formerly the M2B System), an e-book system that displays digital teaching resources and provides such functional tools as highlighting and annotation (Ogata et al., 2017); the use of the functional tools is logged and can be collected and used for analysis. In our previous studies, we found that some learning behaviors were related to the reflection activity on the BookRoll system (e.g., highlighting confusing content, then removing the highlights after the content is understood) and proved to be related to some cognitive strategies, such as rehearsal strategies (Chen et al., 2019), or some metacognitive strategies, such as social regulation strategies (Chen et al., 2020).

The Metaboard consists of two parts, a reading path overview and a detailed view. First, the Metaboard visualizes the overview of a user's reading path on all the pages of the digital learning materials, noting the time spent on each page and showing rough information about the numbers of highlight markers or memo annotations added on each page, separated into in-class and out-of-class use. The Metaboard gathers information required by the learners (Roberts et al., 2017), including the statistical record of the use of BookRoll's system tools by page and previews of pages where the tools were used.

An important part of the self-evaluation process in SRL is evaluating learning performance or behaviors in comparison with standard and others' performance (Belfiore & Hornyak, 1998). It is difficult to determine the best criteria or standards to use for learning behaviors for learners, but the Metaboard shows learners an anonymized comparison of their use of BookRoll's tools with those of the class average. The intended effect of this visual comparison was to motivate learners to use this dashboard consistently to help them reflect on and make sense of their learning behaviors; we reasoned that their curiosity about the differences between their behaviors and others' behaviors would motivate them to visit the Metaboard more frequently.

The following paragraphs describe the functional details of the Metaboard.

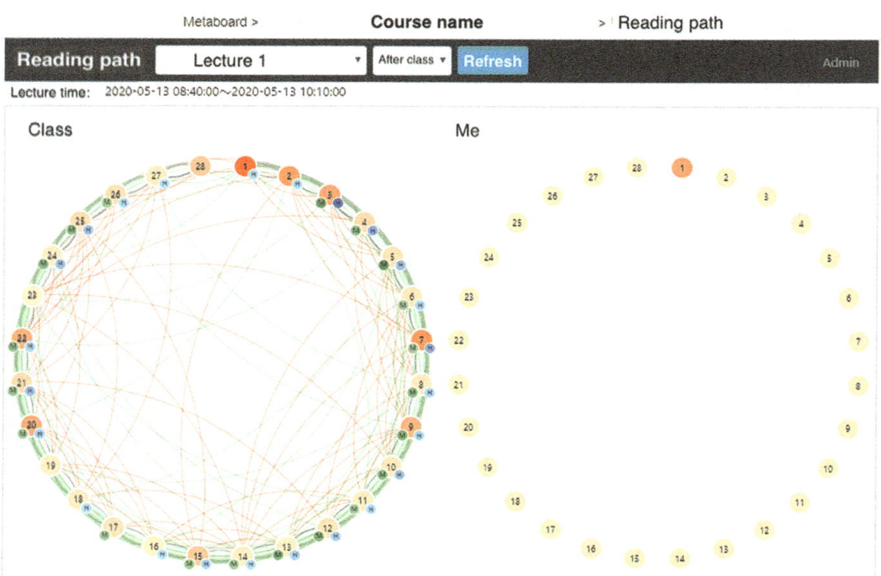

Fig. 8.1 Reading path overview of the Metaboard

8.3.2 *Reading Path Overview*

Figure 8.1 shows an example of a Metaboard graph of a reading path overview, including the reading duration for each page and whether highlights or memos were added and for which pages. Learners have the option of viewing the learning materials by course and see both in-class or out-of-class usage data.

We arranged the nodes representing all the pages in a circle and connected them with links that indicate the "from–to" relationships in the reading path—that is, the pages re-read or referred back to. The color intensity of a node with a page number indicates the reading duration for that page (the darker the color, the more time the learners spent on that page); the thickness of a link indicates the number of times the page was re-visited (a thicker link means more learners were reading on the same path); and the color of the links shows the different direction of the reading path (light gray: turn to the next page, dark gray: turn to the previous page, mint: jump forward, orange: jump backward). The smaller circles attached to a page node (i.e., circles with the letter H and M) means that learners added highlight markers (H) or memo annotations (M) on that page, and the color intensity indicates the total number of uses of those tools on the page.

When a page node is clicked, the node and the "from–to" links related to this page will be highlighted (cyan: from the selected page, magenta: toward the selected page); while other nodes and links will be presented with higher transparency (Fig. 8.2).

To help learners to compare their information with the class average information easily, the Metaboard shows a pair of graphs for the reading path overview and a detailed view.

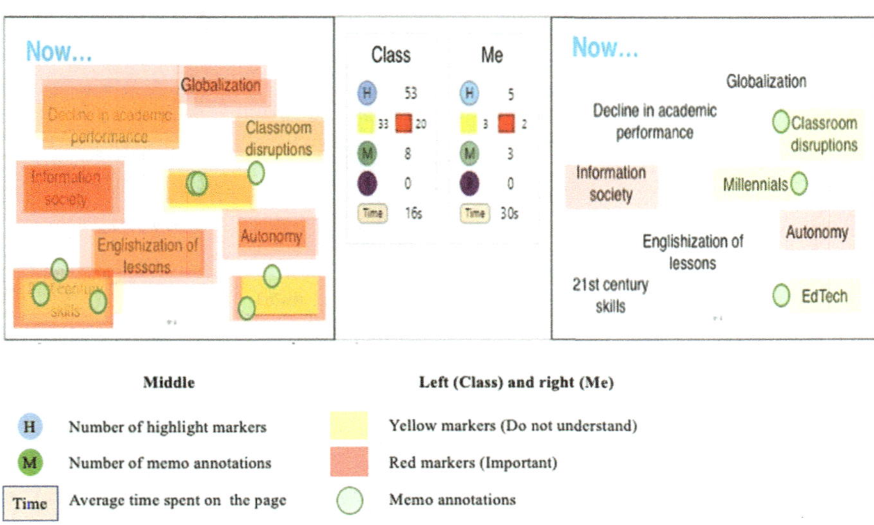

Fig. 8.2 Reading path overview shown on the Metaboard (example: p. 24)

8.3.2.1 Detailed View

As shown in Fig. 8.3, when learners click the page node, the detailed information is provided in the statistical information area in the middle of the Metaboard, including the specific numbers of the highlight markers and memo annotations and the reading time durations for the page. The left and right sides of the dashboard display what contents were highlighted and the memo symbols used. The dashboard shows individual information and class average information to allow comparison and to remind learners which parts of the content they did not understand fully or they ignored. The aim is to encourage them to revise their cognitive strategy selection and use it to improve their learning behaviors.

8.3.3 Prototype of the Metaboard

The initial prototype of the Metaboard focused on the realization of the data-processing flows and visualization techniques to gauge their effectiveness in generating the expected visualizations for the formative experiments. We developed a data processing module using Python 3.0 and a web-based JavaScript visualization module based on D3.js (the Data-Driven Documents JavaScript library); we integrated these two modules into a learning dashboard prototype. The prototype was a plugin for the learning management system (LMS) used by Kyushu University for educational technology research. The users of the LMS were able to access the learning dashboard from their course pages and browse the reading path graphs (their own and those of the whole class) for each BookRoll text in the course.

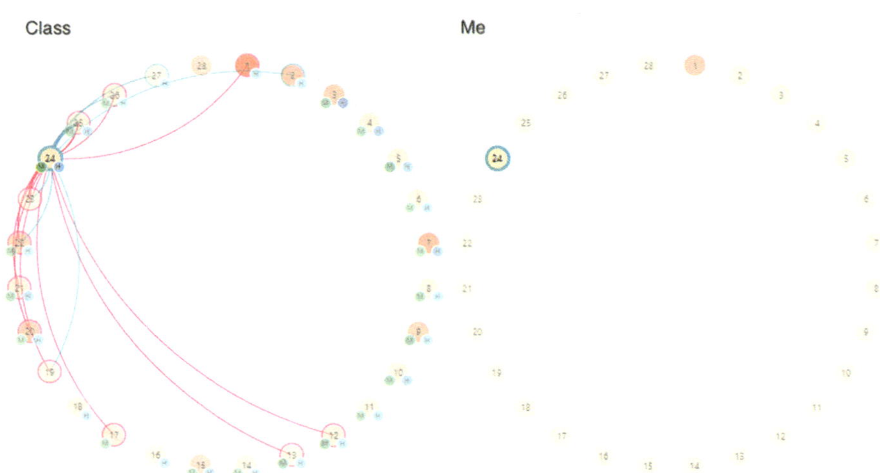

Fig. 8.3 Detailed view of the Metaboard

8.3.3.1 Data Processing Module

We developed this module to query the necessary log records from the BookRoll database and process the raw data to generate something that could be used by the visualization module to render the graphics. The data sources to be processed by this module mainly included the following:

- Information about the web sessions: the identities of the users, courses, and teaching materials, the requested range, etc.
- Information from the LMS: mainly the time periods of the classes of each course and the list of students registered to the courses
- Raw data from the BookRoll database: operational event logs and learner-generated content

In consideration of the processing load and storage costs, we strategically used offline processing for the preprocesses and to handle the data that would be used frequently, and we used online processing for the data that would be used less frequently or incidentally requested. The offline processes could be conducted regularly at off-peak times. These were the typical data processes:

- Preprocesses: Screen and group the data fetched from the operational event log data table in the BookRoll database according to the teaching text, user ID, log types, etc.
- Page navigation sequence construction: From the timestamp-based log records, generate the sequences of page numbers and time spent on the pages.
- Reading path generation: Using each page navigation sequence, sum the total time spent on each page for the nodes in the reading path graph; collect all the links between the different pages from each pair of two neighboring items in the sequence; count the number of links with the same source page and target page; append the numbers of the highlight markers and memo annotations to the nodes (i.e., pages); store the set of the nodes and links of a reading path with the start and end times.

8.3.3.2 Web-Based Visualization Module

We developed the web container of the graphics for the Metaboard with Flask, a microweb framework written in Python 3.0. We developed the visualization functions with JavaScript based on D3.js and wrapped them into an independent module as a.js file. The visualization module provides programming interfaces for the binding data to generate the reading path graphs and learner-generated content overviews and to set up the visual variables, including font and symbol sizes and colors. The module outputs the graphs in the SVG (scalable vector graphics) format for display in the web container. The module also handles and delivers some of the interactive events. For example, all the mouse-click events on the graphics of nodes, links, accessories, and symbols of learner-generated content are raised to the upper layers (e.g., the web container of the graphic). This adds scalability potential to the module to enable future functionality for the learning dashboard and connectivity with other LMS plugins.

8.3.4 Formative Evaluation of the Metaboard and Future Work

8.3.4.1 Participants

Later in the design cycle, before the Metaboard would be used in instructional practice, we conducted a formative study to test the perceptions of using the Metaboard with four participants (two male and two female students), and used the results for the design and revision in the next phase. The participants were all undergraduate students attending Kyushu University in Japan: one senior from the School of Engineering, two seniors from the Interdisciplinary Department, and one sophomore from the School of Science. All four had taken the Education Fundamentals course before; the learning materials used in this evaluation were taken from that course. The duration of the evaluation was 120 min.

8.3.4.2 Procedure

The evaluation experiment consisted of three parts. In part one, the students were required to complete the Metacognitive Awareness Inventory (MAI) (Harrison & Vallin, 2018) as a pre-MAI to assess their prior awareness of metacognition (5 min). The MAI consists of two dimensions, *knowledge of cognition* and *regulation of cognition* and contains 19 items.

In part two, we described the use of the Metaboard, then asked the four participants to read a set of learning materials on BookRoll while using the Metaboard (100 minutes). During the reading, the students were provided with the access to the Metaboard so they could see their reading paths, specific information about the reading processes, and comparisons of their reading behaviors with the class average.

In part three, the participants completed the LAD success questionnaire (Park & Jo, 2019) as a post-LAD questionnaire to evaluate the effectiveness of the Metaboard (15 minutes). The post-LAD success questionnaire consisted of five dimensions—*visual attraction*, *usability*, *understanding level*, *perceived usefulness*, and *behavioral changes*—and contained 28 items. The questionnaire provided a free-form text space where the participants could describe their perceptions and reflections on the experience of using the Metaboard based on three guiding questions: (1) What do you think are the advantages of this dashboard? (2) What do you think needs to be revised or added into this dashboard? (3) On what occasion would you be willing to use this dashboard consistently for your learning?

8.3.4.3 Results and Discussion

The MAI to assess the participants' awareness of metacognition before using the Metaboard showed differences among the four participants. In the results relating to the participants' knowledge of cognition (see Table 8.1 and Fig. 8.4), student D had the highest score (4.17), and none of the other three (students A, B, and C) got scores over 4.00. (The maximum possible score was 3.75.) In the results relating to the participants' regulation of cognition (see Table 8.2 and Fig. 8.5), students A, B, and C got the comparatively higher scores (over 4.00), but student D's score was under 4.00. These results indicated that among the four participants, students A, B, and C had comparatively low knowledge of cognition and high awareness of the regulation of cognition; conversely, student D had the most knowledge of metacognition but the least understanding of the regulation of cognition.

The LAD success questionnaire that the participants completed after using the Metaboard measured their perceptions of the dashboard. Most of the scores were over 4.00 for all five dimensions, with two exceptions: student A rated the Metaboard's *visual attraction* as 3.57, and student B rated the Metaboard's *perceived usefulness* as 2.75. (The maximum possible score for each dimension was 5.00.)

According to Park and Jo (2019), *visual attraction* refers to whether a dashboard contains useful information presented as visual elements in a concise, direct, and clear form. In the free-form text section, student A's reasons for the low *visual attraction* score was that it was "difficult to adjust the graphs (of the reading path overview)," the dashboard didn't fit on the screen.

Student B gave low ratings for the *perceived usefulness* of the Metaboard, which refers to whether a dashboard displays essential information and information that will help users achieve specific learning objectives and support them in monitoring their goal-related activities (Park & Jo, 2019). In the free-form text section, student B expressed a strong desire to see additional information, not just the simple reading path overview graph and the reminder function to support users' learning goals.

In the pre-MAI assessment, both students A and B had scored higher on regulation of cognition than knowledge of cognition. In light of these scores and their post-LAD Metaboard reviews, we could infer that users who have a high awareness of regulating their own learning and the motivation to do so but who lack sufficient knowledge of cognition (e.g., what strategies to use, how to use them in specific situations, etc.) are likely to want more information (feedback) and to have strong

Table 8.1 The average score of each dimension in the MAI (pre-MAI)

	Knowledge of cognition	Regulation of cognition
Student A	3.33	4.20
Student B	3.58	4.30
Student C	3.75	4.03
Student D	4.17	3.67
Mean (SD)	3.71 (0.35)	4.05 (0.28)

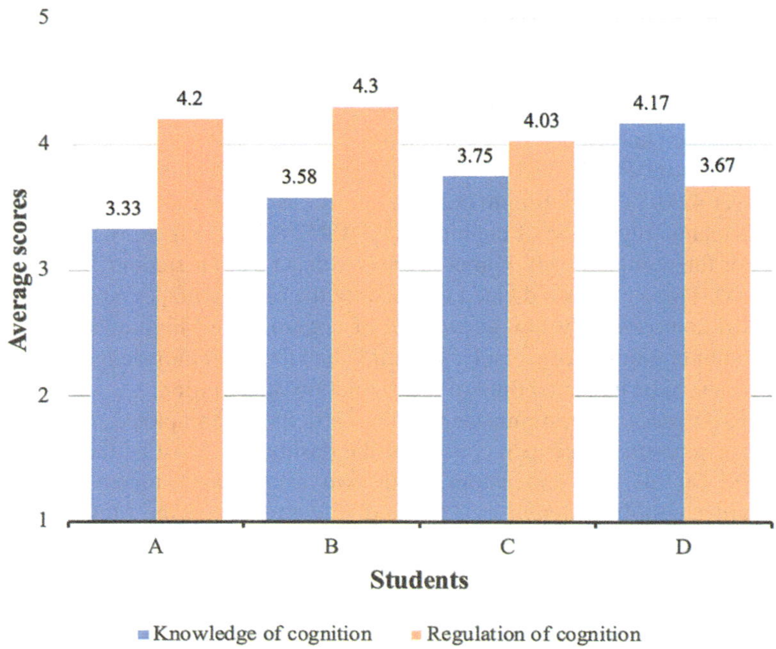

Fig. 8.4 The results of MAI (pre-MAI)

Table 8.2 The average score for each dimension in the LAD success questionnaire (post-LAD questionnaire)

	Visual attraction	Usability	Understanding level	Perceived usefulness	Behavioral changes
Student A	3.57	4.25	4.50	4.00	4.11
Student B	4.86	4.50	4.50	2.75	4.22
Student C	4.71	4.50	5.00	4.75	4.78
Student D	4.71	4.75	4.75	4.50	4.78
Mean (SD)	4.46 (0.60)	4.50 (0.20)	4.69 (0.24)	4.00 (0.89)	4.47 (0.36)

preferences for how the information is visualized. Although the small sample size makes generalization unwise, it seems that such users want more functions, simply presented, that will help them achieve their learning goals. In this study, a formative evaluation was conducted to identify the important factors which should be considered when designing the LAD. However, due to the small size of the sample size, it was difficult to access their levels of metacognition awareness in the larger group. It was also difficult to clarify the effects of students' prior metacognition awareness on the use of the Metaboard. Therefore, the further evaluation of how students use the Metaboard during their learning should be conducted.

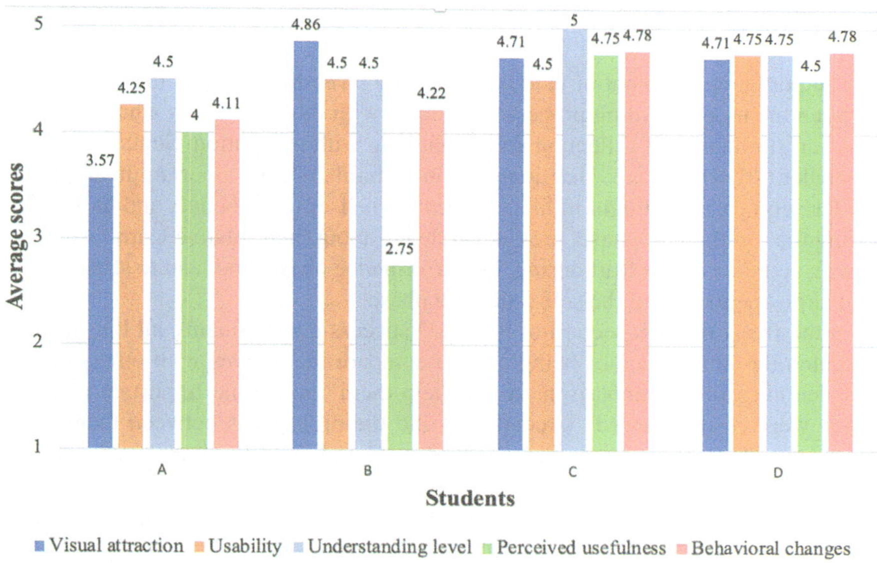

Fig. 8.5 The results of the LAD success questionnaire (post-LAD questionnaire)

8.3.4.4 Important Factors in Designing the Metaboard

Based on previous studies (Park & Jo, 2019; Roberts et al., 2017; Schumacher & Ifenthaler, 2018; Teasley, 2017) and the analysis of the free-form text answers in the LAD success questionnaire, the three most important factors to consider when designing a LAD are these: (1) appropriate graphic representations; (2) comparison functions; (3) monitoring functions that relate to goal attainment; and (4) consistent feedback.

Appropriate Graphic Representations

In their commentary on the Metaboard's *visual attraction*, the participants thought that the graphic representations on the Metaboard were appropriate, approving of the "simple design without redundant graphs" (student B), "clear color representations for different meanings" (student A), and "clear relations between two pages visualized by the links" (student C).

However, they also proposed some additional functions, such as a "categorizing function for the pages according to the time spent" (student B) and "more clear explanation for the different colors" (student C).

As Roberts et al. (2017) indicated in their study, learners prefer customizable dashboards that they can adapt to fit their needs and preferences, which suggests that multiple, diverse users should be involved in the LAD design phase.

Comparison Functions

One important component of the *understanding level* dimension is the comparison between the user's learning processes or status with those of others since comparisons can stimulate self-reflection and might help with motivation. Schumacher and Ifenthaler (2018) indicated that comparison with others was important in all phases of SRL: comparisons might influence learners' self-efficacy beliefs and motivation during the forethought phase; provide feedback about their relative learning performance, activities, and effort during the performance phase; and lead to strategy and goal adjustments during the self-reflection phase.

In the free-form text section of the LAD success questionnaire, all the students indicated the importance of being able to compare themselves with others during their learning since comparisons would help them "notice the important contents which were easily ignored" (student A), find the differences between themselves and others, "reflect on the contents" (student C and D), and "improve the understanding" (student B). Student B also proposed the addition of a "keyword searching function to see what keywords were frequently searched by others." However, student B also worried that seeing comparisons could "strengthen his pressure and reduce the motivation for learning." Roberts et al. (2017) and Teasley (2017) received similar feedback, which suggests that users' participation in the anonymized comparisons should be voluntary and that displaying the comparisons should be optional.

Monitoring Functions that Relate to Goal Attainment

In the *perceived usefulness* dimension, one of the important purposes of the LAD is to visualize essential information, especially information related to learning goals, to help users monitor their activities.

In LAD success questionnaire, student D indicated that the Metaboard's ability to distinguish between in-class and out-of-class activity was useful because it enabled users to "easily…reflect on the contents related to teachers' instruction" and the users' own learning activities. The participants offered many suggestions for additional functions. Student B proposed a "remind function" that would help learners remain aware of their learning plan or goals in the forethought phase and support their time management in the performance phase (Schumacher & Ifenthaler, 2018). Student D proposed a "function to link the reflection and the new learning goals setting," something that was also proposed by Schumacher and Ifenthaler (2018) as an important LA feature for "revision of previous learning content" to activate learners' basic prior knowledge and help them make strategic plans.

Consistent Feedback

The *behavioral changes* dimension covers learners' adaptations after using the LAD. They include such behavioral changes as focusing on different information, feeling more motivated to learn, making plans to achieve specific learning goals, and actively working to improve their academic achievement, and honing their self-management skills. In the LAD success questionnaire, the participants indicated some behavioral changes that they thought they would be likely to make after using the Metaboard to view their own learning processes and comparisons with others' processes. Among these were "reflect on and revise my learning pattern with reference to others' learning patterns" (student B) and "make sense of the contents which I ignored but not by others" (student D).

However, to encourage and support users' positive behavioral changes and to promote more consistent use of a LAD (in this case, the Metaboard), it is necessary to provide consistent feedback (Roberts et al., 2017; Schumacher & Ifenthaler, 2018; Teasley, 2017). In this study, students expressed a desire for explanations of the numbers, functional tools, and why certain contents attracted more attention.

8.4 Conclusions and Future Work

SRL can help learners achieve their academic goals by applying their cognitive learning strategies to specific tasks in academic contexts. In SRL, through metacognition, students monitor and reflect on their own learning, and this can improve their self-confidence, satisfaction, and motivation. Considering the difficulties of observing learners' processes and measuring learners' metacognition, researchers have examined the effectiveness of using a LAD to do this. In this paper, we described a prototype version of the LAD we developed (the Metaboard) to observe and measure users' metacognitive activities. After developing the prototype Metaboard, we tested its use with four participants using learning digital learning materials, then asked them to evaluate the LAD.

After using the Metaboard, the participants reported that it was a useful tool for learning digital learning materials. Two of the four students had a better grasp of the regulation of cognition than of knowledge of cognition, and they were the two who wanted the Metaboard to provide more information with improved visualization, and they proposed additional functions related to learning goal attainment.

The free-form post-LAD commentary on the Metaboard revealed that the participants considered four categories of elements to be important: (1) appropriate graphic representations (e.g., simple design, clear criteria for what was represented); (2) comparison functions (e.g., most frequently searched keywords, anonymized multiuser comparisons to protect privacy); (3) monitoring functions closely related to learning goal attainment (e.g., reminder functions, links between current and new learning goals); and (4) consistent feedback to help them make sense of the results and improve their self-reflection. Most of the user feedback was consistent with that gathered in previous studies, but some comments were specific to the features and functions of the Metaboard.

This study had some limitations. First, the sample size was small and fairly homogenous. Future studies should use a much bigger and more diverse subject population. Second, the LAD tested was a prototype version of the Metaboard during its design and development phase, which provided an initial formative study on its effectiveness. Follow-up studies should be done after the proposed revisions to the design and development have been implemented, and this version should be used in a practical evaluation experiment with a larger sample size.

Acknowledgments This work was supported by a JST AIP, Grant No. JPMJCR19U1, and JSPS KAKENHI grant number JP19H01716, JP20K19939, Japan.

References

Ahn, J., Campos, F., Hays, M., & DiGiacomo, D. (2019). Designing in context: Reaching beyond usability in learning analytics dashboard design. *Journal of Learning Analytics, 6*(2), 70–85. https://doi.org/10.18608/jla.2019.62.5

Baker, R., & Siemens, G. (2014). Educational data mining and learning analytics. In R. K. Sawyer (Ed.), *The Cambridge handbook of the learning sciences* (2nd ed., pp. 253–272). Cambridge: Cambridge University Press.

Belfiore, P. J., & Hornyak, R. S. (1998). Operant theory and application to self-monitoring in adolescents. In D. H. Schunk & B. J. Zimmerman (Eds.), *Self-regulated learning: From teaching to self-reflective practice* (pp. 184–202). New York: The Guilford Press.

Bodily, R., Ikahihifo, T. K., Mackley, B., & Graham, C. R. (2018). The design, development, and implementation of student-facing learning analytics dashboards. *Journal of Computing in Higher Education, 30*(3), 572–598. https://doi.org/10.1007/s12528-018-9186-0

Chen, L., Goda, Y., Shimada, A., & Yamada, M. (2020). Effects of in-class and out-of-class learning behaviors on learning performance and self-regulated learning awareness. In *Companion Proceedings 10th International Conference on Learning Analytics & Knowledge (LAK20)* (pp. 5–8). Association for Computing Machinery.

Chen, L., Inoue, K., Goda, Y., Okubo, F., Taniguchi, Y., Oi, M., … Yamada, M. (2020). Exploring factors that influence collaborative problem solving awareness in science education. *Technology, Knowledge and Learning, 25*, 337–366. https://doi.org/10.1007/s10758-020-09436-8

Chen, L., Lu, M., Goda, Y., & Yamada, M. (2019). Design of learning analytics dashboard supporting metacognition. *Proceedings of 16th International Conference Cognition and Exploratory Learning in Digital Age (CELDA 2019)* (pp. 175–182). IADIS Press.

Chen, L., Yoshimatsu, N., Goda, Y., Okubo, F., Taniguchi, Y., Oi, M., … Yamada, M. (2019). Direction of collaborative problem solving-based STEM learning by learning analytics approach. *Research and Practice in Technology Enhanced Learning, 14*(1), 1–28. https://doi.org/10.1186/s41039-019-0119-y

Chiu, M. M., & Fujita, N. (2014). Statistical discourse analysis: A method for modeling online discussion processes. *Journal of Learning Analytics, 1*(3), 61. https://doi.org/10.18608/jla.2014.13.5

Dinsmore, D. L., Alexander, P. A., & Loughlin, S. M. (2008). Focusing the conceptual lens on metacognition, self-regulation, and self-regulated learning. *Educational Psychology Review, 20*(4), 391–409. https://doi.org/10.1007/s10648-008-9083-6

Flavell, J. H. (1979). Metacognition and cognitive monitoring: A new area of cognitive–developmental inquiry. *American Psychologist, 34*(10), 906–911. https://doi.org/10.1037/0003-066X.34.10.906

Fincham, E., Gašević, D., Jovanović, J., & Pardo, A. (2019). From study tactics to learning strategies: An analytical method for extracting interpretable representations. *IEEE Transactions on Learning Technologies, 12*(1), 59–72.

Harrison, G. M., & Vallin, L. M. (2018). Evaluating the metacognitive awareness inventory using empirical factor-structure evidence. *Metacognition and Learning, 13*(1), 15–38. https://doi.org/10.1007/s11409-017-9176-z

Hofer, B. K., Yu, S. L., & Pintrich, P. R. (1998). Teaching college students to be self-regulated learners. In D. H. Schunk & B. J. Zimmerman (Eds.), *Self-regulated learning: From teaching to self-reflective practice* (pp. 57–85). Guilford Press.

Jacobs, J. E., & Paris, S. G. (1987). Children's metacognition about reading: Issues in definition, measurement, and instruction. *Educational Psychologist, 22*(3–4), 255–278. https://doi.org/10.1080/00461520.1987.9653052

Kim, J., Jo, I.-H., & Park, Y. (2016). Effects of learning analytics dashboard: Analyzing the relations among dashboard utilization, satisfaction, and learning achievement. *Asia Pacific Education Review, 17*(1), 13–24. https://doi.org/10.1007/s12564-015-9403-8

Klerkx, J., Verbert, K., & Duval, E. (2017). Learning analytics dashboards. In C. Lang, G. Siemens, A. Wise, & D. Gašević (Eds.), *The handbook of learning analytics* (pp. 143–150). New York: Society for Learning Analytics Research (SoLAR). https://doi.org/10.18608/hla17

Lan, W. Y. (1998). Teaching self-monitoring skills in statistics. In D. H. Schunk & B. J. Zimmerman (Eds.), *Self-regulated learning: From teaching to self-reflective practice* (pp. 86–105). New York: Guilford Press.

Lee, D., Watson, S. L., & Watson, W. R. (2019). Systematic literature review on self-regulated learning in massive open online courses. *Australasian Journal of Educational Technology, 35*(1) https://doi.org/10.14742/ajet.3749

Lu, M., Chen, L., Goda, Y., Shimada, A., Yamada, M. (2020a). Development of a learning dashboard prototype supporting meta-cognition for students. In *Companion proceedings of the 10th international conference on learning analytics & knowledge (LAK20)* (pp. 104–106). Association for Computing Machinery.

Lu, M., Chen, L., Goda, Y., Shimada, A., & Yamada, M. (2020b). Visualizing studying activities for a learning dashboard supporting meta-cognition for students. In N. Streitz & S. Konomi (Eds.), *Distributed, ambient and pervasive interactions. HCII 2020. Lecture notes in computer science, 12203* (pp. 569–580). Cham: Springer.

McDowell, L. D. (2019). The roles of motivation and metacognition in producing self-regulated learners of college physical science: A review of empirical studies. *International Journal of Science Education, 41*(17), 2524–2541. https://doi.org/10.1080/09500693.2019.1689584

Ogata, H., Taniguchi, Y., Suehiro, D., Shimada, A., Oi, M., Okubo, F., Yamada, M., & Kojima, K. (2017). M2B system: A digital learning platform for traditional classrooms in university. In S. Shehata & J. P.-L. Tan (Eds.), *Practitioner track proceedings of the 7th international learning analytics & knowledge conference (LAK 2017)* (pp. 155–162). Paper presented at Simon Fraser University, Vancouver, Canada.

Park, Y., & Jo, I.-H. (2019). Factors that affect the success of learning analytics dashboards. *Educational Technology Research and Development, 67*(6), 1547–1571. https://doi.org/10.1007/s11423-019-09693-0

Pintrich, P. R. (1999). The role of motivation in promoting and sustaining self-regulated learning. *International Journal of Educational Research, 31*(6), 459–470. https://doi.org/10.1016/S0883-0355(99)00015-4

Pintrich, P. R., Smith, D. A. F., Garcia, T., & McKeachie, W. J. (1993). Reliability and predictive validity of the motivated strategies for learning questionnaire (MSLQ). *Educational and Psychological Measurement, 53*(3), 801–813. https://doi.org/10.1177/0013164493053003024

Roberts, L. D., Howell, J. A., & Seaman, K. (2017). Give me a customizable dashboard: Personalized learning analytics dashboards in higher education. *Technology, Knowledge and Learning, 22*(3), 317–333. https://doi.org/10.1007/s10758-017-9316-1

Rodrigues, R. L., Ramos, J. L. C., Silva, J. C. S., Dourado, R. A., & Gomes, A. S. (2019). Forecasting students' performance through self-regulated learning behavioral analysis. *International Journal of Distance Education Technologies (IJDET), 17*(3), 52–74. https://doi.org/10.4018/IJDET.2019070104

Schumacher, C., & Ifenthaler, D. (2018). Features students really expect from learning analytics. *Computers in Human Behavior, 78*, 397–407. https://doi.org/10.1016/j.chb.2017.06.030

Schunk, D. H. (1982). Progress self-monitoring: Effects on children's self-efficacy and achievement. *The Journal of Experimental Education, 51*(2), 89–93. https://doi.org/10.1080/00220973.1982.11011845

Schwendimann, B. A., Rodríguez-Triana, M. J., Vozniuk, A., Prieto, L. P., Boroujeni, M. S., Holzer, A., … Dillenbourg, P. (2016). Perceiving learning at a glance: A systematic literature review of learning dashboard research. *IEEE Transactions on Learning Technologies, 10*(1), 30–41. https://doi.org/10.1109/TLT.2016.2599522

Shimada, A., Okubo, F., Yin, C., Kojima, K., Yamada, M., & Ogata, H. (2015, July). Informal learning behavior analysis using action logs and slide features in e-textbooks. In *2015 IEEE 15th international conference on advanced learning technologies* (pp. 116–117). IEEE Computer Society.

Teasley, S. D. (2017). Student facing dashboards: One size fits all? *Technology, Knowledge and Learning, 22*(3), 377–384. https://doi.org/10.1007/s10758-017-9314-3

Thiede, K. W., Griffin, T. D., Wiley, J., & Redford, J. S. (2009). Metacognitive monitoring during and after reading. In D. J. Hacker, J. Dunlosky, & A. C. Graesser (Eds.), *Handbook of metacognition in education* (pp. 97–118). Abingdon: Routledge.

Tobias, S., & Everson, H. T. (2009). The importance of knowing what you know: A knowledge monitoring framework for studying metacognition in education. In D. J. Hacker, J. Dunlosky, & A. C. Graesser (Eds.), *Handbook of metacognition in education* (pp. 107–127). Abingdon: Routledge.

van Leeuwen, A. (2015). Learning analytics to support teachers during synchronous CSCL: Balancing between overview and overload. *Journal of Learning Analytics, 2*(2), 138–162. https://doi.org/10.18608/jla.2015.22.11

Verbert, K., Govaerts, S., Duval, E., Santos, J., Van Assche, F., Parra, G., & Klerkx, J. (2014). Learning dashboards: An overview and future research opportunities. *Personal and Ubiquitous Computing, 18*(6), 1499–1514. https://doi.org/10.1007/s00779-013-0751-2

Weinstein, C. E., Acee, T. W., & Jung, J. (2011). Self-regulation and learning strategies. *New Directions for Teaching and Learning, 2011*(126), 45–53. https://doi.org/10.1002/tl.443

Winne, P. H., & Nesbit, J. C. (2009). Supporting self-regulated learning with cognitive tools. In D. J. Hacker, J. Dunlosky, & A. C. Graesser (Eds.), *Handbook of metacognition in education* (pp. 259–277). Abingdon: Routledge.

Winne, P. H., & Stockley, D. B. (1998). Computing technologies as sites for developing self-regulated learning. In D. H. Schunk & B. J. Zimmerman (Eds.), *Self-regulated learning: From teaching to self-reflective practice* (pp. 106–136). New York: Guilford Press.

Yamada, M., Okubo, F., Oi, M., Shimada, A., Kojima, K., & Ogata, H. (2016). Learning analytics in ubiquitous learning environments: Self-regulated learning perspective. In *Proceedings of the 24th International Conference on Computers in Education* (ICCE 2015) (pp. 306–314). Asia-Pacific Society for Computers in Education.

Yamada, M., Shimada, A., Okubo, F., Oi, M., Kojima, K., & Ogata, H. (2017). Learning analytics of the relationships among self-regulated learning, learning behaviors, and learning performance. *Research and Practice in Technology Enhanced Learning, 12*(1), 13. https://doi.org/10.1186/s41039-017-0053-9

Yin, C., Okubo, F., Shimada, A., Oi, M., Hirokawa, S., Yamada, M., Kojima, K., & Ogata, H. (2015). Analyzing the features of learning behaviors of students using e-books. In *Proceedings of the*

23rd International Conference on Computer in Education (ICCE 2015) (pp. 617–626). Asia-Pacific Society for Computers in Education.

Zimmerman, B. J. (1986). Becoming a self-regulated learner: Which are the key sub-processes? *Contemporary Educational Psychology, 11*(4), 307–313. https://doi.org/10.1016/0361-476X(86)90027-5

Zimmerman, B. J. (1989). A social cognitive view of self-regulated academic learning. *Journal of Educational Psychology, 81*(3), 329–339. https://doi.org/10.1037/0022-0663.81.3.329

Zimmerman, B. J. (1998). Developing self-fulfilling cycles of academic regulation: An analysis of exemplary instructional models. In D. H. Schunk & B. J. Zimmerman (Eds.), *Self-regulated learning: From teaching to self-reflective practice* (pp. 1–19). New York: Guilford Press.

Zimmerman, B. J. (2002). Becoming a self-regulated learner: An overview. *Theory Into Practice, 41*(2), 64–70. https://doi.org/10.1207/s15430421tip4102_2

Chapter 9
Diversity as an Advantage: An Analysis of the Demand for Specialized and Social Competencies for STEM Graduates Using Machine Learning

Karin Maurer, Annika Hinze, Heidi Schuhbauer, and Patricia Brockmann

9.1 Introduction

The rise of digitalization rapidly permeates through all aspects of modern life. As a result, the demand for qualified software development professionals continues to increase. This growing demand, paired with a simultaneous shortage of available graduates, exacerbates the widening skills gap between job openings and qualified applicants. In the past, digitalization has often been viewed solely as a threat to those on the losing side of the digital divide. Instead, machine learning methods could be examined as a positive instrument to search for hitherto unseen career opportunities for people from traditionally disadvantaged groups.

Up until now, a number of groups have remained underrepresented among students in technical subjects: first-generation students, people who come from a migration background, single parents, and women. In Germany, the proportion of people who come from a migration background (23%) is significantly higher than the proportion of migrant students enrolled at universities (11%) (BAM, 2011). At the Nuremberg Institute of Technology in Germany, 65% of the students are from non-academic families, approximately 5% have children, 10% are international students and 5% do not have a high school diploma (THN, 2018). The number of female students in the Computer Science Department has actually been declining. During the winter semester of 2018/2019, only 17% of newly enrolled students were female (IN THN, 2018).

From the point of view of these underrepresented groups, future careers in STEM (Science Technology Engineering Math) subjects could offer a path to financial independence and self-empowerment. For potential employers, people from these

K. Maurer · A. Hinze · H. Schuhbauer · P. Brockmann (✉)
Computer Science Department, Nuremberg Institute of Technology, Nurenberg, Germany
e-mail: patricia.brockmann@th-nuernberg.de

D. Ifenthaler et al. (eds.), *Balancing the Tension between Digital Technologies and Learning Sciences*, Cognition and Exploratory Learning in the Digital Age,
https://doi.org/10.1007/978-3-030-65657-7_9

underrepresented groups could offer an untapped source of uniquely qualified talent for hard to fill job openings.

This work considers two research questions:

1. What are the most important specialized skills and social competencies STEM majors need to master for future careers in a digital society?
2. Can machine learning methods be useful in analyzing large quantities of unstructured data about the job market to derive information about which professional competencies will be needed in the near future?

Section 2, Related Work discusses literature related to specialized skills and social competencies which are important in STEM professions, as well as uses of machine learning in educational contexts. Section 3, Methodology describes the setup of the experimental process and the machine learning methods used in this investigation. Section 4, Results presents the results of the analysis and discusses their limitations. Section 5, Conclusions discusses the implications of the results and their relevance for IT graduates from diverse backgrounds. Section 6 describes plans for further research.

9.2 Related Work

9.2.1 Career Goals, Motivation, and Competencies

This section discusses work related to the career goals and motivation of STEM students as well as the competencies students need to learn for their future careers in STEM fields. International, cultural, and gender aspects are also considered.

Liebenberg and Pieterse (2016) investigated the career goals of software development students and professionals in South Africa. They found that both students and professionals valued stability and work/life balance most highly. Professionals additionally expressed the value of creativity. They assert that knowing people's motivation can help to improve recruitment and retention of software developers.

The intercultural competencies necessary to work in global software development teams have been investigated by a number of authors. Beecham et al. (2017) conducted a wide-scale literature review of distributed global software engineering courses. They identified a number of difficulties inherent to working in international software teams, which students need to learn to address: distance, teamwork, soft skills, stakeholders from industry, infrastructure, and distributed software development processes. They categorized various types of distances, such as physical (geographic), time zones, cultural, language, and institutional distances. Other authors, such as Hoda, Ali Babar, Shastri, and Yaqoob (2016) concentrated on the sociocultural capabilities which students need to learn in order to work effectively in global software development teams. They pointed out the importance of overcoming language barriers, different perspectives regarding time, attitudes toward achievement, differences in autonomy and work habits as well as assumptions about national culture. They underline the importance of cross-cultural training. One example of the importance of cultural sensitivity in requirements engineering was reported by Hinze et al. (2018). To develop a medical app aimed at improving the health of Maori communities, sensitive personal data needed to be collected. They stressed the importance of establishing personal relationships in order to create a trusting environment between multi-cultural stakeholders. Ideally, they recommend that one member of the research team should come from the cultural community studied, in order to help build bridges between the two worlds.

A number of authors have analyzed the effect of gender on computational thinking in schools and in the workplace. Budinska and Mayerova (2017) investigated the relationship between computer science concepts and computational thinking, in this case graph tasks. They found that boys were comparatively better at tasks with simple, relatively abstract representation and a larger amount of text, with the goal defined to identify a problem. They found that girls were better at tasks with less text, but with a relatively more complicated representation of structure, with a focus on simple operations on graphs. They concluded that because boys and girls have different methods of acquiring mechanical and abstract thinking, they each need different types of assignments to increase their motivation. Cheryan, Siy, Vichayapai, Drury, and Kim (2011) examined whether role models have an effect on self-confidence. They found that women who interacted with non-stereotypical role models believed they would be more successful in computer science than those who interacted with stereotypical role models. Faulkner (2009) discusses the subtle dynamics which can contribute to a feeling of "belonging" in work relationships. She discusses the importance of informal conversation topics among colleagues, which can make women and other underrepresented groups feel like outsiders. Branz, Pastran Reina, Richter, Waizmann, and Brockmann (2019) used Sentiment Analysis to evaluate how male and female team members interact on software engineering projects. They used statistical and machine learning methods to analyze a large data set from an incident management system to investigate the emotional content of project communication. They found that the types and intensities of sentiments expressed differed considerably between male and female developers.

The literature discussed here illustrates some of the challenges which STEM students will face upon graduation. The question arises as to whether students from underrepresented groups can leverage their backgrounds to make unique contributions to increase the diversity of ideas contributed to STEM teams.

9.2.2 Machine Learning Methods in an Educational Context

Mitchell (2010) describes machine learning as suitable for analyzing large amounts of data and generating knowledge from the data that is not yet available. Kucak, Juricic, and Dambic (2018) performed a comprehensive literature study to categorize the use of machine learning in education. They categorized uses of machine learning into four areas:

1. Grading
2. Improving retention
3. Predicting performance
4. Testing.

They found that machine learning can help to reduce human bias in grading. Machine learning also aided in the early identification of "at risk" students. Retention can be improved by contacting and supporting these students before they fall too far behind. By analyzing the past performance of each student, machine learning algorithms can aid in predicting future performance. Machine learning-based testing delivers frequent feedback to teachers and students about learning progress and can show which additional support may be needed to achieve learning goals.

Blind application of the technology of machine learning alone will not solve all problems, however. The validity and accuracy of the analysis performed is often highly influenced by its context. Ifenthaler and Widanapathirana (2014) investigated the use of support vector machines to validate a learning analytics framework. They found that educational data used for learning analytics are defined quite specifically to the context of each different educational institution. These divergent definitions can be the source of variations in results found when comparing different educational institutions and programs of study.

Furthermore, ethical issues need to be considered when applying machine learning methods to educational institutions. It is often assumed that gaining knowledge of a student's behavior will result in advantages for all of the stakeholders involved (students, instructors, educational institutions). However, the collection of personal data about students presents a number of legal and ethical challenges. Slade and Prinsloo (2013) proposed a socio-critical framework for the use of learning analytics.

9.3 Methodology

One goal of this research is to try to determine which technical and social competencies are most important for future STEM graduates. The second goal is to examine whether machine learning methods can be useful in analyzing large amounts of unstructured data from the labor market. The methods used to investigate these two research questions are described here.

9.3.1 Experimental Procedure

This section describes the setup of the experimental procedure (Fig. 9.1).

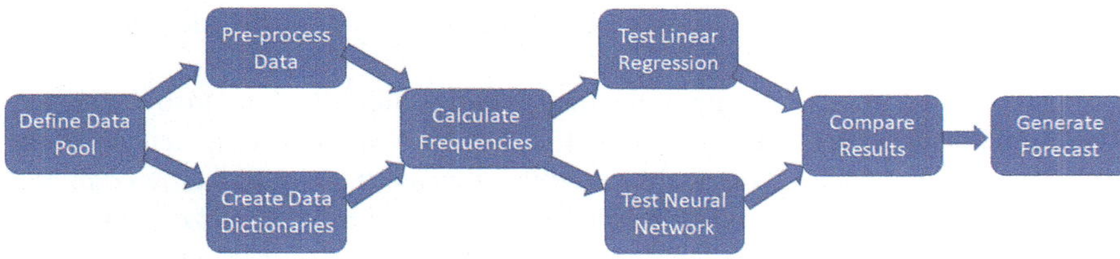

Fig. 9.1 Experimental procedure

The first step is to define an appropriate pool of test data. The higher the relevance and the quality of the data, and the better the data preparation before analysis, the more useful the subsequent results of the analysis will be (Mohri, Rostamizadeh, & Talwalkar, 2012). This study requires a data pool which contains sufficient information about the labor market and from which it can be deduced which vocational competencies may be needed in the near future. One source of information on the competencies required for particular jobs can be found in job advertisements. For this reason, the job advertisements from a job market database maintained by 15 universities in Bavaria, Germany, are used for this research. Since the university job market database also contains job advertisements for non-STEM occupations, a subset of the data was pre-selected. Only job advertisements for STEM occupations were considered for this study.

The next step is to pre-process the data, which can often be a lengthy process. Starting with the selection of the data and continuing up through the evaluation of the results, this pre-processing phase can often take up 60–70% of the total time expenditure (Maurer, 2019). To be able to perform analyses with a large amount of unstructured data, the data must be available in a format which can be analyzed directly by a computer algorithm. Job advertisements are usually formulated as unstructured text, with the required competencies listed as enumerations within this text.

In order to answer the first research question about which specialized skills and social competences are most important, a list of all the competences contained in the job advertisements is required, as well as an aggregation of how often each of these competencies were requested. This enables an assessment of how important a single competency is across the multitude of job advertisements.

The second research question on the competencies which will be needed in the near future requires the generation of a prognosis. This requires an analysis of time series data. To generate future forecasts, the frequency of how often each competence was requested in the past is aggregated for each year.

The pre-processing of the data starts with tokenizing. Words and punctuation marks are separated from each other, so that each word and each character can be considered independently. All characters and numbers that are not alphanumeric and the so-called "stop words" are removed. Stop words are filler words which do not carry meaning, such as "a," "the," "this," or "that." (McNamee & Mayfield, 2004). In job advertisements, company names and standard non-discrimination clauses can also be removed, because they are not relevant for the identification of competencies.

Parallel to the pre-processing of the data, "data dictionaries" are also developed. These data dictionaries contain a complete list of all of the job competencies found so far. The competencies from the data dictionaries are then individually compared with the words from the job advertisements. If the algorithm recognizes that a word from the job advertisements is similar to a word from the dictionary, it is marked as a match. This, however, carries the risk of redundancies, e.g., due to different conjugations used for a word in the job advertisements. In order to avoid redundancies, both the words in the job advertisements and the words in the dictionaries are "stemmed." Stemming means to reduce a word to its root (Agnihotri, Verma, & Tripathi, 2014). For example, "programs" and "programming" would both be reduced to their common stem "program." After stemming the words, the actual information extraction takes place. The analysis is then conducted on this extracted information.

In order to obtain the best possible analysis results, a large amount of historical data is required. (Mohri et al., 2012). It is also necessary to use the same number of data records for every year. Since this work is based on the number of occurrences of a competency, a different number of data sets per year would falsify the result. For example, the university career services database used here has less than 100 records each for the initial years 2003 and 2004. Starting with 2005, each year has approximately 10.000 records. For this reason, only the years 2005 to 2018 were used in this analysis.

Table 9.1 shows an example of the data used to identify which competencies have been requested by potential employers in job advertisements over the past few years. It also includes information on how often each competency was requested in each year. An aggregation of the list per competency enables the determination of which technical and social competencies were important for STEM majors in the past.

In order to determine whether machine learning methods can be useful to analyze large amounts of unstructured data from the labor market, two different methods are tested: linear regression and neural networks. The results of both analyses are then compared in order to evaluate the performance of these two methods for

Table 9.1 Results of the data pre-processing

Record number	Year	Number of occurrences	Competence
1	2018	1114	Experience
2	2017	1029	Experience
3	2013	967	Experience
4	2015	956	Experience
:	:	:	:
:	:	:	:
1794	2013	1	Environmental consciousness
1795	2017	1	Tolerance

this research domain. Each of these two methods are tested to provide forecasts for the years 2017 and 2018. The results of these forecasts are then compared with the actual figures for 2017 and 2018 from the list in Table 9.1.

Each method must first be fed with historical data. This procedure is called "training," because the models "learn" by adapting their behavior to the historical data (Mitchell, 2010). Thus, new knowledge can be generated from past data. When training a machine learning model, it is important to use as much training data as possible, because the more training data you use, the more accurate the prediction result will be. Therefore, the years 2005 to 2016 are used as training data in this experiment.

After completion of the training phase, both the linear regression and the neural network generate forecast predictions for the years 2017 and 2018. This data from 2017 and 2018, which is used to check the model's accuracy, is called test data. A comparison of the forecast predictions with the actual figures from 2017 and 2018 makes it possible to compare the accuracy of linear regression versus the neural network for this forecast. The model which delivers better test accuracy is then used for the actual future forecast for 2020.

9.3.2 *Experimental Methods*

This subsection describes the two competing analytical methods selected to compare their performance: linear regression and a neural network. In this study, linear regression was selected as an example of a well-accepted, computationally-lean statistical method. Neural networks were selected as a second analytical method because they are one of the most commonly used methods of modern machine learning. Linear regression serves as a baseline method to measure whether more modern, "exotic" methods, such as neural networks, actually deliver better results. The utilization of a more complex method which consumes more computational resources can only be justified if it delivers superior results to faster, more efficient statistical methods.

Linear regression is a classical statistical method that mathematically investigates the relationship between two variables and uses this mathematical correlation to predict future values (Teschl & Teschl, 2007). One variable, the scalar response, is dependent on the other independent variable (Raschka, Mirjalili, & Lorenzen, 2018). A number of authors have expressed differences of opinion as to whether linear regression actually qualifies as machine learning. Goodfellow et al. describe linear regression as a machine learning method (Goodfellow, Bengio, & Courville, 2016). In contrast, Dangeti (2017) differentiates between the two: "Statistical modeling is a formalization of relationships between variables in the data in the form of mathematical equations," while "Machine Learning is an algorithm that can learn from data without relying on rules-based programming."

In the case of job advertisements, the number of requests for a certain competency depends on the year in which they were made. Linear regression is a method of supervised learning which calculates continuous values using a regression line (Raschka et al., 2018). The future values can be read off from the extrapolation of the regression line. Since linear regression is a model that provides continuous values, the years 2005 to 2016 can be used as training data without further adjustments. The years 2017 and 2018 can be used as test data. The regression line resulting from the training and test data can be used to read the forecast data and compare its prediction accuracy with the test data.

The second method selected for evaluation is a neural network, which mimics the behavior of the human brain. A neural network consists of neurons that work in parallel and send information to each other via weighted, directed connections (Hecht-Nielsen, 1989). The algorithm is first fed with input values. The subsequent output values are determined by a number of hidden layers, which consist of different weightings of the connections between individual neurons in each layer. Supervised learning utilizes a type of neural networks which must first be trained. This means that the neural network is first fed with input values. During the backpropagation phase, the output produced is compared to the correct output. If the network produced the correct output, the weights between neurons which led to this correct output are strengthened. Weights between neurons which led to incorrect outputs are decreased. During each training pass, the algorithm adapts the weights of the connections between individual neurons to improve network behavior.

Once the network has been trained, it can be used to make the actual predictions. Since the neural network learns iteratively and uses data from previous years, the network must first be trained. In this study, the neural network was trained in two steps. First, the data from years 2005 to 2014 are used as input values. The data from 2016 and 2017 are used to verify the values forecast. Next, the data from years 2006

Table 9.2 Partitioning of the training and test data for the neural network

Year	Training Data x	Training Data y	Test Data x	Test Data y
2005				
2006				
2007				
2008				
2009				
2010				
2011				
2012				
2013				
2014				
2015				
2016				
2017				
2018				

to 2015 are used as input values and the data from 2017 and 2018 serve to verify the forecast values generated by the neural network. Table 9.2 illustrates the distribution of test and training data for the neural network. Both the training data and the test data are offset by 1 year so that the algorithm can learn from slightly different data sets. After both the regression model and the neural network models have been trained, the forecast values for 2017 and 2018 are compared with the actual values for 2017 and 2018. Based on the percentage difference between the forecast values and the actual values, the better machine learning method is selected to generate the actual forecast prognosis.

9.4 Results

9.4.1 Comparison of Method Performance

Figure 9.2 shows an example of the initial results. The left side shows the results of the linear regression; the right side shows the results of the neural network. Both diagrams are based on their performance in analyzing one specific competence, "independence." The green dots are the actual figures, the black dots the forecast figures for the years 2017 and 2018. Figure 9.2 shows that for the competence "independence," the neural network worked better than the linear regression.

A total of 12 test runs (one test run for each of the 12 competencies) were carried out for linear regression and for the neural network. Accuracy is defined as the number of correct predictions divided by the total number of predictions. The evaluation of all of the results from 24 test runs per method (12 test runs each for the years 2017 and 2018) are shown in Fig. 9.3.

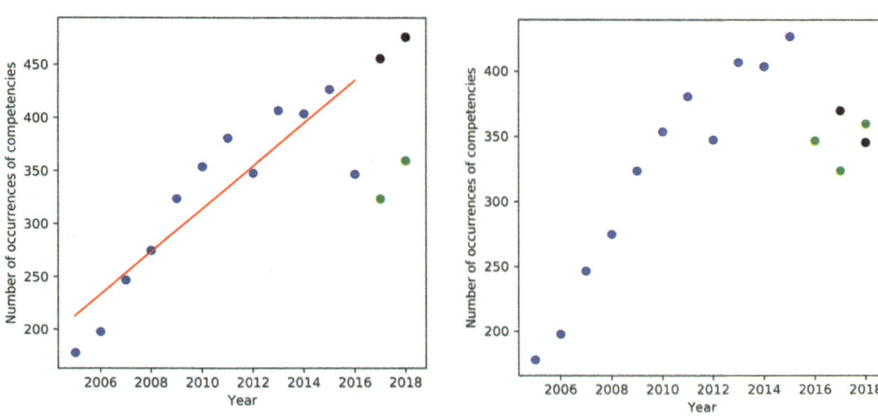

Linear Regression Neural Network

Fig. 9.2 Comparison of forecasts for the competence "independence"

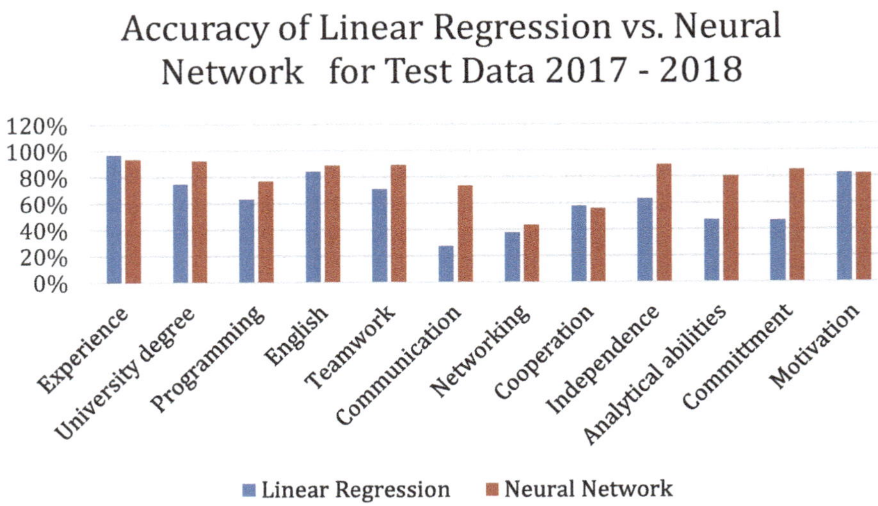

Fig. 9.3 Comparison of accuracy of linear regression to neural network

Table 9.3 Forecast of most important specialized skills and social competencies

Specialized skills		Social competencies	
Experience	12,200	Teamwork	5214
University degree	9199	Independence	4574
Programming (Excel)	3731	Analytical abilities	3060
English	2781	Commitment	2882

The neural network delivered higher accuracy in 18 of 24 cases. Overall on average, the neural network had a significantly higher average accuracy of 79.6% than linear regression 63.3%. For this reason, the neural network was chosen as the preferred method to forecast future competencies.

9.4.2 *Frequency of Demand for Career Competencies*

After selecting the neural network as the forecast method, it was possible to address the first research question:

1. What are the most important specialized skills and social competencies STEM majors need to master for future careers in a digital society?
   Table 9.3 shows the four most important specialized skills and social competencies that STEM graduates should master for their future careers. The most commonly requested specialized skill is work experience. Completion of a university degree came in second. Specialized skills in programming or technical tools such as Excel comes in a distant third. The fourth specialized skill identified as

one of the most important is the mastery of English as a second language. The most important social skills identified are teamwork, followed by independence. The third most important social competency is analytical ability. The final important competency listed for a future career in a digital society is commitment.

After identifying the most important technical skills and social competencies, the second research question can now be addressed:

2. Can machine learning methods be useful in analyzing large quantities of unstructured data about the job market to derive information about which professional competencies will be needed in the near future?

Figure 9.4 shows a comparison of the changes in demand for technical and social competencies. The upper line shows the number of social competencies, the lower line the number of technical competencies desired by future employers. Already at the beginning of the time series in 2005, the required number of social competencies is higher than that of technical competencies. This continues through 2018.

In addition, it can be seen in Fig. 9.4 that the demand for technical competencies has apparently stagnated since 2016. In contrast, Fig. 9.4 shows that the demand for social competencies has been rising steadily since 2016. Although some may have assumed that only technical know-how is important in STEM occupations, Fig. 9.4 also shows that social skills have become increasingly more important in the occupational environment than technical skills.

After discussing the historical development of technical and social competencies, the forecast for the near future will now be presented. Table 9.4 presents the concrete forecast frequencies for 2020. These are compared to the actual frequencies available from 2018. Since the neural network used here can only be executed

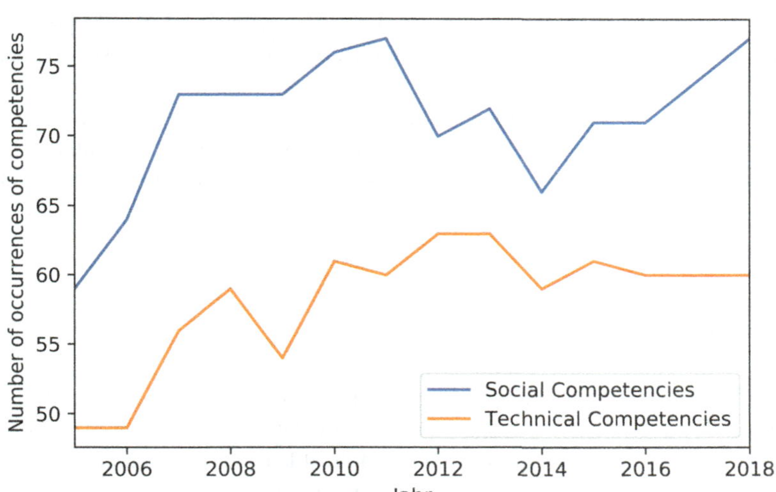

Fig. 9.4 Comparison of social competencies vs. technical skills (Maurer, 2019)

Table 9.4 Comparison of actual frequencies in 2018 to forecast in 2020

Competency	Frequency 2018	Forecast 2020
Experience	1114	1201
University degree	801	838
Programming (Excel)	207	195
English	170	164
Teamwork	359	358
Communication	105	101
Networking	14	11
Cooperation	5	5
Independence	360	379
Analytical abilities	207	205
Commitment	177	173
Motivation	212	219

for one competency at a time, 12 competencies were selected for which a future forecast should be prepared (Maurer, 2019).

Table 9.4 shows that two of the specialized skills, namely "experience" and "university degree," are expected to increase in importance, while the specialized skills "Programming (Excel)" and "English" are forecast to become less important. The neural network forecasts that demand will remain relatively unchanged for the social competencies "teamwork," "communication," "cooperation," "analytical ability" and "commitment." The demand for the competency "networking" is expected to decrease slightly, while an increase in demand for two social competencies "independence" and "motivation" is predicted.

In summary, work experience and the completion of a university degree are predicted to remain the most desirable core competencies. In addition, independence and the ability to work in teams will also be necessary in future STEM professions. In addition, intrinsic motivation remains advantageous, followed closely by analytical skills. However, technical skills, such as programming languages, will continue to be important. As globalization continues to progress, competency in English and communication skills will become increasingly indispensable. Adroitness in networking will also be essential. Finally, graduates seeking their first entry-level jobs should show a high ability to cooperate.

9.4.3 Limitations

The significance of these results is limited by the specificity of the data pool. Because job advertisements from a job market database for German universities were analyzed, these forecasts may not be applicable to the competencies desired by employers in other countries. This underscores the importance of context specificity, as found by Ifenthaler and Widanapathirana (2014). Certain terms can have highly divergent meanings, depending on the context. The term "excel" can mean to be exceptionally good at a certain subject or it can also be interpreted as competence in utilizing the specific software program, Microsoft Excel.

Another further limitation of this research is that the supervised learning machine learning methods implemented here are trained with historical data from past job advertisements. The forecast algorithms can only make predictions for competencies which are already known. There is no possibility to generate forecasts for new, original competencies which the algorithms have never encountered before.

One potential technical source of undetected errors in results could be due to overfitting of the neural network. Overfitting can occur when a neural network is trained with an insufficient amount of training data. The network adapts its behavior too closely to a specific training data set. As a result, the network may fail to recognize new types of data or may fail to generate reliable future predictions. This source of errors could be minimized by implementing the *k-fold* validation technique (Wong, 2015). *k-fold* validation is a resampling technique used to evaluate machine learning models with a limited data sample. The data is randomly divided into a fixed number of *k* sample groups. For each group, the other remaining groups are used as training data and the data from the current group is used as test data. This cycle is repeated *b* times, for each of the *k* groups. Another possible solution to the problem of overfitting would be to use the dropout technique suggested by Svrivastava, Hinton, Krizhevsky, Sutskever, & Salakhutdinov (2014). During the training phase, randomly selected neurons, along with their connections, are dropped. This prevents the neural network from over-adapting to a specific set of data.

9.5 Conclusions

In conclusion, this study has been able to answer two research questions:

1. What are the most important specialized skills and social competencies STEM majors need to master for future careers in a digital society?

   The most widely sought specialized skills found were: work experience, a university degree, proficiency in programming (Excel) and proficiency in English as a second language. As this is an analysis for graduates from STEM studies at universities, most students will be able to demonstrate this competence upon graduation. A mastery of programming cannot be completely guaranteed for all STEM majors. However, the ability to at least use Excel as a basic analytical tool was often required. For STEM graduates at universities in Germany, English proficiency is a skill which should be readily mastered, as most European universities require courses in English as a second language.

   The most widely sought social competencies were: the ability to work in teams, independence, analytical abilities and commitment. These soft skills can be promoted by student-centered learning methods, such as problem-based learning, in which project-based group work is carried out in teams (Bell, 2010). Analytical ability is further promoted in courses such as "Software Engineering," where, for example, customer requirements have to be analyzed and structured. The second research question could also be answered:

2. Can machine learning methods be useful in analyzing large quantities of unstructured data about the job market to derive information about which professional competencies will be needed in the near future?

   Machine learning methods have been demonstrated to be useful in analyzing large amounts of unstructured data in a career services database. The machine learning methods forecast that the following professional competencies would become increasingly important in the near future: work experience, a university degree, independence, teamwork, motivation, and analytical ability. The neural network model tested here performed superior to linear regression. Both methods used here focused on historical data. Identification of hitherto unknown competencies was not addressed. Problems with overfitting can be addressed by implementing *k-fold* validation crossover and dropout techniques.

9.6 Future Work

These results lead to a question for future research: To what extent can students from underrepresented groups fulfill these competencies? The forecast increase in future demand for social skills could be viewed as a potential advantage for underrepresented groups. Underrepresented groups may have an advantage especially in acquiring soft skills. Although many technical skills can be learned through intensive rote memorization, social skills can often only be acquired through a lifetime of experience. Since students from underrepresented groups have often had to master a large number of soft skills in the course of their lives, e.g., as a minority learning to integrate into majority groups, they may have a potential advantage here.

Graduates from underrepresented groups can bring unique advantages to software development teams. They can help to increase the diversity of perspectives examined. Ilumoka (2012) discusses the importance of diversity in engineering teams. Especially during the requirements engineering phase, non-technical skills, such as intercultural communication and foreign language abilities, can be of exceptional value for multi-national teams or for stakeholders in foreign countries. During the development phase, cooperation, team-building, and conflict management skills can prove vital for the success of a software project.

Students from underrepresented groups may have to develop an earlier sense of self-reliance. Since they cannot rely on academic or financial support from their families, they may be forced to become more independent than the average student. They may lack the network of personal relationships necessary to find internships with prestigious companies during their studies. In order to integrate into a culturally foreign environment, not only a great deal of communication talent and commitment, but also personal initiative is required. This could enable first-generation students, students with a migration background, single parents or women in STEM to develop higher levels of social competency and resiliency.

This work is part of a larger research project, named "DiaMINT." The goal of the project is to recruit, support, and retain students from underrepresented groups in STEM subjects. DiaMINT covers each phase of the student customer journey, from the initial information gathering phase, through application, admission, orientation, internships, exams, theses, and entry into the job market (Schuhbauer, Brockmann, & Mustafayev, 2020).

Acknowledgments The authors would like to thank the anonymous reviewers for their constructive comments which greatly improved the quality of this work. This research project, "DiaMINT," was supported by a grant from the Staedtler Foundation.

References

Agnihotri, D., Verma, K. & Tripathi, P. (2014). Pattern and cluster mining on text data. In *2014 fourth international conference on communication systems and network technologies*, Bhopal, India (pp. 428–432).

BAM (2011). Bundesanstalt fuer Migration und Fluechtlinge, Bestandsaufnahme und Vernetzung. In *Vernetzungsworkshop: Integration von Studierenden mit Migrationshintergrund an deutschen Hochschulen*.

Beecham, S., Clear, T., Barr, J., Daniels, M., Oudshoorn, M., & Noll, J. (2017). Preparing tomorrow's software engineers for work in a global environment. *IEEE Software, 1*(34), 9–12.

Bell, S. (2010). Project-based learning for the 21st century: Skills for the future. *The Clearing House: A Journal of Educational Strategies,* Issues and Ideas, *83*(2), 39–43.

Branz, L., Pastran Reina, L., Richter, J., Waizmann, B. & Brockmann, P. (2019). Sentiment analysis of male and female developer comments: Exploring gender influence on emotional expressions in software engineering projects. In *GE@ICSE19 second workshop on gender equality in software engineering*. Montreal, Canada.

Budinska, L. & Mayerova, K. (2017). Graph tasks in Bebras contest—what does it have to do with gender? In *CSERC 17 computer science education research conference*. Helsinki, Finland.

Cheryan, S., Siy, J., Vichayapai, M., Drury, B., & Kim, S. (2011). Do female and male role models who embody STEM stereotypes hinder women's anticipated success in STEM? *Social Psychological and Personality Science, 2*(6), 656–664.

Dangeti, P. (2017). *Statistics for machine learning*. Birmingham: Packt Publishing.

Faulkner, W. (2009). Doing gender in engineering workplace cultures. Observations from the field. *Engineering Studies, 1*(1), 3–18.

Goodfellow, I., Bengio, Y., & Courville, A. (2016). *Deep learning*. Cambridge: MIT Press.

Hecht-Nielsen, R. (1989). Theory of the backpropagation neural network. In *Proceedings of the International Joint Conference on Neural Networks* (vol. 1, pp. 593–611). IEEE.

Hinze, A., Timpany, C., Bowen, J., Chang, C., Starkey, N., & Elder, H. (2018). Collecting sensitive personal data in a multi-cultural environment. In *HCI 2018 Proceedings of the international human computer interaction conference*. Las Vegas, Nevada, USA.

Hoda, R., Ali Babar, M., Shastri, Y., & Yaqoob, H. (2016). Socio-cultural challenges in global software engineering education. *IEEE Transactions on Education, 3*(60), 173–182.

Ifenthaler, D., & Widanapathirana, C. (2014). Development and validation of a learning analytics framework: Two case studies using support vector machines. *Tech Know Learn, 19*, 221–240.
Ilumoka, A. (2012). Strategies for overcoming barriers to women and minorities in STEM. In *IEEE 2nd integrated STEM education conference*, Ewing, New Jersey, USA (pp. 1–4).
Kucak, D., Juricic, V. & Dambic, G. (2018). Machine learning in education—a survey of current research trends. In *Proceedings of the 29th AAAM international symposium* (pp. 406–410).
Liebenberg, J. & Pieterse, V. (2016). Career goals of software development professionals and software development students. In *CSERC 16 computer science education research conference* (pp. 22–28). Pretoria
Maurer, K. (2019). *Kompetenzentwicklung in MINT-Berufen: Ermittlung zukünftig notwendiger Kompetenzen mittels Machine Learning* (Master's thesis). Nuremberg Institute of Technology, Nuremberg.
McNamee, P., & Mayfield, J. (2004). Character *N*-Gram Tokenization for European language text retrieval. *Information Retrieval, 7*, 73–97.
Mitchell, T. M. (2010). *Machine learning. McGraw-Hill series in computer science*. New York, NY: McGraw-Hill.
Mohri, M., Rostamizadeh, A., & Talwalkar, A. (2012). *Foundations of machine learning. Adaptive computation and machine learning*. Cambridge: MIT Press.
Raschka, S., Mirjalili, V., & Lorenzen, K. (2018). *Machine Learning mit Python und Scikit-learn und TensorFlow: Das umfassende Praxis-Handbuch für Data Science, Deep Learning und Predictive Analytics* (2., aktualisierte und erweiterte Auflage). Frechen: mitp Verlags GmbH & Co. KG.
Schuhbauer, H., Brockmann, P., Mustafayev, T. (2020) Mapping the students' journey to develop student-centered tools. In *IEEE educon global engineering education conference*, Porto 27–30 April, 2020, pp. 61–65.
Slade, S., & Prinsloo, P. (2013). Learning analytics ethical issues and dilemmas. *American Behavioral Scientist, 57*, 1510–1529.
Svrivastava, N., Hinton, G., Krizhevsky, A., Sutskever, I., & Salakhutdinov, R. (2014). Dropout: A simple way to prevent neural networks from overfitting. *Journal of Machine Learning Research, 15*, 1929–1958.
Teschl, G., & Teschl, S. (2007). *Mathematik für Informatiker: Band 2: Analysis und Statistik* (2. Auflage). *eXamen.press*. Berlin, Heidelberg: Springer.
Wong, T. (2015). Performance evaluation of classification algorithms by k-fold and leave-one-out cross validation. *Pattern Recognition, 48*, 2839–2846.
Technical University of Applied Sciences Nuermberg. (2018). Diversity an der Hochschule. Retrieved April 26, 2019, from https://www.th-nuernberg.de/hochschule-region/strategie-und-profil/hochschule-der-vielfalt

Chapter 10
Student Perceptions of Virtual Reality in Higher Education

Tebogo John Matome and Mmaki Jantjies

10.1 Introduction

In recent years, higher education has been viewed as the key to influencing holistic development (both economic and social), in emerging markets. Matherly, Amin, and Al Nahyan (2017) highlight the positive correlation, between the level of investment put into improving the quality of higher education, and the resultant increase in a nation's economic growth; insinuating that the higher the quality of a country's education system, the better the life of the inhabitants of the said country. Lane and Johnstone (2012) also make mention of the ongoing "great brain race" between various countries. They accredit the intensity of the race to the different nations' realization of the role played by higher education, and the level of skilled human capital it produced, when attempting to influence innovation, and economic development.

It is undeniable that the introduction, and evolution of technology, has changed the way in which all industries operate. Accordingly, the education system has not been overlooked by the effects of digitization and globalization. Many have researched the integration of different advanced learning technologies within the higher education sector. These include the study of challenges and prospects of attaining an online higher education within a South African university (Letseka & Pitsoe, 2014), the study of student perceptions regarding the implementation and use of virtual classrooms (Cakiroglu, 2014), and using artificial intelligence techniques, to influence interactive and adaptive, online education systems (Almohammadi, Hagras, & Alghazzawi, 2017). Resultantly, the addition of

T. J. Matome (✉) · M. Jantjies
Department of Information Systems, University of the Western Cape, Cape Town, South Africa
e-mail: 3366229@myuwc.ac.za; mjantjies@uwc.ac.za

D. Ifenthaler et al. (eds.), *Balancing the Tension between Digital Technologies and Learning Sciences*, Cognition and Exploratory Learning in the Digital Age,
https://doi.org/10.1007/978-3-030-65657-7_10

technology, to education, has resulted in positive outcomes on various issues faced by the sector, over the years. Furthermore, Passig (2015) mentions how advanced educational technologies, with their advanced interfaces, generate an accelerated enhancement in a wide range of skills that the natural educational environment alone cannot account for.

However, it has also been found that the implementation of online learning technologies is not easily always accessible to different demographic groups. Cohen, Bancilhon, and Grace (2018) stress that the realization of a digitally connected society is highly dependent on, and cannot be disassociated from, the need for social and economic inclusion. Thus, introducing advanced technologies requires an understanding of the student body context, especially if the electronic learning requirements will extend beyond the campus environment.

In terms of the study's population, South African higher education institutions face variances in the population access to resources. This is as a result of the historic background, socio-economic characteristics, and the presence of a digital divide within the country's different races and groupings. For that reason, this study was conducted in a South African higher educational institution.

Nyahodza and Higgs (2017) use statistics from a 2016 World Bank survey (which concludes that members of South Africa's population, who fall within the lower half of the countries income bracket, account for a mere 8% of the country's income)—to support their argument that a majority of these individuals are unable to afford educational resources, which also include computers and an Internet access. Furthermore, a study by Hersh and Mouroutsou (2015) identifies income and language as the primary factors, defining the access to learning technologies. This is important to note, in the context of South Africa, where the apartheid era has resulted in a long-lasting influence on economic disparity. Nevertheless, when considering the different technologies that can be used in higher education, Virtual Reality (VR), amongst other emerging technologies, provides an opportunity to bridge the resource gap for students allowing them to gain experience in their learning outcome using technology. This can be especially useful in resource constrained environments. Lu, Li, Chen, Kim, and Serikawa (2018) describe VR as a computer system-generated simulation, which creates a virtual environment that can be interacted with, in a realistic manner.

Taking into account that traditional learning takes place in the form of real-time interactions between learners and facilitators, Cakiroglu (2014) makes mention of the implementation of VR-created simulations and assessments, allowing facilitators (teachers and lecturers) to test a learner's ability to react to real-life scenarios, based on their course of learning. This also allows students in high-risk fields (such as surgeons and engineers) to immerse themselves in realistic simulations, without the risks associated with making mistakes during real-life operations, as well as without having physical access to such resources. It is, nonetheless, also important to acknowledge the advantages and developments that may arise from the introduction of VR within education. For example, the growth in the popularity of VR may be the driving force, facilitating an increase in access to local higher education, across the different socio-economic groups, of South Africa. These possible advantages provide a rational reasoning for the necessity of investigating a possible solution, to a better interaction with learning content, for registered students of South African higher education institutions.

10.1.1 Research Questions & Objectives

Unterhalter et al. (2018) acknowledge that although the recent student protests in South Africa shine a light on access and funding problems, these protests are not isolated events. Furthermore, they expand on the notion that these events uncover deep-rooted challenges within the higher education system, stemming from an unresolved colonial legacy. One must therefore consider that, due to the large economic disparity amongst the different socio-economic brackets of South Africa, some students find it more difficult than others to access the higher education system as well as related resources. This research therefore aimed to investigate the main research question of, how can VR be used, to enable access to learning resources for higher education students, in South Africa? This was supported by four research sub-questions, giving direction to the research, and ultimately satisfying predetermined objectives. The research delved into what the main factors contributing to the lack of access to Higher Education in South Africa are, before further reviewing past literature to answer the question of what VR solutions have already been implemented in higher education, across the world. Thirdly, the research investigated how a VR-augmented higher education system affects a diverse student population, before the final research question aimed to identify the different ways in which VR can be used to support students' equitable access to higher education learning content.

The sub-questions will help to satisfy the following objectives, and ultimately answer the primary research. The first objective being to identify the main forms of VR, which have been used to support education in universities. Secondly, the research aimed to determine how VR can be used to support students, in order to interact with their institution's learning content adequately and less-effortlessly. The final objective was to analyze the effects, as well as other advantages, of introducing a VR-augmented higher education system, within a diverse student population.

10.2 Literature Review

10.2.1 Technology in Higher Education

With the advent of technology, higher education qualifications have become more revered than in before. So much so that various countries have invested heavily in improving the quality of higher education, as it directly correlates to a nation's economic development (Matherly et al., 2017). Matherly et al. (2017) identified an increase in salaries, enhanced employment opportunities, better mobility, a longer life expectancy, and an improved quality of life as some of the direct benefits of higher education qualifications. This supports the notion that the importance of university or higher education is ever increasing, in an increasingly competitive global economy. Winters (2011) further notes that the main indicators and influence on a nation's human capital capability, is the presence of HE institutions in the area of dwelling.

However, in order to gauge the success of implementing technology in higher education, one must understand the knowledge and associated challenges of the technologies' primary users. These users include higher education students, as well as the teachers (and lecturers) of institutions. Jantjies and Joy (2016) highlight the cultural and linguistic diversity of students, as a pivotal factor in the introduction of technology, in the education system of South Africa. This is a result of the relatively large number of official languages and cultures recognized in the country, compared to other nations. It is also important to consider the readiness and willingness of students, to accept the introduction of technology-aided learning. Chaka and Govender (2017) found that Nigerian higher education learners were receptive to the idea of a technology-aided learning system, with their perceptions of technology driving this acceptance rate. One can therefore assume that, should South African learners have similar perceptions of technology, the adoption of a blended learning system would result in a successful implementation. A Western Cape based study, by Chigona, Chigona, and Davids (2014), investigated the factors motivating educators to use ICTs in the province's disadvantaged areas. These factors were identified as the individual expectations of, the resultant feeling of achievement from, and the responsibilities associated with the use of ICTs in the aforementioned areas. As a result, this clarifies the need to motivate, inspire, and train educators on the use of ICTs in South African institutions.

10.2.2 Introducing Virtual Reality in Higher Education

VR enables the simulation of virtual environments through software applications to mimic the real-world environment. Hardware enabling users to gain access to VR applications are either head mounted gear such as the oculus rift and the HTC Vive, or mobile devices supported through a head mounted casing. As VR hardware becomes more affordable, educational institutions are finding ways to enhance learning by supporting the immersive learning experience afforded by VR (Dută et al., 2011; Freina & Ott, 2015). Furthermore, a CAVE (Cave Automatic Virtual Environment) which is a 4 or 5 screened cube "room" can be used to allow between 2 and 8 participants to get an immersive 3D experience of a learning task.

A CAVE projects on the surrounding screens allowing surrounding 3D visuals of the content being projected. It is also supported by surround sound (Leder, Horlitz, Puschmann, Wittstock, & Schütz, 2019). While CAVES usually entail high implementation costs, they in turn provide simultaneous access to multiple students, as more than two students can use a CAVE at once. While there is a growing number of accessible VR hardware, there is a need for more developments of contextual and multilingual applications for various study fields. Many of the existing learning applications such as DentSim designed to support restorative dentistry education and The Geneva System developed to teach dental anatomy (Dutã et al., 2011) were developed for a specific context. This provides an opportunity for African universities to develop applications which are specific to their fields as well considering contextual issues such as the availability of content in multiple languages.

However, it is also important to acknowledge that with the advances in technology, there are various high-powered applications of VR, which have been implemented within the modern-day higher education curricula. For example, Al Awadhi et al. (2018) make mention of the Titan of Spare application, which immerses students into a virtual realm of the solar system, allowing them to discover and better study the solar system's planets, from the comfort of their lecture halls. King et al. (2018) also elaborate on the CILVRS (Collaborative Immersive Learning Virtual Reality Series) project of Bournemouth University, in the United Kingdom. This simulation, in which medical students take on the role of a medical practitioner, awaiting the arrival of their patient, provides students with the opportunity to learn from their mistakes (virtually) without jeopardizing the wellbeing of real patients.

10.3 Research Methodology

10.3.1 Design & Methodology

A mixed-method survey, which is described by Creswell and Plano Clark (2007) as the use of both qualitative and quantitative data sourcing, in a single study, was used to conduct this study. A self-administered questionnaire, which was development and loosely guided by the 13 principles of questionnaire construction, was made available to the student population, using a simple random sampling approach. Responses from 81 ($N = 81$) registered students, completed the survey. The data was analyzed using a thematic approach. This is described by Clarke and Braun (2017) as the identification and analysis of responses, in order to deduce substantial patterns within a qualitative data set, ultimately classifying them as themes. The formulation of themes from the data, gave the researcher an opportunity to identify patterns and similarities in the students' responses, by finding commonalities in the problems identified, as well as the suggestions provided.

The use and analysis of both qualitative and quantitative methods also allowed for the neutralizing of the weaknesses in one method, by using the strengths of the other method. However, a mixed-method analysis also allows a researcher to enhance the strengths of one method, with the strengths of the other (Creswell, 2009). Furthermore, Tobergte and Curtis (2013) describes triangulation as the effort of combining both types of data that have been used within the mixed-method process, in the attempt to use one set to corroborate the other.

As a result of the triangulation, the trends found within the qualitative analysis were used to try and provide a sense of justification/context to the trends found within the different themes of the qualitative data.

10.4 Findings

10.4.1 Identified Themes

After an initial analysis of the data, three themes were identified and constructed, to better understand the correlation between the different sample responses. These themes were further refined and unpacked in line with the research findings.

These identified themes are:

1. Student access to different content platforms, provided by the institution.
2. Student usage and knowledge of technologies in higher education.
3. Students' perceptions of a VR augmented in higher education system.

The first two themes allowed for a better understanding of the third research sub-question, regarding How a VR-augmented higher education system would affect a diverse student population. In the first theme, we sought to understand the current experiences of online learning for students. Furthermore, the third theme was formulated to unpack the fourth research sub-question, which aimed to elaborate on the different ways in which VR can be used to support students' equitable access to higher education learning content.

10.4.2 Results According to Themes

10.4.2.1 Access to Different Content Platforms, Provided by the Institution

In this theme, the study sought to understand the current digital learning devices and resources, which the students' access, and can be used to support VR. While the university provided full Internet access to students and staff, 26% of participants stated that the main issue they faced with Internet access, was the challenges of a periodic slow Internet connection. Other reasons stated included the lack of access to substantive online journal article portals, with 9% highlighting this as a challenge to them. 16% of the participants reflected on the inability to access the institution's online student portal (Learning Management Systems) and learning platform, at certain times of the day, due to high usage. Almost half of the students displayed in Fig. 10.1 did not experience problems when accessing current online learning

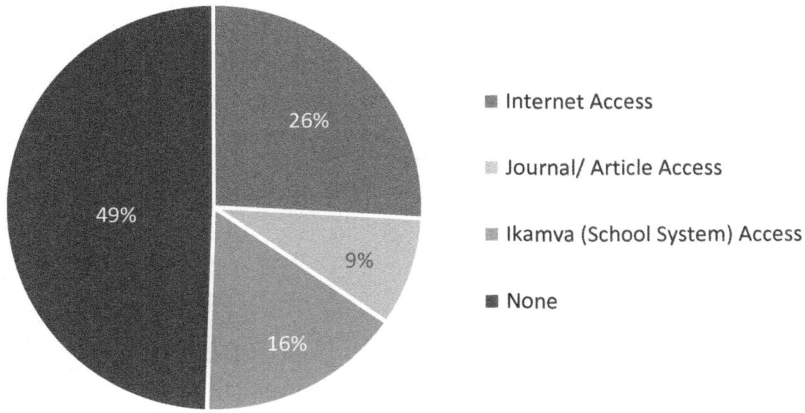

Fig. 10.1 Problems faced when accessing online learning content

Table 10.1 Access to learning platforms on and off campus

		Desktop	Mobile	Tablet	Total
Residence	Off campus	14	6	2	22
	On campus	47	11	1	59
Total		61	17	3	**81**

resources, as the campus had a good network support as well as technician and e-learning walk-in support centers. To understand how VR will be accessed by the student population, the research investigated the types of devices used, to access learning content.

Table 10.1 reflects that almost two-thirds of the students accessed technology on campus and relied on the university computer labs. This provides a reflection on the type of VR applications, as well as tasks, that could be given to students.

10.4.2.2 Student Usage and Knowledge of Technologies in Higher Education

This theme sought to understand the different platforms used by students to interact with their course content. In Fig. 10.2, 16% of participants stated that their most used applications were Microsoft Office Suite or similar software. Further followed by a 20% use of Email, the institution's Online Library, and the university student Learning Management System services, respectively. Furthermore, 20% of students reflected not having had any use of technology in their learning process. Contrastingly, 8% of students perceived the current technology infrastructure in the institution as being either unsatisfactory or mediocre. In aiming to understand student's current knowledge and use of VR technology, the study also found that more than 60% of the students have neither knowledge of, nor experience with, VR.

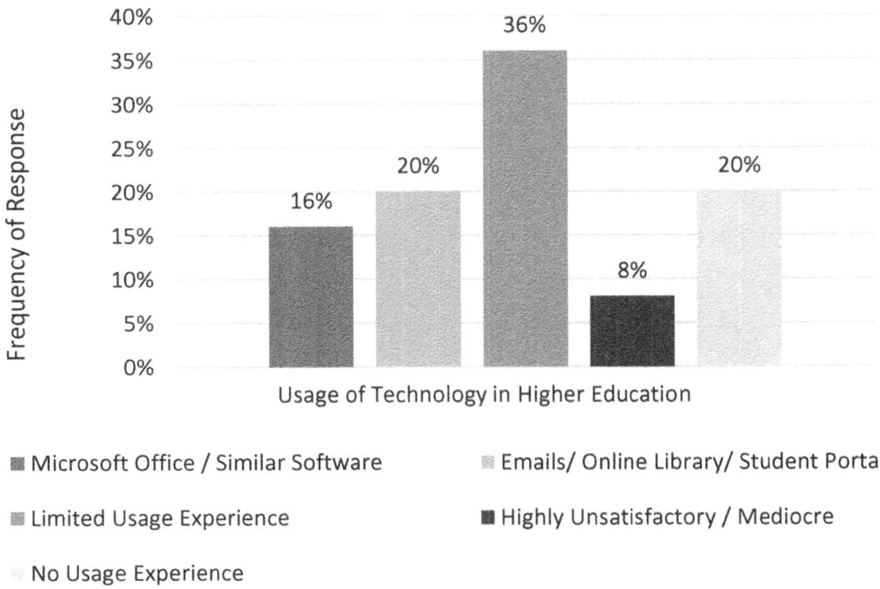

Fig. 10.2 Participant usage of technology in higher education

However, 12% of the students did state that they had experienced VR, with participants highlighting their knowledge for its use in gaming. Additionally, 26% of the students also stated that they only had some knowledge of VR, but had no previous interaction with it. The students reflected that this knowledge of VR was gained from either self-study, or first year Information Systems lectures. The results in the study reflected on the student population's lack of previous interaction with any VR applications, especially in the learning context. VR was often known to them as a gaming tool. The results also reflected the importance of students' training support for VR, as a small percentage of them were currently not expected to use technology in their tasks.

10.4.2.3 Students Expectations of a VR-Augmented Higher Education System

Regarding what students would expect from a VR supported digital learning system, the results in Fig. 10.3 reflect that 15% of students expected VR to provide them with realistic experiences of learning. Examples of these realistic experiences included visiting nature reserves or foreign destinations, from the comfort of their home. Five percent of the students expect that a VR supported learning experience will have a negative impact on their learning experience. Just over a fifth of the students expect VR to aid in education and learning capabilities of the population, both in formal institutions as well as through private/individual use.

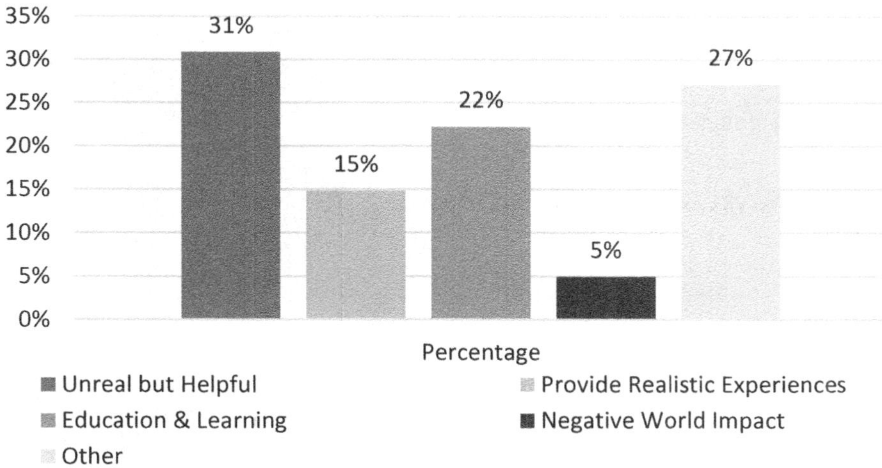

Fig. 10.3 Student expectations of the introduction of virtual reality

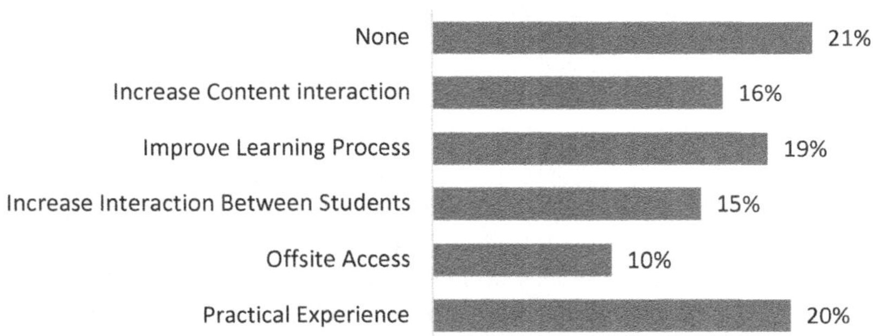

Fig. 10.4 Student's perceived advantages of virtual reality in higher education

Regarding students' perceptions on the advantages of integrating VR within Higher education, as reflected in Fig. 10.4, the study found that while 22% of the respondents were unable to find any advantages of VR in their education, 20% of the students stated that it would afford them the opportunity of practical experience, as opposed to relying solely on the theory they are taught. A further 19% identified a possible increase in their learning and knowledge retention process, with 15% perceiving this innovation as one that will increase interaction between students, as well as lecturers. 10% of the students believe that the ability to access and better interact with course content, when not on the university campus will be the greatest advantage of implementing VR. Additionally, the remaining 16% of respondents highlight an increased interaction with course content as the main advantage associated with a VR-augmented Higher Education system.

In terms of the different ways that students believe the implementation of VR can be made inclusive, the following findings were recorded. In Fig. 10.5, 22% of

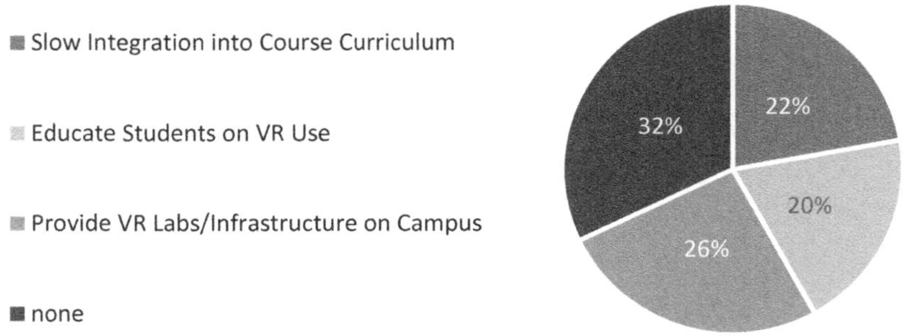

Fig. 10.5 Student suggestions on how to make VR implementation inclusive to all students

students believe that slowly integrating the technology into their curriculum would be the best way to influence equitable use of the technology. However, while a fifth of the respondents highlight the necessity of educating students on the effective use of VR in higher education, a quarter of the respondents felt that the best way to promote equitable access to the technology is through the building and provision of VR-dedicated labs (and infrastructure), on their campus.

10.5 Discussion

One aspect of the research aimed to find the main issues faced by students, when interacting with the different online platforms of their institution. This was analyzed within the theme of Student Access to Different content platforms, provided by the institution. The most prevalent issues were the lack of access to a stable internet connection, as well as the inability to access their online learning management system during certain periods of the day. Resultantly, with institutions having a limited number of computer labs (and computers), compared to the overall student population, most students would be required to use personal computers and laptops to interact with their online content.

If taken in the context of Nyahodza and Higgs (2017)—who stated that a large fraction of South African students are unable to afford educational resources, such as personal computers—this lack of access to a workspace, and adequate internet connection poses a challenge to VR, within higher education learning. In cases where students do not own smartphones, the lecturers would be expected to provide VR platforms or provide alternative solutions, which consider the resources that students have access to. Dedicated VR supportive labs or access to mobile devices, smartphones or a CAVE would be important in ensuring that students can effectively access and use VR in their education. The results also reflected the importance of the existing e-learning center to support VR systems training support which was currently effective in supporting other e-learning training services.

Regarding the theme of "Student Usage and Knowledge of Technologies in Higher Education" the study reflected that many of the students residing on campus make use of the institution's Wi-Fi services, and thus mainly relied on the university network services. Furthermore, students residing off campus would be required to carry the cost of an increase in the bandwidth required to efficiently use VR. The role of free Wi-Fi in supporting current e-learning activities reflected the importance of VR software which was also available offline. The study further reflected that academics using VR should be considerate of the role of the Internet in the e-learning process, and thus cannot solely rely on VR applications requiring access to the Internet, especially with tasks required beyond the lecturer room (Blignaut, Els, & Howie, 2010).

Furthermore, with a combined total of 56% of students having reported that they had either limited, or no, experience with technology usage in higher education, this reflected on literature by Walker and Mkwananzi (2015), highlighting that students that are enrolled in higher education institutions, filter in from positions of extreme inequality. When discussed in conjunction with other findings, where 60% of the students reported that they had no knowledge or experience of VR, extensive training and education on the usage of the VR technology was found to be important in its successful implementation.

When investigating the theme of "Students' expectations of a VR-Augmented Higher Education System," many of the participants had no prior knowledge of VR and its use in education. With regards to the advantages expected from a VR-augmented Higher Education, a combined 26% of the respondents cited perceived advantages in possible interaction and collaboration between students, as well as with lectures, as a result of better accessibility and interaction with course content. This realization correlated with Viberg and Grönlund (2013), who stated that the introduction of VR would change the traditional delivery of, and access to, learning content; by overcoming the spatial-temporal characteristic of modern-day formal education. Furthermore, Viberg and Grönlund (2013) also made mention of the advancements afforded by VR, resulting in habitual online meetings and classrooms, as well as the opportunity to integrate mobile learning into curricula, to further overcome these spatial-temporal factors. Considering the increase in mobile technology adoption (relative to that of personal computers) in developing countries, with mobiles being identified as the second largest platform usage for accessing online learning content—tailoring the introduction of VR to a more mobile-based platform would deliver the most benefits, and a greater realization of the aforementioned advantages.

In terms of promoting inclusivity of, and equitable access to, VR in the higher education system, the findings identified the need for dedicated VR labs and infrastructure (provided by the institution) as being the most imperative factor to widespread adoption and usage. In reference to Letseka and Pitsoe (2014) who stated that education delivery models of global institutions are moving toward one that is cost-effective, while not foregoing service quality, this finding would highlight the need for South African institutions to transition toward integrating and procuring emerging technologies, within their respective campuses. The findings reflect on the notion of Hersh and Mouroutsou (2015) who state the main hindrances of access to these emerging technologies include the cost of, and lack of funding for, the maintenance of emerging technologies, as well as a need for training and support for academics. Future use of VR application in higher education could thus enhance learning effectively through both infrastructure and human resource support to ensure its success.

10.6 Recommendations

10.6.1 Training

10.6.1.1 Student/Instructor/Institutional Readiness

Considering the data collected from the respondents, there is an evident need for some sort of training before the introduction of a VR-education system. This will also need to be coupled with a slow integration of the new system, into the traditional learning system, in order to gradually introduce faculty-specific applications, that will increase the learner's efficiency.

Students will need to be provided with training and base-level education on the use of VR, as well as its advantages in their respective spheres of learning and work environments. This will allow for an increased user buy-in, which is imperative to the introduction of any new technology. Ultimately, it will also ensure that learners are more willing to adopt and use VR in their field of learning, from the initial introduction.

On the other hand, it is also important to not assume that all lecturers and instructors have knowledge about VR (and more importantly VR in their field of teaching). Institutions will need to train lecturers and instructors to use, maintain and effectively incorporate Course-specific applications into their teachings.

In terms of infrastructure readiness, not all South African institutions currently have either the infrastructure or necessary financial requirements to facilitate a large-scale introduction of a VR-education system. However, as the use of immersive technology is slowly embedding itself into everyday life, it would be beneficial for institutions to begin gradually investing in and introducing the necessary infrastructure and applications into the lives of their students.

This could be in the form of a Immersive Campus Tour application, A faculty/course demonstrative application for learners (before they commit to a course), or even a series of immersive videos where the student can experience attending an institution's sporting event, or stadia. Such implementations will allow students to gradually adjust to virtual experiences on campus and look forward to using similar applications in their respective courses and learning journeys.

10.6.2 Feasibility Studies

10.6.2.1 Student

It is also worth noting that not all learners will have the financial abilities to afford some of the hardware associated with the use of VR in a learning environment, or even a remote learning experience. In an attempt to remedy the latter (during the recent COVID-19 pandemic) institutions started supplying students in need with computers and monthly data packages. This was done in the aim of facilitating more effective online learning, for students who may not have been able to initially afford it.

With an increase in the number of applications being developed for mid-range and lower tier smartphones, a similar pilot project could be rolled out (on a smaller scale). This would entail the provision of cost-effective smartphones, as well as VR boxes (such as Google cardboard), to select students in specific faculties. This would facilitate the need to acclimatize to VR-learning, as a first attempt, and allow institutions to readjust their introduction strategy accordingly, after a predetermined period of time.

10.7 Conclusion

There has been a growing increase in the use of VR technologies within the global education system. These technologies play an important role in supporting teaching and learning within education. Furthermore, they are able to provide immersive learning experiences particularly in subjects requiring experience in the learning process.

In addition to increasing the accessibility of online interaction with an institution's learning content, as well as providing practical experience and knowledge of one's respective field of study; VR in higher education will provide facilitators with increased avenues of disseminating and assessing course-related content, in a way that will be easier to remember for the students. In introducing such advanced technologies, basic technology adoption considerations need to be made by any educational institutions considering the context of study. With the data and internet, prices of South Africa being identified as some the most expensive in the world (in comparison to other developing countries) this would affect the experiences of the bandwidth-heavy nature of VR usage and would limit the holistic adoption of the technology across the entire student population. Furthermore, in South Africa, learners are more likely to have access to mobile phones compared to desktop computers, motivating innovations which are largely mobile based. These limitations, in one way or another, may have had an influence on the outcome of the study; and could be mitigated in future research. The study thus reflects that VR technology can play an important role in education when important contextual adoption considerations are made.

Acknowledgments I would like to acknowledge Prof. M. Janjties for her continued support and motivation over the years. I would also like to thank Mr. Rodrique George, of the University of the Western Cape's EMS faculty for having always been the voice of reason, and for sparking my now fruitful love for research. Ultimately, my deepest and heartfelt gratitude is owed to my parents and sister, as well as my partner, for always believing in my abilities, and for constantly pushing me to strive for greater heights. Their contribution to the fruition of this research may have not been a direct one, but is as important as any other.

References

Al Awadhi, S., Al Habib, N., Al-Murad, D., Al Deei, F., Al Houti, M., Beyrouthy, T., & Al-Kork, S. (2018). Interactive virtual reality educational application. *Advances in Science, Technology and Engineering Systems Journal, 3*(4), 72–82.

Almohammadi, K., Hagras, H., & Alghazzawi, D. (2017). A survey of artificial intelligence techniques employed for adaptive educational systems within E-learning platforms. *Journal of Artificial Intelligence and Soft Computing Research, 7*(1), 47–64. https://doi.org/10.1515/Jaiscr-2017-0004

Blignaut, S., Els, C., & Howie, S. (2010). Contextualizing South Africa's participation in the SITES 2006 module. *South African Journal of Education, 30*(October 2004), 555–570.

Cakiroglu, U. (2014). Evaluating students' perspectives about virtual classrooms with regard to seven principles of good practice. *South African Journal of Education, 34*(2), 1–19. https://doi.org/10.15700/201412071201

Chaka, J. G., & Govender, I. (2017). Students' perceptions and readiness towards mobile learning in colleges of education: A Nigerian perspective. *South African Journal of Education, 37*(1), 1–12.

Chigona, A., Chigona, W., & Davids, Z. (2014). Educators' motivation on integration of ICTs into pedagogy: Case of disadvantaged areas. *South African Journal of Education, 34*(3), 1–8. https://doi.org/10.15700/201409161051

Clarke, V., & Braun, V. (2017). Thematic analysis. *Journal of Positive Psychology., 12*, 297. https://doi.org/10.1080/17439760.2016.1262613

Cohen, J., Bancilhon, J. M., & Grace, T. (2018). Digitally connected living and quality of life: An analysis of the Gauteng City-region, South Africa. *Electronic Journal of Information Systems in Developing Countries, 84*(1), 1–12. https://doi.org/10.1002/Isd2.12010

Creswell, J. W. (2009). *Research design: Qualitative, quantitative, and mixed methods approaches* (4th ed.). Thousand Oaks: Sage. Retrieved from http://www.ceil-conicet.gov.ar/wp-content/uploads/2015/10/creswell-cap-10.pdf

Creswell, J. W., & Plano Clark, V. L. (2007). *Designing and conducting mixed methods research*. Thousand Oaks: Sage Publications.

Dutã, M., Amariei, C. I., Bogdan, C. M., Popovici, D. M., Ionescu, N., & Nuca, C. I. (2011). An overview of virtual and augmented reality in dental education. *Journal of Oral Health and Dental Management, 10*(1), 42–49. Retrieved from http://scholar.google.co.za/scholar_url?url=https://pdfs.semanticscholar.org/0ca2/E4b2aa03f327ec58f496e56116ed27bfedc9.pdf&Hl=En&Sa=X&Scisig=Aagbfm14jb5sjyvmpvknblcpwmc4xdf7nw&Nossl=1&Oi=scholarr

Freina, L., & Ott, M.. 2015. A literature review on immersive virtual reality in education: State of the art and perspectives. In eLearning and Software for Education (ELSE). Romania. Retrieved from https://www.researchgate.net/publication/280566372_a_literature_review_on_immersive_virtual_reality_in_education_state_of_the_art_and_perspectives

Hersh, M. A., & Mouroutsou, S. (2015). Learning technology and disability: Overcoming barriers to inclusion: Evidence from a multi-country study. *IFAC-Papersonline, 48*(24), 83–88. https://doi.org/10.1016/J.Ifacol.2015.12.061

Jantjies, M., & Joy, M. (2016). Lessons learnt from teachers' perspectives on mobile learning in South Africa with cultural and linguistic constraints. *South African Journal of Education, 36*(3), 1–10. https://doi.org/10.15700/Saje.V36n3a1274

King, D., Tee, S., Falconer, L., Angell, C., Holley, D., & Mills, A. (2018). Virtual health education: Scaling practice to transform student learning. *Nurse Education Today, 71*(December), 7–9. https://doi.org/10.1016/J.Nedt.2018.08.002

Lane, J. E., & Johnstone, D. B. (Eds.). (2012). *Universities and colleges as economic drivers: Measuring higher education's role in economic development*. New York: State University of New York Press.

Leder, J., Horlitz, T., Puschmann, P., Wittstock, V., & Schütz, A. (2019). Comparing immersive virtual reality and Powerpoint as methods for delivering safety training: Impacts on risk perception, learning, and decision making. *Safety Science, 111*, 271–286.

Letseka, M., & Pitsoe, V. (2014). The challenges and prospects of access to higher education at UNISA. *Studies in Higher Education, 39*(10), 1942–1954. https://doi.org/10.1080/03075079.2013.823933

Lu, H., Li, Y., Chen, M., Kim, H., & Serikawa, S. (2018). Brain intelligence: Go beyond artificial intelligence. *Mobile Networks and Applications, 23*(2), 368–375. https://doi.org/10.1007/S11036-017-0932-8

Matherly, L. L., Amin, N., & Al Nahyan, S. S. K. (2017). The impact of generation and socio-economic status on the value of higher education in the UAE: A longitudinal study. *International Journal of Educational Development, 55*(May), 1–10. https://doi.org/10.1016/J.Ijedudev.2017.04.002

Nyahodza, L., & Higgs, R. (2017). Towards bridging the digital divide in post-apartheid South Africa: A case of a historically disadvantaged University in Cape Town. *South African Journal of Libraries and Information Science, 83*(1), 39–48. https://doi.org/10.7553/83-1-1645

Passig, D. (2015). Revisiting the Flynn effect through 3D immersive virtual reality (IVR). *Computers and Education, 88*, 327–342. https://doi.org/10.1016/J.Compedu.2015.05.008

Tobergte, D. R., & Curtis, S. (2013). Social research methods. *Journal of Chemical Information and Modelling, 53*(9), 1689–1699. https://doi.org/10.1017/CBO9781107415324.004

Unterhalter, E., Allais, S., Howell, C., Mccowan, T., Morley, L., Oanda, I., & Oketch, M. 2018. Conceptualising higher education and the public good in Ghana, Kenya, Nigeria, and South Africa. In *Conference: Comparative and International Education Society (CIES)*, No. March: 1–32. Retrieved from http://discovery.ucl.ac.uk/10050089/

Viberg, O., & Grönlund, Å. (2013). Computers & education cross-cultural analysis of users' attitudes toward the use of Mobile devices in second and foreign language learning in higher education: A case from Sweden and China. *Computers & Education, 69*, 169–180. https://doi.org/10.1016/J.Compedu.2013.07.014

Walker, M., & Mkwananzi, F. (2015). Challenges in accessing higher education: A case study of marginalised young people in one South African informal settlement. *International Journal of Educational Development, 40*, 40–49. https://doi.org/10.1016/J.Ijedudev.2014.11.010

Winters, J. V. (2011). Human capital, higher education institutions, and quality of life. *Regional Science and Urban Economics, 41*(5), 446–454. https://doi.org/10.1016/J.Regsciurbeco.2011.03.001

Chapter 11
Open Distance Learning and Immersive Technologies: A Literature Analysis

Afika Ntaba and Mmaki Jantjies

11.1 Introduction

Distance learning (DL) is categorized as an educational form of instruction that is focused on pedagogy, technology and educational design methodologies that effectively impart education to students (Al-Arimi, 2014). Distance education has been traditionally known to refer to correspondence courses that utilize postal services to transport learning materials related to the course (UNISA, 2018). In recent times DL can be described as asynchronous or synchronous, where the former relates to students engaging with their academic material when it is most convenient to them (e.g. viewing videotaped lectures) and the latter, when the students interact with their lecturers in real-time, through technology such as teleconferencing (Baukal, 2010).

There are various reasons as to why individuals make the choice to enrol in distance learning institutions as opposed to traditional institutions of higher learning. It is imperative to investigate these reasons as they will assist in forming the basis of understanding the student experience discussed in this chapter. An account by Simpson and Anderson (2012) states that the initial distance learning students primarily comprised of women, who were motivated by the inadequate services being provided by formal institutions additionally, by the added pressure of having to work to pay for the tuition. The motive for seeking DL higher education has expanded since the first distance learner experience. Several reasons relate to the need for an educational framework that is not limited to a physical lecture room, is affordable, and fundamentally, has a high degree of flexibility (Rodrigues, Affonso, Quinelato, & Montiel, 2014). Distance learners seek to make better use of their

A. Ntaba (✉) · M. Jantjies
University of the Western Cape, Cape Town, South Africa
e-mail: 3428062@myuwc.ac.za

D. Ifenthaler et al. (eds.), *Balancing the Tension between Digital Technologies and Learning Sciences*, Cognition and Exploratory Learning in the Digital Age,
https://doi.org/10.1007/978-3-030-65657-7_11

time, have geographic independence and also often require a flexible schedule (Rodrigues et al., 2014).

These motives are validated by benefits such as an efficient use of time, increased accessibility to a wide range of diverse students, flexible learning hours, and the possibility of being able to tailor academic content to the individual learner (Leszczyński et al., 2017). DL is, therefore, favoured by learners who would be incapable of attending face-to-face lectures on a daily basis, due to their other responsibilities and/or lack of resources.

DL has afforded nonconventional students an opportunity to study and additionally assisted in reshaping traditional education. Yasmin (2013) reported how DL had been valuable in reducing numerous barriers to traditional learning, specifically inaccessibility due to geographical locations, previous subpar educational completion and financial restrictions. Since the introduction of DL as an alternative means of formal education, there have been a number of significant transformations in not only the provision of education in distance education but also the learning experience by students. Although DL presents learners with an opportunity to engage in an environment free of physical interaction (Van Antwerpen, 2016), it is not without its challenges. Pozdnyakova and Pozdnyakov (2017) speculate that these challenges are predominantly related to the lack of corporeal co-presence between the student and lecturer or tutor. Craft, Dalton, and Grant (2010) substantiates this perspective by reflecting on the absence of physical interaction, due to temporal separation, which may invoke feelings of isolation in the student. Craft et al. (2010) further elaborates that this lack of interaction lessens the value of the learning experience.

11.1.1 Research Questions and Objectives

Over the years there has been an increase in the number of courses offered online as well as the number of students enrolled for these courses. However, distance learning students still face several challenges with the administrative and the learning processes throughout their studies. Distance learning institutions' lack of interaction has led to distance learners as having been characterized as having feelings of isolation (Zaborova & Markova, 2016 cited in Markova, Glazkova, & Zaborova, 2017). Mbatha (2013) recommends consistent communication and interaction, especially in the presentation of course material, in distance learning. Although learner support services exist in such institutions (face-to-face tutorials, video and satellite broadcasting, and counselling services), these services are not always effective in addressing such challenges. In view of the above, it is essential to both ascertain and identify these challenges, in order to proceed to examine possible apparatuses that can be employed in enhancing the ODL learning experience. This study will thus only focus on the learning experiences of ODL learning students and how immersive technologies can be used to address these challenges.

Various technology tools have been employed in education, with the goal of enhancing and increasing the quality of the learning experience. In the modern world, learning institutions have looked to employ technology tools that will further stimulate active learner engagement with academic content. Immersive learning technology tools (which involve the use of gaming, simulations and AR and VR) have largely been used in education in most recent years in order to provide immersive learning experiences.

This chapter aims to answer the primary research question; How can open and distance learning be enhanced using virtual and augmented reality technologies?

Three research objectives have been outlined in order to answer the primary research question. The first objective is to investigate the challenges that are currently being experienced in distance learning. Second, to determine what ICT tools are currently being used to support distance learning. Third objective, to investigate how augmented and virtual reality is currently being used to enhance higher education. A systematic literature review will be employed to address these objectives. The study builds on work that support the use of modern technologies in the enhancement of DL courses. In a study conducted by Mawn, Carrico, Charuk, Stote, and Lawrence (2011), the author states that the study is evidence that field- experiments can be incorporated in the DL science coursework. "Distance learning opportunities such as these can enable students to increase their science content knowledge while also developing scientific process skills, all while doing so on their own schedules and from varied locations (Mawn et al., 2011, p. 145)".

11.2 Methodology

This study made use of a qualitative research methodology through a systematic literature review. A total amount of 40 papers were chosen for this chapter. The Research papers analysed in this study included the following words "distance education", "immersive learning", "distance learning challenges", "practical modules distance learning", "distance learning experience", "distance learning South Africa", "distance learning technology", "virtual reality higher learning", "augmented reality higher learning". The selection of keywords was lastly followed by an inclusion and exclusion criteria.

The extraction and analysis of data was guided by the research objectives that have been set out. The study utilized a thematic analysis in order to extract and analyses data. The themes used were based on the research objectives that were outlined. These outlined objectives were classified as distance learning challenges; distance learning technology tools, and immersive learning. Within the first and second objective, three themes were identified; psychological challenges, teaching style and interaction. These themes assisted in the extracting of relevant data in the articles, addressing the objectives of the study and essentially highlighting and validating the statement of the research problem. Additionally, through the review of articles, the findings revealed another theme "Towards immersive technology use in ODL" which highlights practical ODL courses. The third objective aimed to articulate how immersive learning has been utilized in higher education while highlighting the both the positive and negative aspects of its adoption. Initially, the themes were categorized as Immersive learning challenges and immersive learning advantages. Subsequently, through further scrutiny these themes were further dissected into areas of use, learning outcomes, learner engagement, and adoption challenges.

11.3 Systematic Literature Review Thematic

This section of the chapter is in the form of a systematic literature review. The review is in the form of a thematic discussion where the most prevalent themes that were found when reviewing the literature are highlighted.

The first section will discuss distance learning, the challenges that are faced within the distance learning experience. The second section will discuss Immersive learning within higher education and subsequently the adoption challenges within immersive learning.

11.3.1 Distance Learning

Attending a traditional institution of higher learning is a privilege not all can have. The reasons may vary from person to person, these reasons are usually what ultimately leads individuals to look to distance learning as a suitable alternative for obtaining an education. Lentell (2012, p. 25) defines distance learning as “the totality of arrangements made by a university for a student cohort that is separated geographically from its teachers and services”. It generally involves the use of a range of instructional methods such as print, video, and multi way communication systems for students and instructors that are bounded by physical separation (Bitter & Gregory, 2002).

11.3.1.1 Psychological Challenges

Distance learning institutions have a diverse group of students that use various learning styles. Craft et al. (2010) cites this factor as one that may hinder the process of providing education to DL students, as DL has a tendency of employing a “one size fits all” method of teaching. Case in point aural learners may be disadvantaged as they would most likely fair better in traditional lecture rooms, while visual learners are more likely to prefer making use of the internet. Craft et al. (2010) is of the opinion that there should be a greater focus on seeking different methods of instructing course material, as the current method risks creating intellectual isolation in learners. Simpson (2013) reinforces this notion by maintaining that evidence revealed that DL instructors concentrate on the delivery of course material instead of the learning experience of students. Moreover Simpson (2013) elaborates by drawing a parallel between the aforementioned factors and the dissatisfying graduation rate in DL institutions, when contrasted with traditional institutions of higher learning. Despite these concerns, advocates for distance learning insist that DL is capable of being as efficient as traditional learning; it may even surpass it (Markova et al., 2017). Critics of distance learning Zaborova & Markova (2016) highlight a trend that occurs in learners in virtual environments, reporting that students display feelings of confusion, isolation and frustration. Distance learning students described their learning experience as being disappointing, filled with anxiety, constant tiredness, pressure, marginalization, and relegation (Gravani, 2015). Distance learning's largest hindrance is the psychological issues related to the provision of course material, knowledge attainment and general engagement between both the lecturer and the learner Koutsoupidou (2014). Additionally, Koutsoupidou (2014) further details that isolation could possibly develop into an off-putting factor for students' interest in the course and may even serve as justification for them dropping out.

11.3.1.2 Distance Learning Instruction Tools

Distance learning has been one of the main driving forces that advocated for the continued use of information and communication technologies (ICT) in institutions of higher education (Markova et al., 2017). This is evident as Koutsoupidou (2014) reports on the importance of digital tools in the command of various courses in DL. Technological devices from printed letters, radio, compact disc players, television and video to the modern internet-based learning, are various examples of how technology has always shaped the engagement between learners, educators and the academic content that is taught in ODL settings (Guri-Rosenblit, 2009). The rapid improvement of online course can be credited to technology and the internet as they have been able to replace traditional face-to-face interaction. DL is reliant on modern-day methods of instruction, which employ of technological resources as a medium of communication between lecturer and students Rodrigues et al. (2014). DL students have a vast array of technological resources at their disposal; university learning management systems, e-libraries, e-database, online- textbooks, video conferencing facilities, email and printed material (Markova et al., 2017). However, as is common in traditional institutions of learning is the same for distance learning institution, there is not as equal distribution of education therefore several of these resources are unavailable in a number of institutions. Alternatively, in other institutions teaching and the provision of course material is carried out through the use of tutorial classes, e-learning, and audio course material. DL lecturers commonly utilize online platforms for lectures, tests, exams and course materials but rarely employed these platforms for interactive training methods (Markova et al., 2017). The enhancement of the quality of student learning in DL lies with the improvement of the instruction level and also the need for innovative teaching methods instead of a replication of traditional methods. Dede (2014) maintains that the rapid growth of technology will have a considerable effect on distance learning education.

Martín-Gutiérrez, Fabiani, Benesova, Meneses, and Mora (2015) supports this by reporting there has been a keen interest amongst educators, researchers, and pedagogues on utilizing new visualization methods to enhance current education frameworks. Modern technology that currently exhibits the most potential is virtual and augmented reality (AR) (Table 11.1).

11.3.2 Immersive Learning

Immersive technology is technology that tries to imitate the real world though the use of a digital or computer-generated world. An immersive experience places a user in a simulation where the user acknowledges the virtual components as a portion of the entire world, thus actually becoming less aware that their surrounding components do not exist in the physical reality.

"Immersive technology comprises virtual reality (VR), augmented reality (AR), and mixed reality (MR), all of which merge the physical and digital worlds to create a unique customer experience. AR will naturally morph into MR, while VR will incorporate elements of MR" (Soper, 2020).

11.3.2.1 Areas of Use

The evolution of the internet prompted a surge in the use of online teaching programmes and the use of technology in education. Technology tools such as AR make it possible to interlink real-world aspects by capturing them using a camera with multimedia factors, for instance "text, images, video or three-dimensional models and animations" (Martín-Gutiérrez et al., 2015). Therefore, such technologies display prospects of possibly aiding in the challenges experienced by the DL learners.

Table 11.1 Summary of distance learning section

Main section	Subsection	Key aspects
Distance learning	Psychological challenges	DL students tend to display feelings of confusion, isolation and frustration. The psychological issues are related to the provision of course material, knowledge attainment and general engagement between both the lecturer and the learner. The "one size fits all" teaching method employed by DL educators is another contributing factor
	Distance learning instruction tools	DL institutions make use of a wide variety of instructional tools. DL students have at their disposal resources such as university learning management systems, e-libraries, e-database, online-textbooks, video conferencing facilities, email and printed material. Lecturers commonly utilize online platforms for lectures, tests, exams and course materials

Herrington, Reeves, and Oliver (2007) reported a substantial amount of various institutions as having incorporated simulations into their educational environment. Dede (2009) reported that augmented reality assists students engage in realistic exploration of the real world. Johnson, Smith, Levine, and Haywood (2010) noted that the most common use of augmented reality was to annotate spaces that already exist through an overlay of location-based information. Klopfer and Squire (2008) demonstrated, how the use of AR allowed students to encounter scientific phenomena that impossible to experience in the real world (e.g. chemical reactions). Liu et al. (2007) launched numerous AR systems that form part of this purpose; through discoveries in AR, learners were able to see the virtual solar system on the classroom table or to visualize the process of photosynthesis. Wu et al. (2013) highlights that the benefits offered by augmented reality was also realized by educators. Teachers found that the advantage of augmented reality through the use of 3D imagery and were of the belief that augmented reality made inaccessible subject matter more accessible to students.

Realistic simulations were commonly used in education settings that are deemed to have "high stakes". These sectors comprise of space training, medical education, and piloting. Rizov and Rizova (2015) substantiates this by drawing attention to the possibility for augmented reality to be beneficial when employed in various disciplines, such as medicine, education, and architecture. "Alien Contact!", a scenario-based simulation, was employed to enhance mathematical thinking capabilities (Dunleavy, Dede, & Mitchell, 2009; Mitchell, 2011). Gamification and role-play-based AR has been implemented to increase motivation and a sense of realisticness in medical science (Rosenbaum, Klopfer, & Perry, 2007). The above-mentioned is indicative of the potential that immersive learning has in the enhancing of the ODL learning experience. This additionally, is in conjunction with the argument made by Mawn et al. (2011) which emphasized the necessity to not only enhance theoretical knowledge but to also hone the learning of practical skills. The use of AR in education may be relatively new; however, there are several AR applications that may be utilized in various learning contexts.

Virtual reality similarly to augmented reality, has been employed in various educational contexts. Flint and Stewart (2010) implemented a virtual exercise in a microbiology course. From the exercise it was deduced that students benefited from the virtual exercise, the exercise also met the outlined objectives additionally, it was relatively inexpensive for the university. Despite these findings, Flint and Stewart (2010) did not endorse the use of virtual laboratories in isolation as a replacement for traditional laboratories. Within their findings they placed a large emphasis on student's prior knowledge of laboratory techniques.

The use of virtual laboratories to supplement learning has been studied by several other researchers. Dalgarno, Bishop, Adlong, and Bedgood Jr (2009) proposed a plan to familiarize distance chemistry learners with a chemistry laboratory. Students supported this plan, stating it would aid in their comprehension of a traditional laboratory setting. A more unique direction was taken by Koretsky, Kelly, and Gummer (2011) as engineering students were exposed to virtual laboratory settings and practices that were reported to portray industry more realistically than the traditional laboratory could. The varying factor in the study conducted by Koretsky et al. (2011), was that it was not accompanied by a physical laboratory counterpart. Koretsky et al. (2011, p. 567) concluded that "virtual laboratories can facilitate a broader experience for students". Koretsky et al. (2011) further elaborated that the different types of experiments (physical vs. virtual) guide a students' perception of their learning towards various factors, e.g. "laboratory procedures in traditional laboratories versus critical thinking and higher-order cognition in virtual laboratories".

11.3.2.2 Learning Outcomes

Learning outcomes form the basis of the knowledge and skills that students should acquire after the completion of a module. These outcomes are pertinent in interpreting whether teaching methods have been effective. A study conducted by Rizov and Rizova (2015) yielded positive results in the use of augmented reality as an instruction tool. Rizov and Rizova (2015) reported that it was beneficial to utilize technology tools that portray the modern world because it aided in maintaining student's interests. Herrington et al. (2007, p. 13) reported "the task is the crucial component of immersion and engagement in higher-order learning". When appropriate technologies are employed as cognitive instruments to discover solutions to difficult problems, the learning obligation is shifted to the student as opposed to the engineer of the virtual world (Herrington et al., 2007). Herrington et al. (2007) considers this learning method to be reflective. Therefore, the use of virtual worlds could prove to be useful in tackling the DL challenge stressed by Simpson (2013); of educators concentrating on the distribution of content as opposed to the learning experience of students. It is crucial to view technology tools as an aid to content delivery, learner engagement and meeting learning outcomes as Lee, Sergueeva, Catangui, and Kandaurova (2017) reports in instances where virtual reality mainly is mainly utilized for the distribution of content, only a minor benefit exists in terms of content absorption. Lee et al. (2017) further details that despite this, virtual reality should not be view as being ineffective.

Martín-Gutiérrez et al. (2015) undertook a study that centred on the following teaching contexts; the use of electrical equipment in the laboratory, examination and interpretation of illustrations for reviewing installations, and autonomous learning of course work. Motivated students were reported to have had a positive academic performance. The positive outcomes and feedback from students from the use of new technology correlated with a study by Leszczyński et al. (2017) of an emergency medicine DL course. Leszczyński et al. (2017) found that students were excited about the prospect of employing of modern methods of education, they valued the substantive quality and innovation offered by these new tools and were able to swiftly adapt to the new technologies.

The use of immersive technology in education has been found to enrich the skills that students are expected to obtain. Patiar, Ma, Kensbock, and Cox (2017) study asserted that the VFT (Virtual Field Trip) enabled learners to improve both their cognitive and attitudinal skills. Moreover, the VFT tool aided learners' personal development and presented opportunities for independent knowledge seekers by altering learners' experience of reality. Webster (2014) examined a US Army traditional lecture supplemented with immersive virtual reality-based multimedia teaching, in terms of attaining declarative knowledge. The findings indicated that the VR tool could offer high time-compressed training, it was also able to be customized to the learners' knowledge level, and could allow frequent repetitions that were needed in order to improve mastery (Webster, 2014). Immersive technologies be viewed as a suitable measure to address a DL experience challenge reference by Craft et al. (2010) where a "one size fits all" teaching approach is utilized by educators, which results in a negative impact in a student' ability to grasp information. Rizov and Rizova (2015) corroborated this as the authors reported that educators expressed that AR succeeded in decreasing the amount of time spent attempting to assist students in assimilating information.

11.3.2.3 Learner Engagement

Learner engagement is necessary in the efficient delivery of the academic content because students who take in interest in the course material tend to find in the learning process beneficial. Rizov and Rizova (2015) performed a test on learners based on their knowledge of engineering graphics after having used AR aides. The findings revealed that after educators utilized augmented reality in aiding with the exhibiting of space objects, a growth in the results could be seen. Furthermore, Rizov and Rizova (2015) noted a benefit that physical interaction between students and the application contributed further to the learning of various geometric shapes in the cosmos. Contrastingly a study conducted by Dyrberg et al. (2017, p. 358) aimed to examine "a pilot study on student attitude, motivation and self-efficacy when using the virtual laboratory programme Labster". Dyrberg et al. (2017) elaborated that learners found less value in interacting in the virtual laboratory in comparison to participating in a traditional laboratory. Dyrberg et al. (2017) reported that students deemed the cases to be less engaging, motivating and useful in contrast to performing real-life laboratory work. This also contradicts Flint and Stewart (2010) whose study found that learners in a microbiology course experienced increased enjoyment with a virtual laboratory as opposed to a traditional laboratory. The results by Dyrberg et al. (2017) do, however, support the findings by Dalgarno et al. (2009) who reported that virtual laboratories could be employed when familiarizing students with virtual laboratory before the official lectures. Furthermore, Dyrberg et al. (2017) reported that students had notably greater self-confidence when performing laboratory experiments.

11.3.2.4 Adoption Challenges

Differing opinions exist amongst researchers regarding the practicality of employing immersive technologies in education. Despite numerous positive reviews, a select amount of researcher still remain doubtful of immersive learning. However, their uncertainty is not unfounded as Saleem et al. (2017) drew attention to a significant health issues that related to the extended use of these devices (immersive and wearable devices). Saleem et al. (2017) specified that the health issues are linked these devices were to skin allergies, rashes, etc. furthermore the weight of the device had also been a cause for concern since it was not considered to be of a size that it could easily carried and be mobile. The cost of immersive learning simulations was significantly high, therefore resulting in their limited use (Herrington et al., 2007).

Conversely, Martín-Gutiérrez et al. (2015) argued for augmented reality's affordability and its ability to present course work in more interesting manner learners. Saleem et al. (2017) The IDC (a firm providing global market intelligence) estimated in 2014 that the need for immersive and wearable devices would increase up to 112 million units by 2018, with Google glasses being the most expensive device ranging around 1500$ (Saleem et al., 2017). Saleem et al. (2017) particular countries would view these as affordable prices, however, it would be possible that the rest of the world the rest of the world may not be able to afford these prices, particularly developing countries.

Access could possibly be another major drawback in the application of virtual worlds in the classroom (Ellaway, Dewhurst, & Cumming, 2003). Ellaway et al. (2003) expands on this claim expressing that although the internet has become prevalent in all parts of modern-day society, there are notable obstacles that could impact the potential users when gaining access to the system, e.g. the lack of networked computing equipment. Ellaway et al. (2003) cited staff development as an additional restriction to the application of a virtual environment in education. In order to facilitate system development, the staff will require technical training on the system. Saleem et al. (2017) expounded on access in the context of connectivity as the research revealed that the devices need a constant data update that will also require an Internet connection. This presents a challenge especially in developing and several developed countries due to the countries' inability to supply all its citizens with adequate Internet. Saleem et al. (2017, p. 692) reports that in 2013 "a mere 42% of the global population had internet access. When narrowed down to the continents 13% of the population in North America had no access, 28% in Australia, 30% in Europe, 48% in Latin America, 52% in the Middle East, 64% in Asia, and 74% in Africa still did not have access to the Internet".

Bearing these limitations in mind, researchers have still found the use of immersive learning in higher education learning mainly in courses that require practical engagement, as considerably beneficial in improving the learning experience (Baukal, 2010; Güven, 2014; Mawn et al., 2011). Ellaway et al. (2003) nonetheless, expressed that even though virtual learning environments had the capacity to provide exceptional support for the complex, dispersed and integrated practices of medical education, investment, community commitment, support by stakeholders would continue to be the most significant hindrance to implementation. Without these factors, virtual learning environments would be inclined to linger on the periphery, to only be utilized by enthusiasts and early adopters. Porter, Graham, Bodily, and Sandberg (2016) cited the adoption innovation patterns as a factor that could either hinder or facilitate the implementation of an immersive environment in higher education. Porter et al. (2016) provided a rationale on how to positively influence users in order to obtain the latter outcome. Innovators and early adopters could be influenced through the provision of infrastructure and assistance, and in addition to clarifying the justification for implementing blended learning (Porter et al., 2016). The early majority was reported to be influenced by evaluation data. The conclusion of the report on the late majority suggested that adequate training and support in a safe environment would result in their support (Table 11.2).

11.4 Discussion

Cornelius, Medyckyj-Scott, Forrest, Williams, and Mackaness (2008) stated that in the modern world, it would be rare to find a "fulltime student". Instructors were aware of the unavoidable need to meet the modern student's needs; students who work part-time, study from the convenience of their house, and or in their workplace, residence halls, and also those who fall in the gap of being unable to access computing resources. This description supported the statement by Rodrigues et al.

Table 11.2 Summary of immersive learning section

Main section	Subsection	Key aspects
Immersive learning	Areas of use	Immersive technologies such as AR and VR have been employed in various learning areas. It has been highlighted that they have added the most benefit in areas that require practical learning. Areas such as space training, medical education, and piloting
	Learning outcomes	Learning outcomes are pertinent in interpreting whether teaching methods have been effective. Research indicated that immersive learning was effective in maintaining student's interest in coursework as the visual element it adds allows students to physical views some theoretical concepts
	Learner engagement	There are conflicting views on whether immersive technology tools are able to increase learner engagement with their course work. Some students have been found to find virtual simulations beneficial whilst others preferred to use the real-life tools
	Adoption challenges	There are several significant factors that affect the adoption of AR and VR in higher education learning. Some of these were health factors cited with prolonged use of the devices, the high cost associated of purchasing the devices which results in lower demand, and the accessibility of these devices especially in third world countries where a significant amount of citizens do not have internet

(2014) detailing why individuals became distance learners. Despite the advantages presented by distance learners, the distance learners were confronted with numerous challenges that affected their learning experience (Gravani, 2015; Koutsoupidou, 2014; Markova et al., 2017) these challenges included psychological challenges; feelings of isolation, anxiety marginalization, a lack of efficient in-depth content delivery, physical interaction when having a dialogue, rigid course materials and learning methods. Although there were several technology tools such as; e-libraries, e-database, online-textbooks, video conferencing and email (Markova et al., 2017), learners were still found to experience these challenges.

Cornelius et al. (2008) affirmed that a virtual environment presented an opportunity for educators to meet the needs of the modern student while, at the same time, affording students with several, not all, of the components of a real-life environmental experience. Although a virtual environment may not have been able to compare to a real visit to space, a national park, an aquarium, the amazon or a historical site, it did offer students an opportunity to obtain a realistic perspective with reference to the module that they are learning. AR and VR have exhibited several benefits that may assist in enhancing the distance learning experience. The majority of the studies (Rizov & Rizova, 2015; Herrington et al., 2007; Martín-Gutiérrez et al., 2015; Patiar et al., 2017; Webster, 2014) indicated that the adoption of immersive learning in higher education fostered an increase in learner engagement, a better assimilation of course content and a better development of practical skills. Bower, Howe, McCredie, Robinson, and Grover (2014) argued that the use of virtual reality may be of more use to educational courses that rely on the development of practical skills as opposed to modules that were based on content absorption. Studies which focused on courses in ODL that required the development of practical skills (Baukal, 2010; Güven, 2014; Mawn et al., 2011), substantiated the argument by Bower et al. (2014).

The most significant factors that may have potentially impeded the adoption and application of immersive technology in ODL are funding, health factors and accessibility or a lack thereof, in both developed and developing countries. However, recent advancements in mobile computing and operational performance had brought about an increased allocation of resources to the development of mobile AR systems (Bower et al., 2014; Johnson et al., 2010). Augmented reality had, therefore, become more commonly available to the general public, in comparison to previously, when it had been exclusively located in high-end laboratory research and industry. One of the challenges faced by distance learning students is the feeling of isolation, which stems from a lack of interaction and a "blanket" teaching style. Bower et al. (2014) maintained that the more intelligent and advanced augmented reality became, it would possess the ability to alert lecturers and tutors to a pupil's learning needs, possible behavioural concerns, and recommend an applicable course of action in real-time.

It is essential to note that the studies that made use of AR and VR in higher education did not implement immersive tools in isolation. Singh, O'Donoghue, and Betts (2002) maintained that the augmentation of virtuality in higher education learning must be viewed as a tool to enrich teaching and learning and not as a complete replacement of traditional learning. Cornelius et al. (2008) substantiated this notion stating that virtual environments should only be considered as complements or supplements to the existing teaching methods and not as a replacement. This statement also satisfied the main objective of this study which is to answer the primary research question; "How can open and distance learning be enhanced using virtual and augmented reality technologies?" Additionally, Ellaway et al. (2003) reported that when aiming to take full advantage of the educational learning outcomes of immersive learning, it would be essential that the virtual learning environment is aligned with the practices and outcomes of the course. The course must inform the VLEs (Virtual Learning Environments) and VLEs need to be modifiable to meet course outcomes and not to modify them to meet VLE limitations.

11.5 Conclusion

This study investigated how immersive technologies (AR and VR), may be used to enhance the distance learning experience. The study presented a mini systematic literature review in which a thematic discussion was implemented. The themes for the thematic discussion were informed by the three research objectives which were (1) to investigate the challenges that are currently being experienced in distance learning (2) to determine what ICT tools are currently being used to support distance learning (3), to investigate how AR and VR is currently being used to enhance higher education. Literature findings revealed that the most predominant challenges amongst ODL students were psychological challenges, feelings of isolation, anxiety marginalization, a lack of efficient in-depth content delivery, rigid course materials and learning methods. Learners still experienced challenges with the ODL learning experience regardless of having had technology tools that are currently being employed in ODL (e-libraries, e-database, online-textbooks, and video conferencing) to improve the learning experience. Additionally, the literature revealed that there has been several uses of AR and VR in higher education, however, it is more beneficial when employed in courses that require the development of practical skills. Various authors noted that although immersive learning presents various advantages in the learning environment, AR and VR should not be used as a replacement for traditional teaching but rather as a supplement. AR and VR were found to have contributed to learner engagement and enhanced learning outcomes but there are still several significant challenges that may still impede the adoption of immersive learning in ODL. The outlined aims of the study were achieved however, in order to fully answer the primary research question, future studies should be focused on how to address the adoption challenges of AR and VR, especially in developing countries.

References

Al-Arimi, A. M. A. K. (2014). Distance learning. *Procedia-Social and Behavioral Sciences, 152*, 82–88.

Baukal, C. E. (2010). Continuing engineering education through distance learning. *European Journal of Engineering Education, 35*(2), 225–233.

Bitter, J. A., & Gregory, R. J. (2002). Distance learning. *Journal of Social Sciences, 6*(2), 77–83.

Bower, M., Howe, C., McCredie, N., Robinson, A., & Grover, D. (2014). Augmented reality in education–cases, places and potentials. *Educational Media International, 51*(1), 1–15.

Craft, N., Dalton, A., & Grant, M. (2010). Overcoming isolation in distance learning: Building a learning community through time and space. *Journal for Education in the Built Environment, 5*(1), 27–64.

Cornelius, S., Medyckyj-Scott, D., Forrest, D., Williams, A., & Mackaness, W. (2008). The virtual placement: An alternative to the traditional work placement in the geographical sciences? *Journal of Geography in Higher Education, 32*(2), 287–302.

Dalgarno, B., Bishop, A. G., Adlong, W., & Bedgood, D. R., Jr. (2009). Effectiveness of a virtual laboratory as a preparatory resource for distance education chemistry students. *Computers & Education, 53*(3), 853–865.

Dede, C. (2009). Immersive interfaces for engagement and learning. *Science, 323*(5910), 66–69.

Dede, C. J. (2014). The evolution of distance learning. *Journal of Research on Computing in Education, 22*(3), 247–264. https://doi.org/10.1080/08886504.1990.10781919

Dunleavy, M., Dede, C., & Mitchell, R. (2009). Affordances and limitations of immersive participatory augmented reality simulations for teaching and learning. *Journal of Science Education and Technology, 18*(1), 7–22.

Dyrberg, N. R., Treusch, A. H., & Wiegand, C. (2017). Virtual laboratories in science education: Students' motivation and experiences in two tertiary biology courses. *Journal of Biological Education, 51*(4), 358–374.

Ellaway, R., Dewhurst, D., & Cumming, A. (2003). Managing and supporting medical education with a virtual learning environment: The Edinburgh electronic medical curriculum. *Medical Teacher, 25*(4), 372–380.

Flint, S., & Stewart, T. (2010). Food microbiology—Design and testing of a virtual laboratory exercise. *Journal of Food Science Education, 9*(4), 84–89.

Gravani, M. N. (2015). Adult learning in a distance education context: Theoretical and methodological challenges. *International Journal of Lifelong Education, 34*(2), 172–193.

Guri-Rosenblit, S. (2009). Distance education in the digital age: Common misconceptions and challenging tasks. *Journal of Distance Education, 23*(2), 105–122.

Güven, M. (2014). Distance learning as an effective tool for medical interpreting training in Turkey. *Open Learning: The Journal of Open, Distance and e-Learning, 29*(2), 116–130.

Herrington, J., Reeves, T. C., & Oliver, R. (2007). Immersive learning technologies: Realism and online authentic learning. *Journal of Computing in Higher Education, 19*(1), 80–99.

Johnson, L., Smith, R., Levine, A., & Haywood, K. (2010). *The 2010 horizon report: Australia—New Zealand edition*. Austin, TX: T. N. M. Consortium.

Klopfer, E., & Squire, K. (2008). Environmental Detectives—the development of an augmented reality platform for environmental simulations. *Educational technology research and development, 56*(2), 203–228.

Koretsky, M., Kelly, C., & Gummer, E. (2011). Student perceptions of learning in the laboratory: Comparison of industrially situated virtual laboratories to capstone physical laboratories. *Journal of Engineering Education, 100*(3), 540–573.

Koutsoupidou, T. (2014). Online distance learning and music training: Benefits, drawbacks and challenges. *Open Learning: The Journal of Open, Distance and e-Learning, 29*(3), 243–255.

Lee, S. H., Sergueeva, K., Catangui, M., & Kandaurova, M. (2017). Assessing Google cardboard virtual reality as a content delivery system in business classrooms. *Journal of Education for Business, 92*(4), 153–160.

Lentell, H. (2012). Distance learning in British universities: Is it possible? *Open Learning: The Journal of Open, Distance and e-learning, 27*(1), 23–36.

Leszczyński, P., Charuta, A., Gotlib, J., Kołodziejczak, B., Roszak, M., & Zacharuk, T. (2017). Distance learning methods in continuing education of paramedics. *Studies in Logic, Grammar and Rhetoric, 51*(1), 53–70.

Liu, T. Y., Tan, T. H., & Chu, Y. L. (2007). 2D barcode and augmented reality supported English learning system. In 6th IEEE/ACIS International Conference on Computer and Information Science (ICIS 2007) (pp. 5–10). IEEE.

Martín-Gutiérrez, J., Fabiani, P., Benesova, W., Meneses, M. D., & Mora, C. E. (2015). Augmented reality to promote collaborative and autonomous learning in higher education. *Computers in Human Behavior, 51*, 752–761.

Markova, T., Glazkova, I., & Zaborova, E. (2017). Quality issues of online distance learning. *Procedia-Social and Behavioral Sciences, 182*, 685–691.

Mawn, M. V., Carrico, P., Charuk, K., Stote, K. S., & Lawrence, B. (2011). Hands-on and online: Scientific explorations through distance learning. *Open Learning, 26*(2), 135–146.

Mbatha, B. (2013). Beyond Distance and Time Constrictions: Web 2.0 Approaches in Open Distance Learning, the Case of the University of South Africa (UNISA). Mediterranean Journal of Social Sciences

Mitchell, R. (2011). Alien contact!: Exploring teacher implementation of an augmented reality curricular unit. *Journal of Computers in Mathematics and Science Teaching, 30*(3), 271–302.

Patiar, A., Ma, E., Kensbock, S., & Cox, R. (2017). Students' perceptions of quality and satisfaction with virtual field trips of hotels. *Journal of Hospitality and Tourism Management, 31*, 134–141.

Porter, W. W., Graham, C. R., Bodily, R. G., & Sandberg, D. S. (2016). A qualitative analysis of institutional drivers and barriers to blended learning adoption in higher education. *The Internet and Higher Education, 28*, 17–27.

Pozdnyakova, O., & Pozdnyakov, A. (2017). Adult students' problems in the distance learning. *Procedia Engineering, 178*, 243–248.

Rizov, D., & Rizova, D. (2015). Augmented reality as a teaching tool in higher education. *International Journal of Cognitive Research in Science, Engineering Education, 3*(1), 7–16.

Rodrigues, S. J., Affonso, S. A., Quinelato, E., & Montiel, J. M. (2014). Distance learning in undergraduate education: The challenges of building a collaborative environment. *Procedia-Social and Behavioral Sciences, 116*, 3499–3501.

Rosenbaum, E., Klopfer, E., & Perry, J. (2007). On location learning: Authentic applied science with networked augmented realities. *Journal of Science Education and Technology, 16*(1), 31–45.

Saleem, K., Shahzad, B., Orgun, M.A., Al-Muhtadi, J., Rodrigues, J.J. & Zakariah, M. (2017). Design and deployment challenges in immersive and wearable technologies. *Behaviour & Information Technology, 36*(7), 687–698.

Simpson, M., & Anderson, B. (2012). History and heritage in open, flexible and distance education. *Journal of Open, Flexible, and Distance Learning, 16*(2), 1–10.

Simpson, O. (2013). Student retention in distance education: Are we failing our students? *Open Learning: The Journal of Open, Distance and e-Learning, 28*(2), 105–119.

Singh, G., O'Donoghue, J., & Betts, C. (2002). A UK study into the potential effects of virtual education: Does online learning spell an end for on-campus learning? *Behaviour & Information Technology, 21*(3), 223–229.

Soper, J. (2020). Jump right. In 7 ways immersive technology can transform business. Retrieved June 2020, from https://cmo.adobe.com/articles/2017/4/immersive-technology-.html#gs.8n8do8

Unisa.ac.za. (2018). Open Distance Learning (ODL) is a different way of learning. [online] Available at: http://www.unisa.ac.za/sites/corporate/default/Apply-for-admission/Honours-degrees-&-postgraduate-diplomas/ODL,-Unisa-and-postgraduate-studies/Open-Distance-Learning-(ODL)-is-a-different-way-of-learning

van Antwerpen, S. (2016). The quality of teaching and learning of BCom honours degree students at an open distance learning university in South Africa. *Africa Education Review 12*(4):680–695

Webster, R. (2014). Declarative knowledge acquisition in immersive virtual learning environments. *Interactive Learning Environments, 24*(6),1319–1333.

Wu, H. K., Lee, S. W. Y., Chang, H. Y., & Liang, J. C. (2013). Current status, opportunities and challenges of augmented reality in education. *Computers & Education, 62,* 41–49.

Zaborova, E. N., & Markova, T. L. (2016). February. Students as social actors of virtual educational environment. In Actual Issues of Sociology of Culture, Education, Youth and Management: Materials of the All-Russian Scientific Conference with international participation (pp. 392–397).

Chapter 12
Technological, Organisational and Socio-Interactional Affordances in Simulation-Based Collaborative Learning

Kirsi Lainema, Timo Lainema, Kirsi Heinonen, and Raija Hämäläinen

12.1 Introduction

The use of different learning technologies has rapidly increased on all educational fields, and along with it, the study of affordances has gained momentum. The concept 'affordance' was first advanced by Gibson (1979, cited in Salomon, 1993), who used it to refer to the functional properties that determine the possible utility of an object or an environment. Affordances can be defined as action possibilities latent in the object (the learning environment) and dependent on the capabilities of the agent (learners) (Antonenko, Dawson, & Sahay, 2017). Affordances are more than just technical properties of an object, as they represent an action potential that needs to be met with the respective capabilities of the user. A chair represents an everyday case in point. A chair's affordance is its sit-ability, and that it can be used for that purpose by a person who wants and is able to sit. In the context of a digital learning environment, affordances include e.g. view-ability, read-ability and move-ability.

The technological aspects have for long dominated research on digital learning environments, and the socio-interactional and organisational aspects have received less attention. However, we find that it is mandatory to gauge the interplay of the three types of affordances in more detail and develop a better understanding of how digital learning environments can be designed and applied to empower students to utilise their full capacity and all available resources. In doing so, we also re-consider the notion of affordances as resources for computer-supported collaborative learning

K. Lainema (✉) · K. Heinonen · R. Hämäläinen
University of Jyväskylä, Jyväskylä, Finland
e-mail: kirsi.k.lainema@jyu.fi

T. Lainema
University of Turku, Turku, Finland

D. Ifenthaler et al. (eds.), *Balancing the Tension between Digital Technologies and Learning Sciences*, Cognition and Exploratory Learning in the Digital Age,
https://doi.org/10.1007/978-3-030-65657-7_12

(CSCL) (Koschmann, 2012; Park & Song, 2015*; Wang, Fang, & Gu, 2020; Berthelsen & Tannert, 2020).

Analysis of the applicability and usefulness of a technology requires an evaluation of how the affordances of the technology respond to the users' needs and abilities (Antonenko et al., 2017). Some examples of the affordances of social software tools are connectivity and social rapport, and collaborative information discovery and sharing (McLoughlin & Lee, 2007). As an example of the first, technology-based environments support networks of people and facilitate connections between them. These kinds of environments are representatives of what Gee (2004) calls affinity spaces, where people acquire social and communicative skills, and at the same time become engaged in the participatory culture of the environment. In these spaces, learners engage in informal learning and in creative, expressive forms of behaviour and identity seeking while developing a range of digital literacies. One cannot assume that just because social software entails certain affordances, that is all that is required for effective learning. Careful planning and a thorough understanding of the dynamics of these affordances are mandatory (Ke, Pachman, & Dai, 2020; McLoughlin & Lee, 2007). An explicit approach to identifying technological affordances of e-learning tools and the affordance requirements of e-learning tasks should be used to scaffold the learning design process (Bower, 2008; Lin, Hou, & Chang, 2020).

Traditionally, a socio-culturally oriented research perspective on computer-supported collaborative learning (CSCL) is closely associated with affordances (Moate, Hulse, Jahnke, & Owens, 2019). The focus is on group learning and the social context in which collaboration emerges. This presentation is in line with the notion from Arvaja, Salovaara, Häkkinen, and Järvelä (2007), who viewed collaboration as shared knowledge construction, where participants not only cumulatively share knowledge, but where the knowledge construction is jointly built on others' ideas and thoughts (see also Mercer, 2010). The aim is that the activities of the collaborative group are not a collection of individual activities but rather interdependent group processes (e.g. interactions) pursuing a shared conception of a problem (Lin et al., 2020; Roschelle & Teasley, 1995). Furthermore, these shared processes are mediated by the community and social context in which the group work takes place (Stahl, 2012; Stahl & Hakkarainen, 2020).

We argue that the continually increasing amount of resources allocated to the development of educational simulations by educational institutions calls for in-depth studies of affordances. We need to understand how simulations and games can be designed in pedagogically sound ways to empower users to acknowledge the affordances embedded in these environments. Furthermore, we believe that the use students make of online learning environments will very much depend on their attitudes towards these environments and on the perceived affordances. This is also the motivation of our study.

12.2 Affordances

Research of affordances is interdisciplinary, and while it originates from Ecological Psychology (Gibson, 1979), it has found application in Education (Berthelsen & Tannert, 2020; Kirschner, 2002; Kreijns, Kirschner, & Jochems, 2003; Wang et al., 2020; Xue & Churchill, 2019), Information Systems (Majchrzak & Markus, 2012), Organisation Studies and Management disciplines (Baralou & Tsoukas, 2015; Pozzi, Pigni, & Vitari, 2013).

We examine the affordances of a digital learning environment perceived and utilised by dispersed student teams. The novelty of our study lies in incorporating not only the technological aspects of affordances in our analysis but also the socio-interactional and organisational dimensions in our treatment. The use of socio-interactional and organisational affordances plays a key role in how these technologies can be made to work.

As learners engage in a technology-based learning environment, they perceive 'affordances' of objects, defined as the 'acts or behaviors that are afforded or permitted by an object, place, or event' (Michaels & Carello, 1981, p. 17). Affordances are, thus, different from the properties of objects. Affordances are the perceptions of what we can do with the properties of objects. Although affordances can be perceived as preconditions for an activity, they do not imply that a specific activity will occur (Dohn, 2009; Greeno, 1994). As affordances are merely potentials for action, benefitting from them requires that they are triggered (Volkoff & Strong, 2013) or actualised (Strong et al., 2014). Pozzi et al. (2013) recognised four steps in the application of affordances: an affordance exists, the user perceives the affordance, the user actualises the affordance and the actualisation leads to the affordance effect. In our study, we focus predominantly on affordances as the doings in which the actors engage. In doing so, we follow Majchrzak and Markus (2012), who noted that affordances are best phrased in terms of action verbs or gerunds, such as 'share knowledge' or 'information sharing' and involve technological, organisational and social dimensions. McLoughlin and Lee's (2007) list of potential sources of affordances in Pedagogy 2.0 environments (based on the technological possibilities of Web 2.0) allows us to identify various dimensions that create possibilities for affordances in the simulation gaming context:

- **Content**: Simulation gaming-based learning is learner-centred in that the students use the information in the simulation environment to generate analysis and decisions by communicating and collaborating with peers. Thus, students create, share and revise ideas and turn them into actions in the environment. The environment is open to negotiation, and learner input is often inter-disciplinary in nature and blends formal and informal learning.
- **Communication**: Communication is open, peer-to-peer and multi-faceted, and uses multiple media types to achieve relevance and clarity.
- **Process**: Situated, reflective, integrated thinking processes are iterative, dynamic and enquiry-based. Learning tasks can be authentic in nature, learner-driven and designed to enhance experiential learning and enable multiple perspectives.

- **Resources**: Multiple informal and formal sources that are media-rich and often global in reach.
- **Scaffolds**: Support for students comes from a network of peers, teachers, and the learning community.

In our case simulation, the affordances of virtual gaming include the fidelity of a real-world business environment, the ability of remote team members to talk face-to-face with each other in real-time, and the illustration of causal business operations as dynamic processes. These affordances foster the development of a participatory culture with genuine involvement and communication in which the participants are socially connected with one another. Virtual simulation gaming makes use of the affordances of the software tools that enable connectivity, communication, participation and the development of dynamic communities of learning. Affordances in these kinds of environments may stem from many sources. Simulation games can be used to facilitate:

- experiential learning tasks that would be impractical or impossible to undertake in the real world,
- learning tasks that lead to increased intrinsic motivation and engagement,
- learning tasks that lead to improved transfer of knowledge and skills to real situations through contextualisation of learning and
- tasks that lead to richer and/or more effective collaborative learning than is possible in other learning environments.

Next, we discuss the technological and socio-interactional affordances in light of the research literature.

12.2.1 Technological Affordances

The concept of technology affordance refers to an action potential—what an individual or organisation with a particular purpose can do with technology (Majchrzak & Markus, 2012). Technological affordances are not isolated from other issues present in the learning situation and environment. Technologies do not directly cause learning to occur but can afford certain learning tasks that may result in learning or give rise to certain learning benefits. Selwyn (2012) argued that there cannot be a predetermined outcome to the implementation of technologies. Instead, technologies are subjected continually to a series of complex interactions and negotiations with the social and cultural contexts into which they emerge. Steffens et al. (2015) agreed with Selwyn's (2012), stating that techno-centrism focuses too much on the objective capabilities of the technology and too little on the social and contextual aspects of the learning situation. The principal advantage of the more socially nuanced theoretical approach is its capacity for developing a more socially grounded understanding of the realities of educational technology (Selwyn, 2012).

The issue of categorising technological affordances and aligning them with the abilities they afford the users of the technology is seen as essential for analysing the potential utility of educational technologies (Antonenko et al., 2017). Taxonomies and categorisations of affordances in technology-supported learning environments include technological affordances such as accessibility, speed of change, diversity, communication and collaboration, reflexivity, multi-modality and non-linearity, risk, fragility and uncertainty, immediacy, monopolisation and surveillance (Conole & Dyke, 2004); multimodal auditive, linguistic, visual, gestural and spatial affordances presented by composition software (Gall & Breeze, 2007); accessibility, entertainment exchange, information repositories in asynchronous video conferencing tools (Krauskopf, Zahn, & Hesse, 2012); and multimedia access and collection, communication, and representation of thought and knowledge in handheld devices (Song, 2011). Alahuhta, Nordbäck, Sivunen, and Surakka (2014) identified affordances related to enhancing team creativity in virtual worlds. The identified affordances included avatars as graphic self-representations, changing the frame of reference, co-presence, immersion, multimodality, rich visual information, simulation capabilities and supporting tools for creative work. The above taxonomies have some overlap, but they remain context-specific and are therefore of marginal assistance in the pursuit for a more robust tool for analysing technological affordances in collaborative computer-based learning environments. In an attempt to describe affordances based on their physical characteristics and emphasise their functionality, Bower (2008) proposed a methodology for matching the affordance requirements of learning tasks with the technological affordances of ICT tools. Bower's (2008) affordance classification system includes 11 different areas of technological affordances (see Table 12.1). In this paper, we focus predominantly on the actual action possibilities perceived and utilised by the users (usability). Moreover, we analyse the dynamics of technological, organisational and socio-interactional affordances and their combinations as reported by the learners. Expanding our treatment from a predominantly technological view allows for an appreciation of the interplay between and among the various types of affordances.

12.2.2 Organisational Affordances

When analysing the affordances of technological environments, we can and should go beyond the technologies. Scarantino (2003) categorised affordances as being mental, basic physical or non-basic physical. Hartson (2003) recognised cognitive affordances, physical affordances, sensory affordances and functional affordances. However, Kennewell (2001) noted that ICT is just one component of the setting, although it is particularly important because of the special features it can bring to learning, such as access to information and immediate feedback to the learner.

An important aspect particularly in collaborative learning environments is the organisation and coordination of the learners' joint activities. In the context of computer-based learning to organise, collaboration can be defined as a set of

Table 12.1 Identified technological affordances (adapted from Bower, 2008)

Technical affordance	Explanation	Example of how shows in student essays
Media affordances	Input and output	*Being able to call and interact/discuss with other team members while the game is running was what made the game so alive and exciting* (Team 4)
Spatial affordances	Ability to resize, move and place elements	
Temporal affordances	Access anytime anywhere, synchronous versus asynchronous	*I decided to join my team in the morning, at 05:00 UTC+0. At this time a few of my team members already worked on (−).* (Team 12)
Navigation affordances	Capacity to browse other sections of a resource and move back/forward	*(−) A chat with all the team members inside the programme can help us to converse easier because now we had to handle two programmes at the same time.* (Team 3)
Emphasis affordances	Capacity to highlight aspects of resources	*(−) We started tutoring new people by explaining everything that we were doing loudly and showing it directly in the simulation.* (Team 3)
Synthesis affordances	Capacity to combine multiple tools to create a mixed media environment	*We mainly discussed by writing but also had Skype call(s) (−). I prefer not to speak English so writing was ideal for me* (Team 1)
Access-control affordances	Capacity to allow or deny who can operate; capacity to support one–one/many–many contributions	*I (−) expected that production, inventory and sales would've been spread into different pages making it all faster to control and not (−) waiting someone to finish their own tasks* (Team 1)
Technical affordances	Capacity to be used on various platforms, ability to adapt to bandwidth, efficiency of tools	*In my shift was one girl who had a bad internet connection and therefore she couldn't take part in our Skype- conference* (Team 3)
Usability	Intuitiveness, ease of manipulating a tool	*The game was easy to get into (−)* (Team 2)
Reliability	Robustness	*If the game cut out it would automatically re-connect (−)* (Team 4)

activities related to collaborating, discussing or asking questions about the learning tasks (Lampe, Wohn, Vitak, Ellison, & Wash, 2011). In digital and virtual teamwork, organising is an elementary activity which is needed to ensure that collaboration proceeds smoothly and effectively (Hernández-Sellés, Muñoz-Carril, & González-Sanmamed, 2019; Hossain & Wigand, 2004). Thus, the way the teamwork is organised has a fundamental influence on what is learned during the exercise. At first glance, organisational affordances may seem a fuzzy category, as it is sometimes difficult to discern which actions can be understood as organising. In our context, organising is regarded as assembling the available resources to attain order, structure and organisational objectives (BusinessDictionary, 2019). In addition to assembling resources to attain certain objectives, organisational affordances also entail managing the process and the participants, which is accomplished through communication; thus, organisational and socio-interactional affordances are closely linked and intertwined. The organisational affordances may therefore guide students in organising their personal and team efforts and actions and help them achieve the goals of the course. Organisational affordances may also steer students in their choices of technologies and communication tools.

12.2.3 Socio-Interactional Affordances

Social interaction in CSCL environments needs to be purposefully facilitated via technological solutions that allow for synchronous and/or asynchronous communication between the learners. Kreijns, Kirschner and Vermeulen (2013) found that the properties of the CSCL environment function as facilitators for the learners' social interaction and, thus, work as socio-interactional devices. In CSCL environments, collaboration and learning depend largely on interaction, that is, on written and oral communication, information sharing and knowledge creation.

It is generally assumed that participants in a CSCL environment will interact because the environment makes it possible (Kreijns et al., 2003). However, the nature of CSCL environments makes them more salient and critical in respect to social interaction than face-to-face settings. Computer-based collaborative learning environments depend on technology for mediating interactions among students and teachers, whereas face-to-face learning situations provide an unrestrained social context and direct opportunities for interaction (Weidlich & Bastiaens, 2019). The forms of interaction in groups have an impact on mutual knowledge construction and often play a central role in how successful the groups are in their collaborative task (Oliveira, Tinoca, & Pereira, 2011).

Computer-mediated communication systems embedded in CSCL environments prompt what kind of messages are exchanged and how these messages are interpreted (Jeong & Hmelo-Silver, 2016; Norman, 1999). Many CSCL environments are predominantly used for task execution, except for social, off-task communication, which has been found essential for building trust among the team. This tendency is further reinforced by the fact that group members often are unacquainted with each other and have no common history (Kreijns et al., 2003). Thus, many socio-interactional elements affect communication in CSCL environments that need to be taken into account.

Socio-interactional affordances comprise the various synchronous and asynchronous forms of communication: emailing, chatting on Skype or Facebook, and online talk using Voice over Internet Protocol (VoIP) applications. Sociability and the socio-interactional affordances of a learning environment seem to be connected (Kreijns, Kirschner, & Vermeulen, 2013). Some studies suggest that sociable social network solutions (e.g. Facebook) are perceived as being overwhelmingly more suitable for sharing materials and resources, for receiving updates and for overall course interaction than traditional learning management systems such as Moodle or Blackboard (Jong, Lai, Hsia, Lin, & Liao, 2014). Consequently, sociable learning environments are more likely to afford both task-related and informal interaction between students. Furthermore, when social and interactional affordances are perceived, learners are encouraged to engage in activities that are in accordance with these affordances (Kirschner, Strijbos, Kreijns, & Beers, 2004). Learning in collaborative digital environments depends highly on the learners' possibilities to interact and collaborate in the environment. Therefore, it is important to gain more knowledge about socio-interactional affordances.

This study examines what kinds of technological, organisational and socio-interactional affordances students perceive when collaborating in a simulation-based learning environment. Furthermore, this study investigates how these affordances are employed in the collaborative learning task.

12.3 Case Learning Environment, Data and Analysis

12.3.1 Context

The simulation-based learning environment is a clock-driven business simulation (REALGAME; Lainema, 2003; Lainema & Makkonen, 2003) in which events in the simulation game processes evolve hour by hour. Students were placed in teams of 10–13 members (18 teams in total). Teams were recommended to have at least 3–4 participants online at all times during the 14-h simulation exercise. Thus, working in shifts was necessary.

In the simulation, the teams made decisions to manage the information and material flows and in the supply chain (purchases, inventories, production, deliveries) of the simulation company. The clock-driven nature of the simulation required that team members run their simulation companies in synchronous collaboration.

All participants had a real-time view of their simulation companies on their computer screens through a remote connection and were able to make decisions in the simulation. Virtual communication tools, such as VoIP (Skype), chat and email, were used for communication in the teams.

12.3.2 Participants and Data Collection

Data were collected through reflective essays of 177 undergraduate students participating in an online business simulation course in higher education. The students came from 10 universities (Austria, Belgium, China, Estonia, Finland, New Zealand and the US) and represented 38 different nationalities (the biggest ones being Finnish 52 students, New Zealander 52, Austrian 29, Belgian 15, and Chinese 14).

The students were tasked with writing a reflective essay in English after the first simulation game session of the course addressing teamwork, roles, tasks, and virtual collaboration and communication. Most students wrote lengthy descriptions of their gaming experiences in which they explained how their team got organised and how they worked as a team. All students participating in this study were required to sign informed consent forms.

12.3.3 Analysis

Data were analysed via qualitative content analysis using a data-driven analysis approach (e.g. Krippendorff, 2004). The data analysis process was inductive, allowing the analytical categories to emerge from the data rather than attempting to fit the data into existing theoretical categories. In this study, content analysis was applied to the data to answer the research question: what issues/elements in the gaming exercise enabled or hindered the team task?

The analysis entailed careful close reading of the data in iterative rounds. First, two of the authors conducted the qualitative analysis independently. During the reading, the observations were summed up and coded in categories of different types of 'doings'. The findings were mutually discussed, and the analytical categories were further refined to better respond to the aim of the study. Further analyses helped sharpen the focus and yielded three main categories of action potentials: technological, socio-interactional and organisational affordances. The analysis details how participants in the learning simulation perceived and seized the various affordances in the learning environment and how these were intertwined and influenced by each other.

The simulation game exercise consisted of different phases and tasks. First, the participants familiarised themselves individually with the relevant materials and finalised the course pre-assignments. Then, the team members became acquainted with each other and organised the team. At this point, the team work factually began and the participants started to interact with each other and with the learning environment and its elements. The simulation game was run on two separate days (2 weeks in between), and there were team assignments and individual assignments between the gaming days and after the final gaming day. Our analysis focuses on two sets of activities: activities before the first simulation exercise and activities during the first simulation exercise. This type of analysis allows us to gauge the specific nature of the affordances perceived and employed at each stage. The next section presents the results of our analysis.

12.4 Results

We report on the preliminary findings of our analysis on a general level that portrays what kinds of technological, organisational and socio-interactional affordances students perceived when collaborating in the simulation-based learning environment; these affordances were employed in the collaborative learning task. Our analysis indicates that these three types of affordances are intertwined and co-dependent.

12.4.1 Technological Affordances

Our analysis employed Bower's (2008) classification of technological affordances. Table 12.1 illustrates how the different technological affordances are shown in the data.

Many of the technological affordances were related to technological prerequisites. When working in a digital learning environment, some basic requirements need to be met before the gaming can take place. For example, in Bower's (Bower, 2008; see Table 12.1) classification, media and spatial affordances are normally prerequisites for a functional e-learning system. Teams selected different communication technologies for different purposes and different tasks. For example, during gaming, email was found to be clumsy, but before gaming, it was deemed an efficient form of disseminating information. In teams where members shared more personal information, applications like Facebook were used more often than in teams with less personal information sharing. The choice and use of communication technologies played a key role in how the teams got organised. In teams with a poor audio connection, chat or text messaging was the technology of choice. Some teams moved from audio to chat due to problems with audio. Group discussions in audio were sometimes deemed chaotic due to simultaneous talk and delays in broadcast. However, some teams were successful in using audio and found it useful and convenient. The gaming exercise required simultaneous use of multiple technologies, and some teams quickly saw which combinations were most fruitful.

Our analysis revealed that while the overall learning environment was the same for all teams, the teams utilised and combined the various affordances differently and complementarily. Our findings suggest that the way different affordances were combined depended, for example, on the availability and functionality of the technologies, the participants' personal preferences or a mutual team agreement, or the team members' technical skills. Our findings are in line with Faraj and Azad (2012), who suggested that an affordance is a multifaceted relational structure, not just a single attribute, property or functionality of the technology artefact or the actor.

12.4.2 Organisational Affordances

Organising teamwork in a CSCL environment is challenging for multiple reasons: team members do not know each other beforehand; the team task may be loosely defined, especially in cases where the focus is on solving ill-defined problems and creating new knowledge; and the information provided by face-to-face communication is missing in virtual communication. Particularly in an international context, there may be cultural differences and differences in communicative style between the students. Furthermore, students may have uneven levels and combinations of skills and competences.

Much of the organisational work in the simulation gaming exercise was related to securing the availability and timely delivery of resources. Before the simulation exercise, the teams needed to get organised. Shift planning was needed to ensure that there were enough team members online, meaning a minimum of three people at any given time. The teams also needed to decide how to deal with the responsibilities and roles in the game. It was suggested in the game materials that teams choose designated persons for at least three roles: purchasing raw materials, managing the production, and making sales offers and deliveries to customers. In some teams, one of the team members took the initiative to send out a Doodle poll to let the team members indicate when they were available and which roles they felt most comfortable with. Others used different Excel charts or sent emails to each other. Some teams made plans only for the shifts and not the roles.

> *In general, I believe our team was overall very unbalanced, as the roles were not clearly defined (−) and ultimately everybody had something to say to whatever was to be performed as company activity (production, offers, sales, …).* (Team 12)

However, it appeared that the roles needed not to be very precise and carefully planned for the team to function well:

> *I think we had very clear responsibilities and everyone did their best and we supported each other and helped when needed. Of course, because we didn't have a business strategy at all, everything we did was intuitive, so our functions or ways to do things were built up just in time in the game.* (Team 3)

Communication and organisation for the teamwork went hand in hand, and teams with multiple communicative occasions and versatile organisational tools (Doodle, Excel charts, explicit goal setting) were better prepared and oriented to the simulation exercise. In some of the teams, one or two team members even contacted each team member individually to negotiate a suitable shift and role, which was regarded as a welcomed practice:

> *Firstly, the communication and enthusiasm of my team was beyond impeccable in my opinion. As soon as the team lists were released, I had emails from most of the members in my group by the end of that day.* (Team 3)

Some teams had clear leadership, either by self-selection or by mutual agreement. In these teams, the leadership was more established and visible and acknowledged by most team members. Leadership was partly an issue of controversy, as some participants had reservations for strong leadership. In general, however, teams with clear leadership reported more satisfaction and better results.

> *(−) I found the team to be relatively effective, although lacking a leader figure. Because of this, I stepped in and created a Facebook group in which we could communicate quicker than that over email. This was effective, and some team members created a roster where it was outlined what hours each individual was online for, and their duty during that given time.* (Team 15)

Table 12.2 Identified organisational affordances

Organising	Managing	Leading
Organising shifts and tasks (Doodle poll, Excel chart)	Managing one's own task	Pointing out critical areas and initiating discussion
Re-organising shifts and tasks during gaming	'Feeling the pulse'—hearing how others are doing	Setting an agenda for team talks
Gathering information from the team and using it to securing and re-arranging resources	Suggesting what to do next—giving orders	Making projections based on the available data
Ensuring all areas are covered	Compromising through team talk – finding middle ground	Discussing and suggesting strategies
	Managing contacts to collaborative teams	Announcing decisions

Table 12.2 presents the categories of organisational affordances. Our analysis yielded three types of organisational affordances or practices, which partly overlap. The categories are organising, managing and leading. The table also presents the activities in each category.

The most important organisational work before the gaming exercise was to organise the shifts. Teams with enough participants online at any given time were most satisfied with the teamwork. Teams with too few people online found it stressful and chaotic to try to run the simulation company. Our analysis illustrates how organisational affordances are made possible by employing technological affordances, which, in turn, are prerequisites for the whole learning exercise. It is the dynamics of various kinds of affordances and their combinations in the digital learning environment that create the potential for learning.

Decisions in the simulation functions could be executed individually, but intra-team coordination was needed to balance the functions in the simulation company to avoid oversized inventories and bottlenecks along the supply chain. The designated roles of the participants were useful, since the simulation allowed one team member at a time to manage the simulation interface. Teamwork depended on communication and collaboration. Organisational and socio-interactional affordances were both needed and central to the success of the teams. Organising affordances, thus, consisted of combinations of different affordances (see e.g. Sæbø, Federici, & Braccini, 2017).

How organising takes place is essentially a retrospective sensemaking activity (Weick, 1995). Despite the centrality of collaboration as an activity including the organisation and execution of the joint task, little research addresses how collaboration is organised and coordinated or how organising for collaboration is afforded in computer-mediated environments. To date, what organising entails in the context of computer-supported learning environments is unclear, and empirical evidence of how organising takes place in computer-supported environments remains scarce. The lack of research puts further impetus on studying computer-supported collaboration from the affordance perspective in more detail.

12.4.3 Socio-interactional Affordances

Some interactional aspects in the gaming exercise were, similar to technological affordances, prerequisites of communication. For example, gaining access to the relevant information at the right time was imperative; without it, the team members could not function properly. This, in turn, was closely linked with the technology in use. Many teams found email and Skype chat clumsy for rapid communication and chose synchronous VoIP-solutions for talking about pressing issues. The analysis revealed that the socio-interactional affordances fall into four distinctive categories: observing, participating, facilitating and chairing. Some affordances can be placed in multiple categories, but the main difference between the affordances is the level of input and activity.

By alternating between technologies and channels appropriately, different kinds of communicative contributions are relevant for the team task. However, if everyone was disseminating information and nobody was drawing conclusions, the team task and its accomplishment would be compromised. A balanced participation and contribution bring results and increases the team's satisfaction with the team's functioning (Table 12.3).

> *I got so enthusiastic that I watched the game even later in my course because I was so excited about the project. Moreover, I am glad to be a part of such project because it teaches us more than any book about crosscultural and virtual communication.* (Team 12).

When designing learning environments, it is important to acknowledge the role of technological affordances as enablers or hindrances to the learning exercise. The technological affordances can be designed in ways that encourage and facilitate teamwork and interaction and support the development of organisational skills.

Table 12.3 Identified socio-interactional affordances

Observing	Participating	Facilitating	Chairing
Listening to what others are saying	Listening to what others are saying	Listening to what others are saying	Listening to what others are saying
	Acknowledging what others are saying	Encouraging others to speak	Drawing conclusions
	Listening, stepping back	Facilitating the discussion	Making suggestions
	Responding to what others are saying	Repeating what has been said	Announcing decisions
	Disseminating information	Disseminating information	Disseminating information
	Giving feedback	Giving feedback	Giving feedback
		Negotiating, finding middle ground	Negotiating, finding middle ground

12.5 Conclusion

This study examined what kinds of technological, organisational and socio-interactional affordances were identified by the learners collaborating in a simulation-based learning environment. In addition, we investigated how these affordances are employed in the collaborative learning task.

The study shows that a technology must improve interactions between the individual and the environment to be useful (Kaptelinin & Nardi, 2006; Stahl & Hakkarainen, 2020; Tchounikine, 2019). Furthermore, an abundance of affordances does not necessarily mean that they will be perceived or utilised by the actors equally.

Operating in the learning environment required coordinating action to accomplish the team task. Participants evaluated their team success not only in terms of how well the team performed but also in how well the team had worked together and what they had learned. The latter, in our opinion, gives an even better indication of how affordances and their use are connected to learning (see also Jayarathna, Eden, Fielt, & Nili, 2020).

Based on our analysis, we can assume that it is useful to have complementary affordances to allow for individual consideration and to foster motivation and more productive ways of working. This finding is in line with recent research on learning in CSCL environments (e.g. Bonneau & Bourdeau, 2019; Jeong & Hmelo-Silver, 2016).

Teams chose partly different combinations of communication technology. Before the game session, most teams resorted to asynchronous communication technologies, such as email, to better control the flow of information and to produce a record of all communication. During the simulation game, synchronous communication via Skype was found most appropriate by most teams. However, some teams continued to rely on chat and not talk online. For some teams, this choice was motivated by the team members' reluctance to speak English. However, delays in communication lead to misunderstandings, missing information and confusion. Teams with most self-reported motivation and initial success made changes in their use of technology and communication tool according to what took place in the game. These teams also adapted their organisation according to the situation and used appropriate organisational practices to pursue the best possible outcome for the team at any given situation. Our results corroborate the findings of previous research in highlighting the importance of interactional organisation of teamwork for collaborative problem-solving (Perit Çakır, Zemel, & Stahl, 2009). Furthermore, our results have broader implications, as they stress the salience of flexibility, adaptivity and the accommodation of available resources and affordances to the given task as some of the most important skills needed in all areas of life.

Previous studies have found that teamwork in digital environments benefits from abundant communication (Choi & Cho, 2019; Fjeldstad, Snow, Miles, & Lettl, 2012). The data show that teams with less self-reported motivation and successful outcomes had fewer occasions of communication before and during the gaming exercise. These teams also leaned more on asynchronous communication during gaming. The less satisfied teams seemed unable to adjust their team effort or to correct the downfall spiral. By contrast, synchronous communication during the simulation sessions contributed to a more effective teamwork and higher satisfaction. Our results are in line with previous research in suggesting that student-centred learning requires that the learning environment encourages and empowers students to search for information, try different tactics and strategies, test ideas and create new knowledge (Bonneau & Bourdeau, 2019; Martens, Bastiaens, & Kirschner, 2007). However, these potentials need to be carefully and purposefully designed and integrated in the learning environment, as they do not miraculously appear without purposeful planning and effort. Consequently, we find that it is essential to study affordances in more depth and learn how they can be embedded in learning environments to enhance and empower learning. At best, synchronous collaborations facilitate rich learning experiences, such as the one quoted below.

> *(−) This online simulation definitely surpassed by expectation of how much I would learn. Learning how to compromise, learning how to negotiate, learning how to speak up, learning how to manage, and most importantly, learning how to work as a collaborative team through an online virtual world.* (Team 4)

Some limitations of this study also need to be acknowledged. First, the study solely examines three types of affordances: technological, organisational and socio-interactional, omitting any other types of affordances from the analysis. This choice was made to better focus on the selected affordances and to respect the space limitations of the paper. Further analyses could probe into other types of affordances, allowing for a broader view. Second, the study only presents the preliminary results and does not provide a full analysis of the data. However, already the first analyses and their results bring new knowledge of how the selected affordances in a CSCL environment are perceived and employed. Third, the study does not examine the relation between employed affordances and students' learning outcomes. This type of analysis requires different research methods and analyses and was therefore out of the scope of this research. Despite this limitation, the study provides some preliminary conceptions of beneficial and less beneficial aspects regarding collaborative learning in computer-based environments.

This study is advantageous in illustrating how the selected affordances in a CSCL environment are perceived and employed by the learners. It also shows that despite the abundance of affordances in a learning environment, not all affordances are employed equally by all teams, which, in turn, leads to varying outcomes and different perceptions of the success of the teamwork.

A more thorough understanding of the dynamics of affordances can be used to design accessible learning environments and help educators to better understand how the learning process and the use of affordances can be facilitated and supported.

Acknowledgements This research has been supported by The Finnish Work Environment Fund, grant number 190154 Well@DigiWork, and The Academy of Finland, grant number 318095, the Multidisciplinary Research on Learning and Technology profile II of University of Jyväskylä (JYU).

References

Alahuhta, P., Nordbäck, E., Sivunen, A., & Surakka, T. (2014). Fostering team creativity in virtual worlds. *Journal for Virtual Worlds Research, 7*(3).

Antonenko, P., Dawson, K., & Sahay, S. (2017). A framework for aligning needs, abilities and affordances to inform design and practice of educational technologies. *British Journal of Educational Technology, 48*(4), 916–927.

Arvaja, M., Salovaara, H., Häkkinen, P., & Järvelä, S. (2007). Combining individual and group-level perspectives for studying collaborative knowledge construction in context. *Learning and Instruction, 17*(4), 448–459.

Baralou, E., & Tsoukas, H. (2015). How is new organizational knowledge created in a virtual context? An ethnographic study. *Organization Studies, 36*(5), 593–620.

Berthelsen, U. D., & Tannert, M. (2020). Utilizing the affordances of digital learning materials. *L1 Educational Studies in Language and Literature (Special Issue: Danish as L1 in a learning materials perspective)*, 20, 1-23. https://doi.org/10.17239/L1ESLL-2020.20.02.03

Bonneau, C., & Bourdeau, S. (2019). Computer-supported collaboration: Simulation-based training using LEGO®. *Educational Technology Research and Development, 67*(6), 1507–1527.

Bower, M. (2008). Affordance analysis–matching learning tasks with learning technologies. *Educational Media International, 45*(1), 3–15.

BusinessDictionary (2019). Organising. In *BusinessDictionary*. Retrieved June 27, 2019, from http://www.businessdictionary.com/definition/organizing.html

Choi, O. K., & Cho, E. (2019). The mechanism of trust affecting collaboration in virtual teams and the moderating roles of the culture of autonomy and task complexity. *Computers in Human Behavior, 91*, 305–315.

Conole, G., & Dyke, M. (2004). What are the affordances of information and communication technologies? *ALT-J, Research on Learning Technology, 12*(2), 113–124. https://doi.org/10.1080/0968776042000216183

Dohn, N. B. (2009). Affordances revisited: Articulating a Merleau-Pontian view. *International Journal of Computer-Supported Collaborative Learning, 4*(2), 151–170.

Faraj, S., & Azad, B. (2012). The materiality of technology: An affordance perspective. *Materiality and Organizing: Social Interaction in a Technological World*, 237–258. https://doi.org/10.1093/acprof:oso/9780199664054.003.0012

Fjeldstad, Ø. D., Snow, C. C., Miles, R. E., & Lettl, C. (2012). The architecture of collaboration. *Strategic Management Journal, 33*(6), 734–750.

Gall, M., & Breeze, N. (2007). The sub-culture of music and ICT in the classroom. *Technology, Pedagogy and Education, 16*(1), 41–56.

Gee, J. P. (2004). *Situated language and learning: A critique of traditional schooling*. London: Routledge.

Gibson, J. J. (1979). *The ecological approach to visual perception*. Boston: Houghton Mifflin.

Greeno, J. G. (1994). Gibson's affordances. *Psychological Review, 101*(2), 336–342.

Hartson, R. (2003). Cognitive, physical, sensory, and functional affordances in interaction design. *Behaviour & Information Technology, 22*(5), 315–338.

Hernández-Sellés, N., Muñoz-Carril, P. C., & González-Sanmamed, M. (2019). Computer-supported collaborative learning: An analysis of the relationship between interaction, emotional support and online collaborative tools. *Computers & Education, 138*(1), 1–12.

Hossain, L., & Wigand, R. T. (2004). ICT enabled virtual collaboration through trust. *Journal of Computer-Mediated Communication, 10*(1), JCMC1014.

Jayarathna, L., Eden, R., Fielt, E., & Nili, A. (2020). Contextualizing the effective use of social media network for collaborative learning: An affordance perspective. In *Proceedings of the 24th Pacific Asia conference on information systems: PACIS 2020* (p. 118). Association for Information Systems.

Jeong, H., & Hmelo-Silver, C. E. (2016). Seven affordances of computer-supported collaborative learning: How to support collaborative learning? How can technologies help? *Educational Psychologist, 51*(2), 247–265.
Jong, B. S., Lai, C. H., Hsia, Y. T., Lin, T. W., & Liao, Y. S. (2014). An exploration of the potential educational value of Facebook. *Computers in Human Behavior, 32*, 201–211.
Kaptelinin, V., & Nardi, B. (2006). *Acting with technology: Activity theory and interaction design*. Cambridge: MIT Press.
Ke, F., Pachman, M., & Dai, Z. (2020). Investigating educational affordances of virtual reality for simulation-based teaching training with graduate teaching assistants. *Journal of Computing in Higher Education, 32*, 607–627.
Kennewell, S. (2001). Using affordances and constraints to evaluate the use of information and communications technology in teaching and learning. *Journal of Information Technology for Teacher Education, 10*(1–2), 101–116.
Kirschner, P. A. (2002). *Can we support CCSL? Educational, social and technological affordances*. Open Universiteit Nederland.
Kirschner, P., Strijbos, J. W., Kreijns, K., & Beers, P. J. (2004). Designing electronic collaborative learning environments. *Educational Technology Research and Development, 52*(3), 47.
Koschmann, T. (2012). *CSCL: Theory and practice of an emerging paradigm*. Abingdon: Routledge.
Krauskopf, K., Zahn, C., & Hesse, F. W. (2012). Leveraging the affordances of YouTube: The role of pedagogical knowledge and mental models of technology functions for lesson planning with technology. *Computers & Education, 58*(4), 1194–1206.
Kreijns, K., Kirschner, P. A., & Jochems, W. (2003). Identifying the pitfalls for social interaction in computer-supported collaborative learning environments: A review of the research. *Computers in Human Behavior, 19*(3), 335–353.
Kreijns, K., Kirschner, P. A., & Vermeulen, M. (2013). Social aspects of CSCL environments: A research framework. *Educational Psychologist, 48*(4), 229–242.
Krippendorff, K. (2004). *Content analysis. An introduction to its methodology* (2nd ed.). Thousand Oaks: Sage Publications.
Lainema, T. (2003). *Enhancing organizational business process perception: Experiences from constructing and applying a dynamic business simulation game*. Turku School of Economics and Business Administration.
Lainema, T., & Makkonen, P. (2003). Applying constructivist approach to educational business games: Case REALGAME. *Simulation & Gaming, 34*(1), 131–149.
Lampe, C., Wohn, D. Y., Vitak, J., Ellison, N. B., & Wash, R. (2011). Student use of Facebook for organizing collaborative classroom activities. *International Journal of Computer-Supported Collaborative Learning, 6*(3), 329–347.
Lin, P. C., Hou, H. T., & Chang, K. E. (2020). The development of a collaborative problem solving environment that integrates a scaffolding mind tool and simulation-based learning: An analysis of learners' performance and their cognitive process in discussion. *Interactive Learning Environments*, 1–18.
Majchrzak, A., & Markus, M. L. (2012). Technology affordances and constraints in management information systems (MIS). In E. Kessler (Ed.), *Encyclopedia of management theory* (pp. 832–836). Thousand Oaks: Sage Publications.
Martens, R., Bastiaens, T., & Kirschner, P. A. (2007). New learning design in distance education: The impact on student perception and motivation. *Distance Education, 28*(1), 81–93.
McLoughlin, C., & Lee, M. J. W. (2007). Social software and participatory learning: Pedagogical choices with technology affordances in the Web 2.0 era. In *Proceedings ASCILITE*, Singapore 2007. https://researchbank.acu.edu.au/cgi/viewcontent.cgi?article=3049&context=fea_pub
Mercer, N. (2010). The analysis of classroom talk: Methods and methodologies. *The British Journal of Educational Psychology, 80*(1), 1–14.
Michaels, C. F., & Carello, C. (1981). *Direct perception*. Englewood Cliffs, NJ: Prentice Hall.

Moate, J., Hulse, B., Jahnke, H., & Owens, A. (2019). Exploring the material mediation of dialogic space—A qualitative analysis of professional learning in initial teacher education based on reflective sketchbooks. *Thinking Skills and Creativity, 31*, 167–178.
Norman, D. A. (1999). Affordance, conventions, and design. *Interactions, 6*(3), 38–43.
Oliveira, I., Tinoca, L., & Pereira, A. (2011). Online group work patterns: How to promote a successful collaboration. *Computers & Education, 57*(1), 1348–1357.
Park, H., & Song, H. D. (2015). Make e-learning effortless! Impact of a redesigned user interface on usability through the application of an affordance design approach. *Journal of Educational Technology & Society, 18*(3), 185.
Perit Çakır, M., Zemel, A., & Stahl, G. (2009). The joint organization of interaction within a multimodal CSCL medium. *Computer Supported Learning, 4*, 115–149.
Pozzi, G., Pigni, F., & Vitari, C. (2013). Affordance theory in the IS discipline: A review and synthesis of the literature. In *Proceedings of the 20th Americas conference on information systems, AMCIS 2014*.
Roschelle, J., & Teasley, S. (1995). The construction of shared knowledge in collaborative problem solving. In C. O'Malley (Ed.), *Computer supported collaborative learning. NATO ASI series (Series F: Computer and system sciences), vol. 128* (pp. 69–97). Berlin: Springer.
Sæbø, Ø., Federici, T., & Braccini, A. M. (2017). Combining social media affordances for organising collective action. *Information Systems Journal, 30*(4), 699–732.
Salomon, G. (1993). No distribution without individuals' cognition: A dynamic interactional view. In G. Salomon (Ed.), *Distributed cognitions: Psychological and educational considerations* (pp. 111–138). Cambridge: Cambridge University Press.
Scarantino, A. (2003). Affordances explained. *Philosophy of Science, 70*(5), 949–961.
Selwyn, N. (2012). Making sense of young people, education and digital technology: The role of sociological theory. *Oxford Review of Education, 38*(1), 81–96.
Song, Y. (2011). What are the affordances and constraints of handheld devices for learning in higher education. *British Journal of Educational Technology, 42*(6), E163–E166.
Stahl, G. (2012). Ethnomethodologically informed. *International Journal of Computer-Supported Collaborative Learning, 7*(1), 1–10.
Stahl, G., & Hakkarainen, K. (2020). Theories of CSCL. Manuscript in preparation. In U. Cress, C. Rose, A. F. Wise, & J. Oshima (Eds.), *International handbook of computer-supported collaborative learning*. New York, NY: Springer.
Steffens, K., Bannan, B., Dalgarno, B., Bartolomé, A. R., Esteve-González, V., & Cela-Ranilla, J. M. (2015). Recent developments in technology-enhanced learning: A critical assessment. *International Journal of Educational Technology in Higher Education, 12*(2), 73–86.
Strong, D. M., Johnson, S. A., Tulu, B., Trudel, J., Volkoff, O., Pelletier, L. R., … Garber, L. (2014). A theory of organization-EHR affordance actualization. *Journal of the Association for Information Systems, 15*(2), 53. https://doi.org/10.17705/1JAIS.00353
Tchounikine, P. (2019). Learners' agency and CSCL technologies: Towards an emancipatory perspective. *International Journal of Computer-Supported Collaborative Learning, 14*(2), 237–250.
Volkoff, O., & Strong, D. M. (2013). Critical realism and affordances: Theorizing it-associated organizational change processes. *MIS Quarterly, 37*(3), 819–834.
Wang, C., Fang, T., & Gu, Y. (2020). Learning performance and behavioral patterns of online collaborative learning: Impact of cognitive load and affordances of different multimedia. *Computers & Education, 143*, 103683.
Weick, K. E. (1995). *Sensemaking in organizations*. Thousand Oaks: Sage.
Weidlich, J., & Bastiaens, T. J. (2019). Designing sociable online learning environments and enhancing social presence: An affordance enrichment approach. *Computers & Education, 142*, 103622.
Xue, S., & Churchill, D. (2019). A review of empirical studies of affordances and development of a framework for educational adoption of mobile social media. *Educational Technology Research and Development, 67*(5), 1231–1257.

Chapter 13
Enhancement of Experiential Learning in Software Factory Project-Based Course

Muhammad Ovais Ahmad and Kari Liukkunen

13.1 Introduction

In today's digital age, the software industry is moving with a fast pace as highlighted by Marc Andreessen in his essay Why Software is Eating the World (Andreessen, 2011). This shift has also pushed educational institutions to train graduates with various skills and competencies. This shift pushing the universities boundaries related to their offerings in term of courses and programs. The major push is to create software engineering (SE) technology agnostic courses that prepare graduates that fulfill the industry needs and demands. One way to train SE graduates is to provide project-based learning through capstone courses (Erdogmus & Peraire, 2017; Howe, 2010; Walker & Andrew, 2015).

Project-based learning (PBL) is common in various engineering programs and courses such as software engineering and information systems. It helps students to achieve a range of high-level goals and dealt with real life industry-oriented problem or activities. In such project-based courses students play an active role whereas teachers play the role of facilitator and more passive. The students take active responsibility of their learning and have more control on how to solve a given problem. According to Barg and Barg (1999), project-based learning shown increase

M. O. Ahmad (✉)
Department of Mathematics and Computer Science, Karlstad University, Karlstad, Sweden

Faculty of Electronics, Telecommunications and Informatics, Gdansk University of Technology, Gdańsk, Poland

M3S Research Unit, University of Oulu, Oulu, Finland
e-mail: ovais.ahmad@kau.se

K. Liukkunen
M3S Research Unit, University of Oulu, Oulu, Finland

D. Ifenthaler et al. (eds.), *Balancing the Tension between Digital Technologies and Learning Sciences*, Cognition and Exploratory Learning in the Digital Age,
https://doi.org/10.1007/978-3-030-65657-7_13

student motivation and interaction while working on real-world problem and improve social and team-working skills. Along with the technical knowledge the students also require to know the principle, methods, and process while developing software project.

Educational institutions must train students in real-life practical projects where students engage in a collaborative teamwork environment and develop project management skills (Erdogmus & Peraire, 2017; Palacin-Silva, Khakurel, Happonen, Hynninen, & Porras, 2017; Wijnia, Loyens, & Derous, 2011). Universities around the world already include such capstone courses (ACM Joint Task Force, 2014). However, various transversal capabilities—leadership, decision making, negotiation, self-reflection, and the infusion of design thinking—receive little or no attention in these courses.

To provide such competencies, the University of Oulu established the software factory (SWF) learning environment/laboratory in 2012. The Oulu SWF is an infrastructure platform that serves multiple purposes to support SE research, education, and entrepreneurship. The SWF is a test bed for SE ideas and a source for original research on the development of basic scientific software (Ahmad, Liukkunen, & Markkula, 2014). It is an educational vehicle for the university where the artifacts produced in the factory serve to improve learning and provide teaching materials in close collaboration with industry (Ahmad et al., 2014; Fagerholm, Oza, & Münch, 2013). The Oulu SWF is part of a European Union SWF network (Taibi et al., 2016). The aim of the SWF is for students to share their experiences, learn in a collaborative environment, and grow to compete in the fast-growing ICT domain. The Oulu SWF laboratory is equipped with latest computers, software development tools, interactive projector, and wide range of gadgets available for student projects (i.e., smart phones, smart watches, tablets, cameras, and so on).

This chapter describes the design of the SWF project-based course, assessment techniques, student perception, and teaching experiences, and discusses the importance of reflective learning diaries. The paper also aims to identify factors in the SWF learning environment that affect learning, in terms of exploring (1) student achievements in term of skills gained, (2) students' perceptions of the SWF learning environment, as measured with the computer laboratory environment inventory (CLEI) (Newby & Fisher, 1997), and (3) students' attitudes toward the SWF project course, as measured with the attitude toward computers and computing courses questionnaire (ACCC) (Newby & Fisher, 1997).

This chapter is organized as follows: Section 2 sheds light on related work, and Sect. 3 discusses the SWF course under the pedagogic lens and presents the SWF course learning objectives, mode of delivery, overall structure, student team formation, mentoring, and student assessment. Section 4 reports students' perceptions, and Sect. 5 discuss the importance of student reflective learning diaries. Section 6 discusses lessons learned from the teacher perspective in the context of teaching and managing similar project-based courses. Finally, Section 7 concludes the paper and sheds light on future research work and improvement to the SWF course and alike.

13.2 Related Work

Project- and problem-based courses prepare student for real work in the industry (ACM Joint Task Force, 2014; Palacin-Silva et al., 2017). Project-Based Learning (PBL) is based on the constructivist paradigm, which is student-driven learning approach (Bell, 2010; Yilmaz, Yilmaz, & Keser, 2020). PBL supports the development of social and cognitive aspects of learners (Hung, Hwang, & Huang, 2012; Land, 2004; Yilmaz et al., 2020). The students as learner develop the questions, which are investigated under the supervision of teachers. In this process, teachers closely follow every step and approve before the students embark in a direction. Here, the key element is the student choice. According to Bell (2010), *PBL is a key strategy for creating independent thinkers and learners*. Lee, Blackwell, Drake, and Moran (2014) highlighted the important component for PBL is community partnerships, where students collaborate with professionals. Problem-based learning originated in medicine that subsequently adopted in other disciplines (Mann et al., 2020). According to Kolmos and de Graaff (2014) problem-based learning approaches combine the learning approach, the social approach, and the content approach. It is basically student-centered approach which engage them in complex real-life problem with open-ended answers (Chang, Hsiao, Chen, & Tsung-Ta, 2018). According to (Jabarullah & Hussain, 2019) problem-based learning exposed students showed a greater inclination toward deep and strategic learning rather than surface learning. It is necessary for SE students to have hands-on experience and a glimpse of real software industry work during their studies. To prepare students for a software industry, universities around the world offer various capstone courses. The SWF project is based on capstone course concept.

Since 2010, the SWF and SWF-based courses have been offered at various universities around the world, such as the University of Helsinki, the University of Oulu, the University of Eastern Finland, the Free University of Bozen-Bolzano, Tampere University of Technology, the Free University of Cagliari, the Technical University of Madrid, Montana State University, the Catholic University of America, and Bowling Green State University (Ahmad et al., 2014; Chao & Randles, 2009; Fagerholm et al., 2013; Taibi et al., 2016; Tvedt, Tesoriero, & Gary, 2002). These SWFs aim to provide students with practical experience in software development projects and help the students to gain business experience in a collaborative environment, as well as polish their technical expertise. However, research on SWF projects and course curricula is scarce. Most studies report success stories, students experimenting with processes, and positive experiences of students' motivation in such courses or projects. We did not find any studies that reported on SWF course design or mode of delivery or that gave a detailed assessment of students' techniques.

13.3 The Software Factory Course

The SWF laboratory offering a 10 ECTS (290 h of work) advanced-level course for Information Processing Science master's degree program students at the University of Oulu, Finland. The purpose is to expose students to real-life software development projects in a multicultural collaborative environment. The focus is learning by doing—that is, managing authentic, resource-limited project work and integrating the practices of an academic expert in a unique project assigned by a software company. Each year, in the spring semester 15–20 students take the SWF course.

The SWF course is based on various learning theories or approaches—behaviorism, cognitivism, and constructivism—are taken into consideration (Anderson, 2008). The blended learning approach in SWF course goals was to maximize students' learning outcomes from three angles. First, behaviorism school, we observed how teaching staff behavior affects students' learning, e.g., teacher approval of certain items required by the course. In such cases, the teacher acts proactively to respond quickly. In this way, we avoid unnecessary wait times from the students' perspective. Second, cognitivism approach, the students are encouraged to have a mental map of their project and processes. Such encouragement is important, especially in the context of software development. The students need to have a map for a specific goal, which boosts motivation and reduces stress. Third, as constructivists, we do not push students to memorize the concepts taught during the lectures. Various serious games, learning diaries, and discussions enable students to develop their knowledge.

In summary, the SWF course is more inclined toward constructivism due to a student-centered model that focuses on learning by doing in a collaborative environment and problem-based learning. Such a collaborative course and environment have a significant impact on learning (Ahmad et al., 2014; Fagerholm et al., 2013; Khine & Lourdusamy, 2003; Taibi et al., 2016).

13.3.1 SWF Course Learning Outcomes

After completing the course, the students should demonstrate their ability to work on a challenging ICT project. Students learn to acquire and apply professional expertise in the topic of the project. One example of a project is a pathfinder using the Robot Operating System for an autonomous electric car. Students should be able to:

- Act as independent professional members in an ICT project; a team member collectively produces, monitors, and updates the project plan (a project with a fixed deadline and human resources).
- Search up-to-date scientific literature on the subject matter of the project to build professional expertise in the topic and apply to the project work.

- Develop analytical and creative skills for successful completion of the project and monitor and communicate the status (time and human resources used) of the project in real time within the team.
- Develop skills to communicate with customer in a professional context and manage a successful project review with the steering group/project team.
- Report and explain the status (progress, results, and future estimations of the project) to the steering group to support decision making and problem resolution concerning the project's future.
- Work as a project team member with people from different technical and/or cultural backgrounds, produce a realistic outcome in relation to the project deadline and human resources (ok, good, or excellent), and reflect on the relationship between the selected process models (evolutionary, agile, lean, etc.) and management practices followed in the project.

13.3.2 Mode of Delivery

The SWF course adopted blended teaching or mixed-mode instruction to boost collaborative learning (Martyn, 2003). This approach has become popular in SE because it helps develop critical thinking and improves understanding of various concepts (Palacin-Silva et al., 2017). There are four major components of the SWF course: classroom lectures, serious games workshops, weekly customer meetings and monthly progress reports, and individual project work. All SWF project-related communication, materials, and deliverables are stored in the Optima workspace.[1]

Even though the course had strong learning by doing roots, it has evolved more and more toward practical skills and way of working. During the first implementations there was some theoretical content and students did more written documents. Changes were based on the feedback from students and software companies. A typical course execution structure is presented in Fig. 13.1. The SWF course consist of three main parts, lectures, workshops, project, and mentoring meetings.

Introductory lectures and first workshop provide the guiding steps for carrying out the course and discuss relevant course/project information. The students already have a background in project management and software development tools, processes, and techniques. However, in the first three introductory lectures, these concepts are briefly revised. The attractive aspects of the introductory lectures are that they are delivered with the help of software professionals from the local software industry. Oulu is a smart city, and many big ICT companies, such as Nokia, Ericsson, and Bittium, have offices in the city. The invited professionals share their experiences and provide insights into running projects efficiently and adopting a software development and management method or technique based on project needs. Students

[1] Optima is a learning management system used at the University of Oulu, Finland. https://www.discendum.com/references/?q=optima

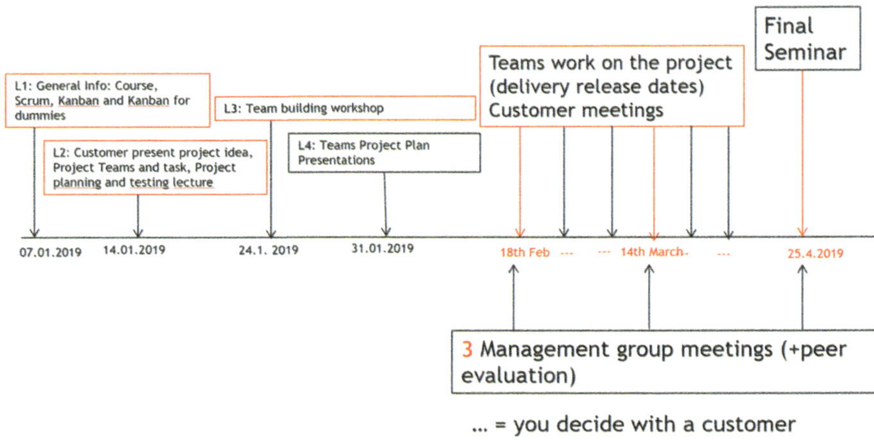

Fig. 13.1 Example of software factory execution plan

are divided into multiple groups during the fourth lecture; each team has three to four members. Additionally, interested software companies are invited to present the project ideas to the student teams. These software companies and entrepreneurs are the real customers of the SWF projects. Such real customer involvement helps students to learn more about the dynamics of software projects and enhance their soft skills. Each team has the opportunity to select a project based on their interest.

Workshops goal is that students present their initial understanding of the project, problems expected, and possible solutions. The students present their project plan, development process, and management practices. The students break down the customer requirements and needs, as well as discussing the delivery of the project to the customer. In this phase, the teaching staff act as facilitators and guide students to successfully implement their projects. The aim of the workshop is to facilitate students through experiential learning to understand group dynamics and the software development methods, tools, and techniques used in the software industry. Various serious games are played, such as the Marshmallow Challenge, Draw Toast, Scrum Simulation with LEGO, and Poker Planning. For example, the Marshmallow Challenge is an instructive design exercise that engages students to work in teams. Such activities help students experience the dynamics of collaborative teamwork, the importance of analyzing each other's perspectives, and iteration planning.

The actual software development starts at the end of the first workshop. The students work on their projects for 3 months. The project team(s) experiment and select development methods, tools, techniques, and related technologies based on the project requirements and customer demand.

Frequent communication inside and outside the team is important. The project team meets the customer based on need, but the course recommends organizing a weekly meeting. Additionally, there are monthly management group meetings, where students present their overall progress, project and team challenges, and the status of the work hours. The students must also write a learning diary once a month. The teaching staff provide formal feedback during management group meetings.

13.3.3 Team Formation and Mentoring

We used a team formation tool called CATME (Hrivnak, 2013), which restricts each team to four or five students. The students' characteristics, such as ethnicity, gender, leadership preferences, specific technology skill level, and relevant knowledge, play a critical role in the formation of diverse teams. The teaching staff separately mentor each team and monitor their performance and dynamics. The teaching staff act as facilitators, help students prioritize their tasks, provide feedback on the development process, and discuss the reflective learning diaries to enhance student learning. To track students' progress and facilitate efficiently, teachers use various techniques, such as daily stand-up meetings, agile retrospectives, burn-down charts, cumulative flow diagrams, and Kanban boards.

13.3.4 Assessment Methods and Criteria

Assessing project work and then grading individual team members is always challenging. It is the teacher's duty to fairly assess each team member and present the criteria clearly and understandably. In the SWF course, we used a rubric-based assessment. It is mandatory for students to attend all the lectures and workshops. The distribution of the SWF course assessment is summarized in Table 13.1.

The peer evaluation and reflective biweekly learning diaries are out of SWF course long journey. Each year, we learnt new things and revised these two evaluation parts. For example, the weekly learning diaries were very exhaustive for the students and hindering the sprint planning and execution. The peer evaluation brings more visibility and transparency to each individual student contribution in the group project. Below we will explain the mentioned two assessment parts in detail.

It is mandatory for every student to write four diary entries during the project. Reflective diaries are core elements of self-regulated learning that promote the development of meta-cognitive strategies (Fulwiler, 1986). Fulwiler described the rationale for introducing reflective diaries as follows: "In the academic world, where we teach students to gain most of their information from reading and listening, we spend too much time telling our students how to see or doing it for them. That's not how I would encourage critical, creative, or independent thinking. Our students have good eyes; let's give them new tools for seeing better: journal writing is, of course, one of those tools."

Peer assessment is a powerful meta-cognitive tool, which has been advocated in various studies (Kaufman, Felder, & Fuller, 2000; Layton & Ohland, 2001; McGourty, Dominick, & Reilly, 1998). According to McGourty et al. (1998): "In a

Table 13.1 Software factory (SWF) course evaluation and grading

Criteria	Points	Criteria
Final Software Product (Group Evaluation)	25	Final evaluation by the customer. The product must fulfill the customer requirements and meet functional and non-functional requirements. Additionally, customer evaluate the groups communication and meeting deadlines
Supporting Documents (Group Evaluation)	25	Supporting documents and evidence regarding the entire development process, including planning, management, implementation process and the planned and actual work hours. The project plan is updated with each sprint, which lasts for 2 weeks
Supervisor Evaluation (Group Evaluation)	10	Supervisor's evaluation of group performance The teaching staff observe each team's progress from the beginning until the final presentation of the project. The teacher provides feedback after every weekly customer meeting and guides the students to improve their preparation for the next steps'. The teachers also consider how the teams prepare their presentations, handle technology during the meetings, and work together internally
Individual Reflective Biweekly Learning Diary	20	Individual three reflective reports (3 × 5 points = 15). Final lesson learned report (5 points)
Peer Evaluation (Individual Evaluation)	20	Peer evaluation of individual performance by other members of the team
Total	100	Final grade

cooperative learning environment, students themselves are often in the best position to provide one another with meaningful feedback regarding both their technical and interpersonal performance." A number of peer assessment tools and advised reducing the possibility of a student intentionally "damaging" his or her peers' scores and ensuring that students do not feel that they are "ratting" on their peers (McGourty et al., 1998; Nicole, Pamela, & Rebecca, 2005; Wilkins, Lawhead, Wilkins, & Lawhead, 2000). In the SWF course, students are required to fill out a form to report aspects of their team members' contribution and behavior characteristics. The Oulu SWF project peer assessment is based on Sanders (Ohland & Layton, 2000; Sanders, 1984). Examples of teamwork characteristic statements include attending team meetings, contributing to the discussion at the meetings, completing tasks on time, and the team member's ability to work with the other team members. The students were asked to score the characteristics using the following scale: Always (2 points), Usually (1), Sometimes (0), Rarely (−1), and Never (−2). Furthermore, students have the opportunity to express their feedback in answer to an open-ended question and report their concerns in detail.

13.4 Student Perception Survey

The objective of survey is to identify factors affect students learning in a SWF and explores the relationships between student achievement, and their perceptions regarding SWF learning environment. The teaching staff conduct a voluntary online survey annually at the end of the SWF course (2012–2018). In 6 years, 50 of 90 students participated in the survey. The students' perceptions of the SWF course and facilities are collected based on the CLEI and the ACCC (Newby & Fisher, 1997) (see Table 13.2). Survey questions use a five-point Likert scale, ranging from 1 (strongly disagree) to 5 (strongly agree). Studying the learning environment (i.e., the SWF laboratory) is one way to explore student perception (Kolb & Kolb, 2005; Newby & Fisher, 1997). The following Table 13.2 shows that the reliability of the factor measurement is high; Cronbach's alpha varies between 0.597 and 0.951. These values show that the CLEI and ACCC constructs are internally consistent and reliably measured.

Newby and Fisher (1997) developed the CLEI to measure students' perceptions of their learning environment. The CLEI has five constructs: student cohesion, open-endedness, integration, technology adequacy, and laboratory availability. In the Oulu SWF context, the laboratory availability construct is not relevant, as the SWF laboratory is assigned to SWF project students 24/7. Furthermore, with the help of the ACCC (Newby & Fisher, 1997), students' attitudes toward computers and computing courses are assessed. The ACCC consists of four constructs: anxiety, enjoyment, usefulness of computers, and usefulness of course. In the Oulu SWF context, we removed the usefulness of computers construct because the targeted students are in the final year of the Information Processing Science master's degree program. These students are proficient in the use of computers.

Students' perceptions about SWF laboratory is measure using the CLEI; their perception is quite positive. The highest mean scores are for "student cohesion" and "open-endedness" (mean = 4.24 and 3.84, respectively). "Technology adequacy" had the lowest mean score (mean = 3.28). The students feel confident and support each other in their project work. The students also have a positive response for "open-endedness" (average mean = 3.84), which might be due to the close industrial collaboration. The SWF project is developed iteratively which helps them to obtain frequent feedback from customer. The students collectively work toward the same goal and seek a solution for the given problem. The students are encouraged to use and put their theoretical knowledge into practice in the SWF project. The teaching staff facilitate throughout the project life. The SWF laboratory is fully equipped

Table 13.2 Student perception survey based on CLEI constructs

Measures	Constructs	M	SD	α
Computer Laboratory Environment Inventory (CLEI)	Student cohesion	4.24	0.63	0.846
	Open-endedness	3.84	0.75	0.694
	Integration	3.41	0.61	0.527
	Technology adequacy	3.28	1.07	0.951

Table 13.3 Student perception survey based on ACCC constructs

Measures	Constructs	M	SD	α
Attitude Toward Computers & Computing Courses Questionnaire (ACCC)	Anxiety	2.64	0.83	0.674
	Enjoyment	3.21	0.55	0.597
	Usefulness of course	3.53	0.72	0.812

with the latest technology, which is very important for executing such student projects. For example, the students have access to the latest computers, various types of tablets, smartphones, and smartwatches. This access is why students provide positive responses to the "technology adequacy" construct (mean = 3.28). This positive perception shows that CLEI constructs play an important role in student learning. Various studies reported a positive association between environmental and attitudinal constructs (Al-Qahtani, 2012; Pyatt & Sims, 2012; Saadon & Liong, 2012).

The ACCC constructs in Table 13.3, show that student have positive attitudes toward the SWF course. The constructs "usefulness of course" and "enjoyment" score the highest (mean = 3.53 and 3.21, respectively). The students found the course useful because they work with a real software company project, develop the product or services based on customers' requirements, and manage and monitor their activities. Further, the students experience first-hand encounters with a real customer, which helps them to learn negotiation skills and develop for future jobs. The "anxiety" construct received the lowest mean score (mean = 2.64), which indicates that the SWF course is exciting and make them bore or create bad experiences. This is also evident in the "enjoyment" construct, which received a mean score of 3.21.

In an open-ended question, the students expressed some of their anxious encounters as:

- *"Groups [one project assigned to one team] have huge gaps and differences because some students are very modest, and they felt shy to say that they are good at programming, and it is also hard to say and to evaluate if one person is good at programming. While the course really works, and I learned so many things from this course, and this is more like a practical course."*
- *"Confusion and sense of competition kept me on my toes for the whole length of the project. I like the course concept, but the variety of cultures among students brings variation in multiple aspects of how the projects flow. After understanding how human the students all are, I was able to forgive and work in a way I felt comfortable with."*
- *"Working in a group where people do not listen or understand what you are talking about when discussing web architectures, object-oriented programming, and other topics makes working in the project depressing."*
- *"When dividing the team, it is better to do it based on the required technical skills distinction, rather than culture differences. It could involve more interaction of the other teams and customers."*

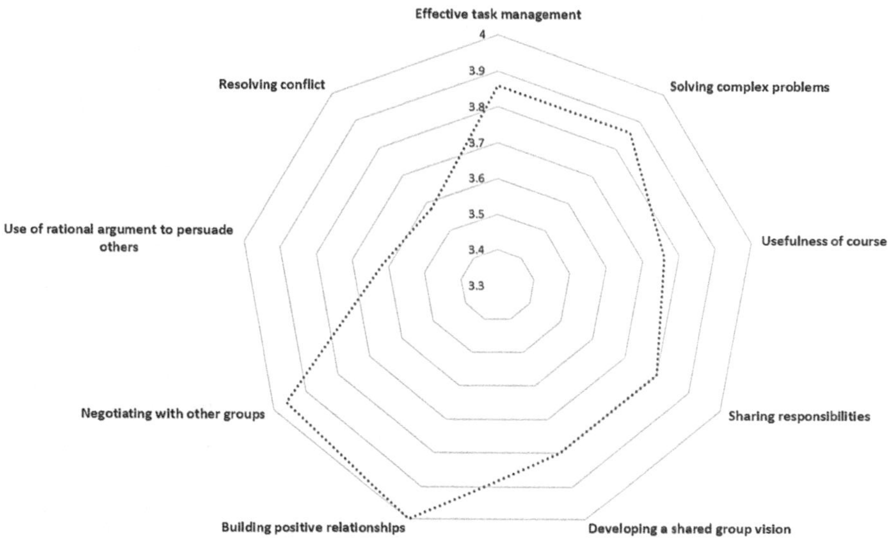

Fig. 13.2 Competencies gained during the SWF project course

- *"It is a 10 credit course, which means it has a big grade, and all the time, I think about not getting a bad grade rather than learning. It has a bad feeling to think that failing this course might really mess up ones study time in the school."*
- *"I think customers should be more involved in the project."*

The students work collaboratively on the SWF project, which helps them gain or improve various competencies as shown in Fig. 13.2. The students rate themselves highly against the achievement of various competencies gained during the SWF project. Building positive relationships with multicultural team members received the highest value, which contributes to developing a shared mental model, managing tasks effectively, solving complex problem, and better negotiating inside and outside the group.

The students expressed their positive experiences and competencies gained during the SWF project as follows:

- *"The [SWF] project allowed me to develop my project management skills as a whole, from planning to scheduling, task allocation to having formal meetings, and collaborating with a steering group. I gained a good insight into what duties a project manager has and what kind of personal traits are needed for the successful management of a project."*
- *"My teamwork and communication skills improved in project."*
- *"I learned about interesting technologies and software tools in usability activities in practice."*
- *"Management has also increased my ability to deal with the different nature of people, how to motivate them and drive or steer them to do the work. Everyone has his own working style to do the work."*

- *"From this course, I found my direction for future work, and I found things that I like to do. Thanks to the teachers, customers, and also my group members!"*

Overall, with these competencies, the students clearly worked collaboratively enable them to generate ideas, solve complex problems, and offer opportunities to form supportive networks in pursuit of improved outcomes. These results are in line with existing literature (Burdett, 2003; Saadon & Liong, 2012).

13.5 Reflective Learning Diaries Importance

Writing reflective diaries is the core element in many medical, nursing, and teacher education programs (Tang, 2002). The reflective diaries can help students to document and redirect their learning to better prepare for challenge position in the software industry. In SWF, reflective learning diaries enable students to see how they can better prepare themselves for the challenging SE profession. Every student writes biweekly learning diary to answer three main questions (*What tasks did you do in this sprint? As a learner, what did you improve or learn compared to the last sprint? What were the issues and challenges you faced in this sprint?*).

It is an opportunity for students to reflect on their individual and group experiences for each sprint. Thus, they can identify their own learning, polish existing skills, and seek improvement opportunities. It also helps students to adapt to individual project needs by understanding software development methods, practices, and tools and their application during software development projects and experiments. For teachers it is provide opportunity to observe each student progress and make the necessary arrangements to assist. It is evident that students reflect on each sprint practice, seek improvement, and adapt development practices based on their experiences. The following are example statements from the students' learning diaries:

- *"We use Trello [online tool;www.trello.com] for task management like Kanban board; our understanding of the Kanban process was minimal. Moving our process to the white board in the Software Factory Lab, we had a deeper course to interact with one another, gather feedback from the supervising teacher, etc. Petty issues, such as missing work-in-progress limits on our board, were quickly raised by the supervising teacher. We thus had to move our work process from Trello to the white board in the Software Factory Lab just to not to repeat ourselves but use a common board and approach."*
- *"Being a multicultural team, sometimes it was hard understanding other people, which affected our work output to an extent. We tried to do the best, however, working schedule of team members was quite problematic. While there is no one to blame, we need to work together, find more consensus, interact on Slack more, be as productive as possible during the few minutes we have together, and persevere in the given task."*

- *"We come from different cultures, different languages, and possess different accents; it was quite difficult for team members to understand each other sometimes during conversations. However, when communicating on Slack, these issues were not present. I believe this was one of the reasons why communication on Slack was more frequent"*

In summary, reflective diaries offer opportunities for students to think critically, look back on the learning activities, help identify what they have learned throughout the practical software development project.

13.6 Teachers' Perspective

As mentioned earlier, course evolved during the years toward more practical direction. It was learned that planning of the project course with real customers in advance has to take count on following issues:

- *Different universities in different areas around the world give very different practical skills set.*
- *Most of the students have not been working in the company environments before or during the studies. So, worker skills and good examples of professional working ways are missing. These are* e.g. *formal meeting procedures and customer presentations.*
- *Project management is not an easy issue for student groups.*

The teaching staff implement the SWF project in relation to the results and adapt teaching techniques to optimize learning outcomes. In this regard, the following lessons were learned:

- *The SWF laboratory involves the local software industry for real projects and customers for the university's SE students. However, non-disclosure agreements must usually be signed. This requirement must be communicated very clearly to students to avoid breaches. Similarly, the message should be communicated to the companies that the students might not be aware of the seriousness of confidentiality, and the companies should be careful when assigning confidential tasks.*
- *The company and real customers are invited to the final project presentation. However, their evaluation should not focus only on the final product. The evaluation criteria should be clearly communicated to these external stakeholders to avoid confusion and promote fair evaluations.*
- *The SWF project course design is very effective for motivating students and plays an active role throughout the project. For example, the SWF project has an almost zero dropout rate, despite requiring intense work over one semester. This is why it is important to include serious games and consider the gamification approach to improve students' motivation and active participation and increase*

*collaboration (*Glover & Glover,2013*;* Sanmugam et al., 2016*;* Sheth, Bell, & Kaiser,2012*).*

- *The SWF project course is also demanding from the teaching staff perspective due to the frequent communication and mentoring. Each SWF project team requires a teaching assistant to provide technical support, monitor their progress continuously, and facilitate throughout the SWF journey. This technique is very effective in a SE project-based environment (*Palacin-Silva et al., 2017*;* Walker & Andrew, 2015*).*
- *Creating balanced teams is a challenge with multicultural and heterogenous students. An unbalanced team with inadequate skills or cultural conflicts can create difficult situations during the long and intensive work period. The teaching staff must proactively oversee the teams' work and communication.*
- *Almost all universities around the world have strong policies regarding the installation of computer laboratory software. Such policies affect students when they need to urgently install software packages. The solution is to install a virtual machine on all students' laboratory computers. This enables students to freely install and update the required software, applications, and tools.*

13.7 Conclusion

This chapter provides a detailed description of a project-based SE graduate course, insight into course delivery, course assessment, peer evaluation, and the use of tool support for forming teams. Additionally, we also documented our 6 years teaching experiences, students' perceptions of the SWF laboratory, and the SWF course.

The SWF laboratory is an innovative learning environment that offers a graduate-level project-based course, where students learn SE processes and building software development products or solutions with real industrial customer. Our SWF experience shows that it is very important to maintain a balance between coding and SE development processes. The students can easily ignore development processes and start focusing on their own individual coding tasks. The main findings from the 6 years SWF course execution is that allow student to experiment with the software developments methods and process. Learning and mastering these processes and worker skills are essential to compete in the competitive job market.

The student survey results show that the majority expressed a positive view in two ways: (1) The SWF course is appreciated as an important course in their master's degree curriculum. The SWF project is a good blend of theoretical and practical training that enhances students' enjoyment, and they find the course useful for achieving the required competencies for future jobs; (2) The SWF laboratory makes it possible to provide opportunities for students to interact with real software industry customers and work collaboratively in a multicultural environment. These findings are aligned with the results of several researchers (Ahmad et al., 2014; Al-Qahtani, 2012; Cico, Jaccheri, Nguyen-Duc, & Zhang, 2020; Pyatt & Sims, 2012).

Fair assessment is a very important and complex activity for teaching staff in such courses. We developed a matrix that considers various aspects of teamwork. Along these lines, free riders and hardworking students in the project can be identified and treated fairly. Peer assessment and individual reflective learning diaries play an important role in encouraging and motivating students in this collaborate course. Reflective learning diaries is an excellent technique to boosts student engagement, assess their epistemological beliefs and conceptions of learning. On the other hand, it provides an opportunity for the teaching staff to create and update strategies for monitoring and regulating learning. For a teaching staff this kind of course requires a good understanding of how SW development projects are executed in company environments. The peer assessment findings are aligned with other studies (Leenknecht et al., 2020; Li, Xiong, Hunter, Guo, & Tywoniw, 2020).

In future work, the SWF course could be run in a geographically distributed context in cooperation with other SWFs and other universities or even as a part of companies' development teams However, this might require much greater technical competencies among teaching support staff and strong collaboration between universities and participating companies. It would be interesting to investigate such courses in different academic and cultural settings.

References

ACM Joint Task Force. (2014). *Software engineering 2014—curriculum guidelines for undergraduate degree programs in software engineering*. Retrieved from https://www.acm.org/binaries/content/assets/education/se2014.pdf

Ahmad, M. O., Liukkunen, K., & Markkula, J. (2014). Student perceptions and attitudes towards the software factory as a learning environment. In IEEE global engineering education conference, EDUCON.

Al-Qahtani, M. F. (2012). Students' perception and attitude towards computer laboratory learning environment. *Educational Research, 3*(4), 402–411.

Anderson, T. (2008). *Theory and practice of online learning*. Athabasca: Athabasca University Press.

Andreessen, M. (2011). Why software is eating the world. *The Wall Street Journal*. Retrieved from http://www.djreprints.com

Barg, M., & Barg, M. (1999). Problem based learning for foundation computer science courses. Basser Department of Computer Science, University of Sydney.

Bell, S. (2010). Project-based learning for the 21st century: Skills for the future. *The Clearing House, 83*(2), 39–43.

Cico, O., Jaccheri, L., Nguyen-Duc, A., & Zhang, H. (2020). Exploring the intersection between software industry and software engineering education—a systematic mapping of software engineering trends. *Journal of Systems and Software, 172*, 110736.

Chao, J., & Randles, M. (2009). Agile software factory for student service learning. In *2009 22nd conference on software engineering education and training* (pp. 34–40). IEEE.

Chang, J. C., Hsiao, Y. D., Chen, S. C., & Tsung-Ta, Y. (2018). Core entrepreneurial competencies of students in departments of electrical engineering and computer sciences (EECS) in universities. *Education + Training, 60*(7/8), 857–872.

Erdogmus, H., & Peraire, C. (2017). Flipping a graduate-level software engineering foundations course. In *39th international conference on software engineering: Software engineering education and training track* (pp. 23–32).

Fagerholm, F., Oza, N., & Münch, J. (2013). A platform for teaching applied distributed software development: The ongoing journey of the Helsinki software factory. In *3rd international workshop on collaborative teaching of globally distributed software development* (pp. 1–5).
Fulwiler, T. (1986). Seeing with journals. *The English Record, 32*(3), 6–9.
Glover, I., & Glover, I. (2013). Play as you learn: gamification as a technique for motivating learners. In *Proceedings of EdMedia 2013—World Conference on Educational Media and Technology*. EdMedia + Innovate Learning. Vol. 2013. AACE.
Howe, S. (2010). Advances in engineering education where are we now? *Statistics on Capstone Courses Nationwide.* Advances in Engineering Education.
Hung, C. M., Hwang, G. J., & Huang, I. (2012). A project-based digital storytelling approach for improving students' learning motivation, problem-solving competence and learning achievement. *Journal of Educational Technology & Society, 15*(4), 368–379.
Hrivnak, G. A. (2013). CATME smarter teamwork (www.CATME.org). *Academy of Management Learning Education, 12*(4), 679–681.
Burdett, J. (2003). Making groups work: University students' perceptions. *International Education Journal, 4*(3), 177–191.
Jabarullah, N. H., & Hussain, H. I. (2019). The effectiveness of problem-based learning in technical and vocational education in Malaysia. *Education+Training, 61*, 552.
Kaufman, D. B., Felder, R. M., & Fuller, H. (2000). Accounting for individual effort in cooperative learning teams. *Journal of Engineering Education, 89*(2), 133–140.
Khine, M. S., & Lourdusamy, A. (2003). Blended learning approach in teacher education: Combining face-to-face instruction, multimedia viewing and online discussion. *British Journal of Educational Technology, 34*(5), 671–675.
Kolb, A. Y., & Kolb, D. A. (2005). *Learning styles and learning spaces: Enhancing experiential learning in higher education.* Source: Academy of Management Learning & Education (vol. 4).
Kolmos, A., & de Graaff, E. (2014). Problem-based and project-based learning in engineering education–merging models. In A. Johri & B. M. Olds (Eds.), *Cambridge handbook of engineering education research (CHEER)* (pp. 141–160). New York: Cambridge University Press.
Layton, R. A., & Ohland, M. W. (2001). Peer ratings revisited: focus on teamwork, not ability. In *Proceedings of the 2001 American Society for Engineering Education Annual Conference & Exposition.*
Land, S. M. (2004). The design and evaluation of a CSCL tool to support reflection and revision of design projects. *Journal of Computing in Higher Education, 16*(1), 68–92.
Lee, J. S., Blackwell, S., Drake, J., & Moran, K. A. (2014). Taking a leap of faith: Redefining teaching and learning in higher education through project-based learning. *Inter-disciplinary Journal of Problem-Based Learning, 8*(2), 2.
Leenknecht, M., Wijnia, L., Köhlen, M., Fryer, L., Rikers, R., & Loyens, S. (2020). Formative assessment as practice: The role of students' motivation. *Assessment & Evaluation in Higher Education*, 1–20.
Li, H., Xiong, Y., Hunter, C. V., Guo, X., & Tywoniw, R. (2020). Does peer assessment promote student learning? A meta-analysis. *Assessment & Evaluation in Higher Education, 45*(2), 193–211.
Martyn, M. (2003). The hybrid online model: Good practice. *Educause Quarterly, 26*(1), 18–23.
Mann, L., Chang, R., Chandrasekaran, S., Coddington, A., Daniel, S., Cook, E., … Dohaney, J. (2020). From problem-based learning to practice-based education: A framework for shaping future engineers. *European Journal of Engineering Education*, 1–21.
McGourty, J., Dominick, P., & Reilly, R. R. (1998). Incorporating student peer review and feedback into the assessment process. In *Frontiers in education conference. Moving from "teacher-centered" to "learner-centered" education.*
Newby, M., & Fisher, D. (1997). An instrument for assessing the learning environment of a computer laboratory. *Journal of Educational Computing Research, 16*(2), 179–190.

Nicole, C., Pamela, D., & Rebecca, S. (2005). Self and peer assessment in software engineering projects. In *Proceedings of the 7th Australasian conference on computing education* (vol. 42, pp. 91–100). John Garratt Publishing.

Ohland, M. W., & Layton, R. A. (2000). Comparing the reliability of two peer evaluation instruments the impact of integration on student's persistence. *View project optimizing student team skill development using evidence-based strategies view project comparing the reliability of two peer evaluation instruments.*

Palacin-Silva, M., Khakurel, J., Happonen, A., Hynninen, T., & Porras, J. (2017). Infusing design thinking into a software engineering capstone course. In *30th conference on software engineering education and training* (pp. 212–221).

Pyatt, K., & Sims, R. (2012). Virtual and physical experimentation in inquiry-based science labs: Attitudes, performance and access. *Journal of Science Education and Technology, 21*(1), 133–147.

Saadon, S., & Liong, C.-Y. (2012). Perception of students on services at the Computer Laboratory: A case study at the School of Mathematical Sciences, Universiti Kebangsaan Malaysia. *Procedia Social and Behavioral Sciences, 59*, 117–124.

Sanders, D. (1984). Managing and evaluating students in a directed project course. *ACM SIGCSE Bulletin, 16*(1), 15–25.

Sanmugam, M., Mohamed, H., Zaid, N. M., Abdullah, Z., Aris, B., & Suhadi, S. M. (2016). *Gamification's role as a learning and assessment tool in education.* International Journal of Knowledge-Based Organizations (Vol. 6, p. 28).

Sheth, S., Bell, J., & Kaiser, G. (2012). Increasing student engagement in software engineering with gamification.

Taibi, D., Lenarduzzi, V., Liukkunen, K., Lunesu, I., Matta, M., Fagerholm, F. Ahmad, M. (2016). "Free" innovation environments: Lessons learned from the software factory initiatives.

Tang, C. (2002). Reflective diaries as a means of facilitating and assessing reflection. In *Proceedings of the 29th HERDSA annual conference* (pp. 7–10).

Tvedt, J. D., Tesoriero, R., & Gary, K. A. (2002). The software factory: An undergraduate computer science curriculum. *Computer Science Education, 12*(1–2), 91–117.

Walker, A. Andrew E. (2015). Essential readings in problem-based learning: Exploring and extending the legacy of Howard S. Barrows.

Wijnia, L., Loyens, S. M. M., & Derous, E. (2011). Investigating effects of problem-based versus lecture-based learning environments on student motivation. *Contemporary Educational Psychology, 36*(2), 101–113.

Wilkins, D. E., Lawhead, P. B., Wilkins, D. E., & Lawhead, P. B. (2000). Evaluating individuals in team projects. *ACM SIGCSE Bulletin, 32*(1), 172–175.

Yilmaz, R., Yilmaz, F. G. K., & Keser, H. (2020). Vertical versus shared e-leadership approach in online project-based learning: A comparison of self-regulated learning skills, motivation and group collaboration processes. *Journal of Computing in Higher Education, 32*, 628–654.

Chapter 14
How to Generate Exercise Questions for Web-Based Investigative Learning

Rei Saito, Akihiro Kashihara, Yoshiki Sato, Miki Hagiwara, and Koichi Ota

14.1 Introduction

Investigative learning is learner-centred, in which learners could investigate any question with abundant information resources in a self-directed way to obtain ample and multidimensional knowledge (Edwards, 2012; Sellwood, 1991). It induces them to self-regulate their own learning process, in which obtained knowledge is individualised (Edwards, 2012). The self-directedness with self-regulation contributes to developing skills in recognising, analysing, and resolving questions (Sellwood, 1991), which meets to the demands for improving competence necessary in today' information society (Raij, 2007).

With the development of information society in recent years, conducting investigative learning on the Web (called Web-based investigative learning) has become popular. On the Web, learners could not only search information, but also investigate various questions with a huge number and variety of Web resources (Land, 2000). The Web-based investigative learning process involves navigating Web resources/pages and constructing their own knowledge (Jonassen, 2000; Land, 2000). However, learners tend to search a limited number of Web resources/pages. As a result, they would construct limited knowledge. In an elaborate investigation about an initial question, learners are expected to widen and deepen the question, which requires them to identify related questions to be further investigated during navigation and knowledge construction with Web resources (Hill & Hannafin,

R. Saito (✉) · A. Kashihara · Y. Sato · M. Hagiwara
The University of Electro-Communication, Chofu, Japan
e-mail: rei.saito@uec.ac.jp

K. Ota
Japan Advanced Institute of Science and Technology, Nomi, Japan

D. Ifenthaler et al. (eds.), *Balancing the Tension between Digital Technologies and Learning Sciences*, Cognition and Exploratory Learning in the Digital Age,
https://doi.org/10.1007/978-3-030-65657-7_14

1997). This corresponds to expanding the initial question into related ones as sub-questions.

In addition, existing Web resources are not always well-structured and reliable for learning. They generally provide learners with no learning scenario implying questions to be investigated and their sequence (Kashihara & Akiyama, 2013). The learners accordingly need to select and navigate the Web resources/pages and integrate the contents learned at the navigated resources/pages (Henze & Nejdl, 2001) to expand a question into the sub-questions by themselves. Such question expansion corresponds to creating a learning scenario, which could play a crucial role in self-regulating navigation and knowledge construction process (Azevedo & Cromley, 2004). Wider and deeper expansion of an initial question would make an investigative learning process more structured and beneficial (Kashihara & Akiyama, 2013).

On the other hand, it is not so easy for learners to create their own scenario in concurrence with navigation and knowledge construction process (Zumbach & Mohraz, 2008). As they tend to pay more attention to navigation and knowledge construction in investigating a question (Hill & Hannafin, 1997), they often miss finding out sub-questions to be further investigated. This brings about an insufficient investigation of the initial question.

In our previous work, we have proposed a model of Web-based investigative learning, and developed a cognitive tool called interactive Learning Scenario Builder (iLSB) as model-based learning environment (Kashihara & Akiyama, 2016a). iLSB provides learners with scaffolding for promoting question expansion as modelled. The results of the case study suggest that iLSB promotes question expansion, and makes learning scenario created by learners wider and deeper. On the other hand, we have also confirmed that learners could not always conduct wider and deeper question expansion even if they use iLSB. It is accordingly necessary to help them acquire skills in Web-based investigative learning (Saito, Sato, Hagiwara, Ota, & Kashihara, 2019).

We are addressing the issue how to develop skills in Web-based investigative learning particularly for novice learners who could not create an appropriate learning scenario on their own. Towards this issue, it is necessary for learners to practise so that they could expand an initial question wider and deeper. Accordingly, the first step to this issue is to provide learners with exercise questions for Web-based investigative learning, which is the novel point in this work.

In this paper, we define an exercise question as the one including not only an initial question but also its sub-questions to be expanded. This paper also proposes a method for generating exercise questions, which extracts keywords representing candidates of sub-questions by means of Linked Open Data (LOD) and Word2vec. Learners are expected to use iLSB to resolve an exercise question, in which they need to select appropriate keywords as sub-questions from the extracted candidates to create a learning scenario. Such exercise contributes to developing skills in question expansion for Web-based investigative learning.

In addition, this paper reports the case studies whose purpose were to evaluate the appropriateness and effectiveness of exercise questions generated by the proposed method. The results suggest that exercise questions are appropriate, and effectively work particularly for novice learners to conduct question expansion.

14.2 Web-Based Investigative Learning

14.2.1 Difficulties in Web-Based Investigative Learning

In Web-based investigative learning, learners are allowed to search for Web resources/pages to investigate any question. In this learning, it is important for learners to construct deeper and wider knowledge as to an initial question by gathering and navigating Web resources, integrating the contents learned in the resources, and expanding the initial question into sub-questions in a self-directed way. In addition, most Web resources do not have any learning scenario such as table of contents in instructional textbook, which implies questions to be investigated and their sequence. Learners accordingly need to create their own scenario in concurrence with navigation and knowledge construction.

On the other hand, it is difficult for learners to concurrently navigate Web resources, construct knowledge and create learning scenario. The more they pay attention to navigation and knowledge construction, the less they pay attention to question expansion. This prevents them from creating appropriate, wider, and deeper learning scenario. In order to address this problem, it is necessary to promote elaborate question expansion (Kashihara & Akiyama, 2016b). In our previous work, we have modelled the process of Web-based investigative learning, and developed iLSB with scaffolding for promoting Web-based investigative learning process as modelled.

14.2.2 Model of Web-Based Investigative Learning

This model consists of three cyclic phases: (a) search for Web resources, (b) navigational learning, and (c) question expansion (Kashihara & Akiyama, 2013; Kinoshita & Kashihara, 2013). In phase (a), learners are expected to use a search engine with a keyword representing an initial question, and to gather Web resources suitable for investigating the question.

In phase (b), they are expected to navigate Web pages in the gathered resources to learn the contents, and to construct their knowledge. In the knowledge construction process, they are also expected to extract keywords representing the contents learned in the pages to make connections among them. Such knowledge construction with navigation is called navigational learning (Kashihara & Akiyama, 2013). In this phase, the learners are required to self-regulate the knowledge construction process, which involves reflecting on whether constructed knowledge is appropriate and sufficient for their learning.

In phase (c), the learners are expected to find out some related sub-questions to be further investigated about the initial question, which could be obtained from knowledge constructed (keywords extracted) in phase (b). This corresponds to expanding the initial question into sub-questions. They are also expected to cyclically investigate each sub-question in the next phase (a) and (b).

These three phases are repeated until no further question expansion occurs. The question expansion results in a tree called question tree, which corresponds to learning scenario. This tree includes part-of relationships between the question and its sub-questions. The root of the tree represents the initial question. Each node also corresponds to a question represented by a keyword (called q-keyword). In this work, q-keyword is used as a substitute for question investigated.

Creating such a tree corresponds to defining the initial question, which specifies what to investigate and how for the initial question. Therefore, learning scenario creation is essential for Web-based investigative learning.

14.2.3 Interactive Learning Scenario Builder (iLSB)

We have developed iLSB as an add-on for Firefox, which intends to scaffold investigative learning process as modelled. iLSB provides a search engine for gathering and selecting Web resources, a keyword repository for navigational learning, and a question tree viewer for question expansion as scaffolding for each phase in the model. Figure 14.1 shows the user interface of iLSB.

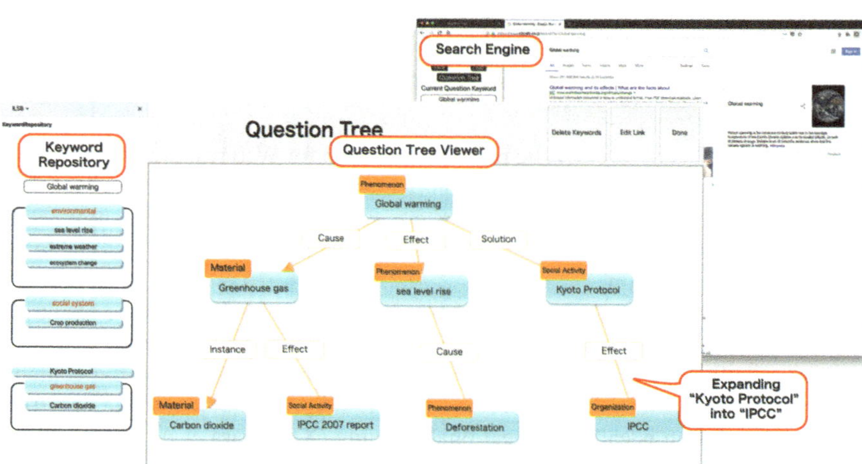

Fig. 14.1 User interface of iLSB

Let us demonstrate how iLSB scaffolds investigative learning process with an example of initial question "*What is global warming?*". First of all, learners are expected to input "*Global warming*" as initial q-keyword into a search engine and collect Web resources suitable for learning about "*Global warming*". Secondly, they are expected to store keywords related to "*Global warming*" such as "*Greenhouse gas*" and "*Kyoto Protocol*" in the keyword repository while they navigate the Web resources/pages. These stored keywords represent the contents learned about the initial question. Then, they are expected to associate the keywords to construct their knowledge in the keyword repository (Kashihara & Akiyama, 2016b).

After that, the learners are expected to find out sub q-keywords such as "*Greenhouse gas*" from the keywords stored in the keyword repository and to add them into the question tree viewer. Such addition corresponds to question expansion. In Fig. 14.1, they expand the initial q-keyword to "*Greenhouse gas*", "*See level rise*", and "*Kyoto protocol*" in the question tree viewer. They are next expected to investigate these sub q-keywords in the same way until new sub q-keywords are not added anymore.

14.2.4 Issue Addressed and Purpose

We have ascertained that some learners still have difficulties in creating their own scenario even with iLSB. It is necessary for such learners to practise question expansion to develop skills in Web-based investigative learning. The first step to the skill development is to provide learners with exercise questions.

The main issue addressed in this paper is how to define exercise questions so that they could conduct model-based investigative learning on the Web. In exercise questions, it is generally necessary to prepare answers as learning scenarios/knowledge to be created/constructed. However, it is difficult to assume them since Web-based investigative learning with the same question could result in diverse learning scenarios/knowledge. Although we do not prepare answers for exercise questions in advance, we have already developed the LOD-based mechanism for diagnosing the appropriateness of learning scenario involving question expansion conducted by learners as another work (Sato, Kashihara, Hasegawa, Ota, & Takaoka, 2019), and ascertained the feasibility.

At the beginning of Web-based investigative learning, we have provided learners only with an initial question so far. However, they often have difficulties in finding out sub-questions to be expanded from the contents learned. In order to develop skills in question expansion, we define an exercise question as the one including not only an initial question but also related sub-question candidates, which make learners more skilful in finding out appropriate sub-questions.

This paper proposes a method for extracting keywords (called c-keywords) as q-keyword candidates to be included in an exercise question from Linked Open Data (LOD), which are related to an initial question. This method generates the exercise question including the initial q-keyword and extracted c-keywords to present to learners. In this method, the c-keywords are divided into positive c-keywords that have strong relationships with the initial question and negative c-keywords that have weak or less proper relationships with the initial question. In resolving the exercise question, learners are expected to select positive c-keywords and avoid selecting negative ones to create their learning scenario. Such selection allows them to develop their skills in proper question expansion.

14.3 Generating Exercise Questions

Let us here describe how to generate exercise questions with LOD and Word2vec.

14.3.1 Linked Open Data (LOD)

LOD is a set of structured data, which interlinks related information on the Web. In this work, we use DBpedia Japanese as LOD (DBpedia Japanese, 2016). The data of DBpedia Japanese are extracted from Japanese Wikipedia, and stored in RDF (Resource Description Framework) format. RDF data is described with triplet: subject, predicate and object. A collection of RDF data also represents a network structure of the triplets.

Data in DBpedia Japanese are extracted by means of query language SPARQL. By sending SPARQL query to DBpedia Japanese, it is possible to extract keywords related to initial q-keyword from the network structure in DBpedia Japanese.

14.3.2 Word2vec

Word2vec is an algorithm, which analyses words in documents to generate word vectors (Ling, Dyer, Black, & Trancoso, 2015). The generated vectors include words and their adjacent words. This algorithm learns weights of adjacent relations between words by means of neural network from a large number of documents as input.

In this work, we first use MeCab (Kudo, 2013) to extract words from documents in Japanese Wikipedia. We second use Word2vec to generate word vectors from the extracted words. The generated vectors are used for calculating cosine similarity between the initial q-keyword and extracted c-keywords from DBpedia Japanese.

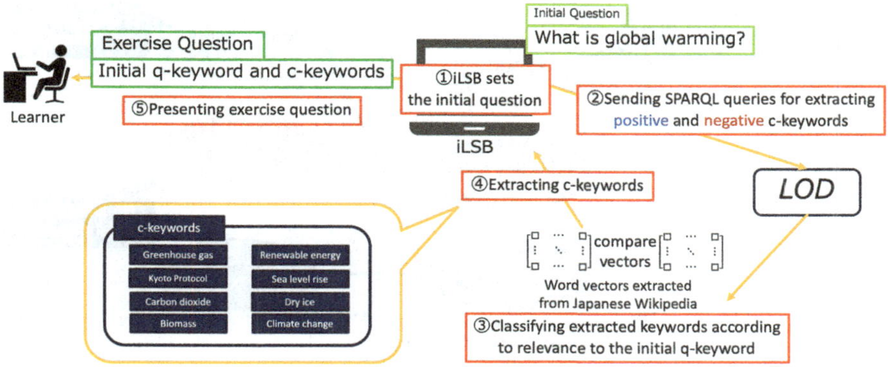

Fig. 14.2 Framework for generating exercise questions

14.3.3 Framework

Figure 14.2 shows the framework for generating an exercise question. The method for generating exercise question is aimed at implementation in iLSB. In the proposed method, exercise question is generated by means of DBpedia Japanese and word vectors prepared in advance by Word2vec. It consists of an initial q-keyword and c-keywords including positive ones highly related to the initial q-keyword and negative one less related to the initial q-keyword.

First of all, iLSB sets up an initial question, and sends SPARQL queries to DBpedia Japanese. The queries allow iLSB to extract related keywords and their relations with the initial q-keyword on DBpedia Japanese to classify as c-keywords according to relevance to the initial q-keyword. It then calculates cosine similarity between the initial q-keyword and all of the classified c-keywords by means of word vectors. C-keywords with higher similarity are decided as positive ones, and c-keywords with lower similarity are decided as negative ones. iLSB next limits the number of positive and negative c-keywords to generate an exercise question, which is presented to learners.

In resolving the exercise question, learners are expected to investigate the initial question by using iLSB and to select sub-questions not from the negative ones but from the positive ones. They are finally expected to create proper learning scenario. Presenting c-keywords in advance intends to help learners select sub-questions to expand the initial question. In addition, the proposed method makes it possible to adjust the difficulty of exercise questions by changing the amount and ratio of positive and negative c-keywords to be presented to learners. Such adjustment allows learners to improve their skills in proper learning scenario creation.

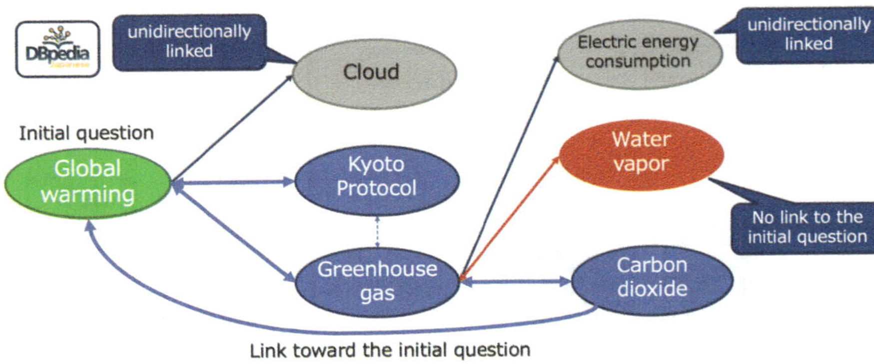

Fig. 14.3 Relations between initial question and c-keywords

```
SELECT DISTINCT ?child
      WHERE
      {
       VALUES ?root_name{"global warming"@ja}
       ?root rdfs:label ?root_name.
       ?root ?p ?child.
       ?child ?p2 ?root.
       FILTER(contains(str(?child),'http://ja.dbpedia.org/resource/'))
      }
SELECT DISTINCT ?grand_child
      WHERE
      {
       VALUES ?root_name{"global warming"@ja}
       VALUES ?child_name{"greenhouse gas"@ja}
       ?root rdfs:label ?root_name.
       ?child rdfs:label ?child_name.
       ?grand_child ?p3 ?child.
       ?child ?p4 ?grand_child.
       ?grand_child ?p5 ?root.
       FILTER(contains(str(?grand_child),'http://ja.dbpedia.org/resource/'))
      }
```

Fig. 14.4 An example of SPARQL queries for extracting positive c-keywords

14.3.4 Extracting c-Keywords with SPARQL Query

In the proposed method, positive and negative c-keywords are classified by means of SPARQL queries, in which the relations between the initial q-keyword and related keywords on DBpedia Japanese are considered. Let us demonstrate the process to extract c-keywords with the example "*What is global warming?*" as initial question. Figure 14.3 shows the links between an initial q-keyword ("*global warming*") and keywords related on DBpedia Japanese.

Since positive c-keywords should be highly related to the initial q-keyword, it is necessary to extract keywords close to the initial q-keyword on DBpedia Japanese. This method accordingly searches and classifies the keywords as positive c-keywords, which have bidirectional links with the initial q-keyword such as "*Kyoto Protocol*" in Fig. 14.3. Figure 14.4 shows an example of SPAQRL queries for extracting such c-keywords close to the initial q-keyword.

It also searches the keywords such as "*carbon dioxide*" and "*water vapor*" that have bidirectional links with the positive c-keywords classified. Among them, the keywords such as "*carbon dioxide*" that have a link with the initial q-keyword are classified as positive c-keywords. But, the keywords such as "*water vapor*" that have no link with the initial q-keyword are not regarded as positive c-keywords.

As for negative c-keywords, on the other hand, we should not select keywords that learners could readily decide as negative ones. This method accordingly searches the keywords like "*water vapor*" in Fig. 14.3 as negative ones via SPARQL queries, which are not so far from the initial q-keyword. In other words, it aims at classifying the keywords as negative ones, which are bidirectionally linked with positive c-keywords classified and have no link with the initial q-keyword.

14.4 Case Studies

Let us next describe the case studies we have conducted for evaluating exercise questions proposed in this paper. Research questions in this study are as follows:

- Whether c-keywords to be presented are appropriate as exercise questions, and
- Whether exercise questions generated contribute to promoting proper learning scenario creation.

To address these, we conducted two case studies. The purpose of first case study was to ascertain the appropriateness of c-keywords and second one was to ascertain the effectiveness of exercise questions.

14.4.1 Appropriateness of c-Keywords

14.4.1.1 Purpose and Procedure

In order to evaluate the appropriateness of exercise questions, we have conducted the first case study. The main purpose of the study was to ascertain whether c-keywords extracted by the proposed method were appropriate as q-keywords to be used/not to be used for expanding an initial question. In this study, three of the authors evaluated each c-keywords as "appropriate" or "inappropriate" by referring to reliable Web resources related to the initial question. The appropriateness was decided by majority of three authors.

In this study, we prepared "What is renewable energy?" (renewable energy as initial q-keyword) and "What is foodborne illness?" (foodborne illness as initial q-keyword) as initial questions. We extracted c-keywords for each initial q-keyword by means of the proposed method described in Sect. 3.4 whose numbers of positive ones were limited to 22 for renewable energy and 35 for foodborne illness, and whose number of negative ones were about half of positive ones (13 for renewable energy and 18 for foodborne illness).

The appropriateness of positive (or negative) c-keywords extracted for each initial question was calculated as the ratio of positive (or negative) c-keywords evaluated as "appropriate" (or "inappropriate") to all the positive (or negative) ones.

14.4.1.2 Results

Figure 14.5 shows the appropriateness of positive and negative c-keywords evaluated for each initial question and for total. As shown in this figure, about 80% of both positive and negative c-keywords were evaluated as appropriate. This suggests that exercise questions generated works as expected.

14.4.1.3 Discussion

Let us discuss inappropriate c-keywords (rated as "inappropriate"). According to Fig. 14.5, the percentages of inappropriate c-keywords were about 20%. Many of positive c-keywords rated as inappropriate represented extreme specifics or company names related to the initial questions. Although these keywords are close to the initial question on DBpedia Japanese, they are evaluated as inappropriate as sub-questions for the initial one.

As for negative c-keywords rated as inappropriate, one example was *mackerel*, which often causes food poisoning. However, this was considered as a negative c-keyword because it had no link with *foodborne illness* on DBpedia Japanese. We think this is the limitation of only using DBpedia Japanese as LOD. Using other LOD with DBpedia Japanese could be a promising solution.

14.4.2 Effectiveness of Exercise Question

14.4.2.1 Purpose and Procedure

We have next conducted the second study whose purpose was to ascertain the effectiveness of exercise questions for helping learners create a proper learning scenario with iLSB. The participants were 14 graduate and undergraduate students in science

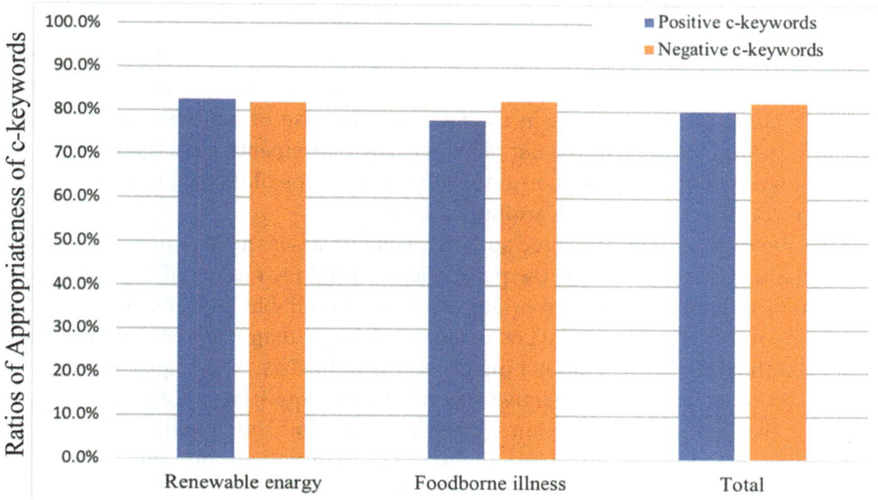

Fig. 14.5 Appropriateness of c-keywords

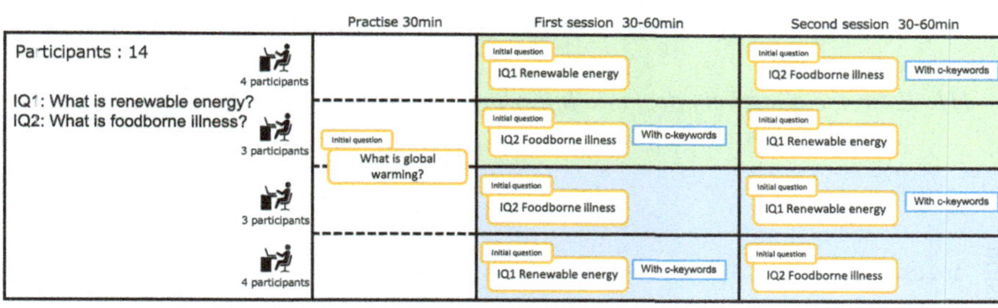

Fig. 14.6 Procedure

and technology. The study was composed of two sessions. In one session, each participant conducted Web-based investigative learning only with an initial question. In the other session, they conducted Web-based investigative learning with an exercise question including an initial question and c-keywords.

The initial questions were, "IQ1: What is renewable energy?" and "IQ:2 What is foodborne illness?" IQ1 was represented as initial q-keyword "renewable energy". IQ2 was also represented as "foodborne illness". We used the same c-keywords for IQ1 and IQ2 that were extracted in the first case study.

Figure 14.6 shows the procedure of the study. The participants were divided into the following four conditions in order to remove the order effect of the initial questions and the presentation of the c-keywords:

(a) Investigating IQ1, then investigating IQ2 with c-keywords,
(b) Investigating IQ2 with c-keywords, then investigating IQ1,
(c) Investigating IQ2, then investigating IQ1 with c-keywords, and

(d) Investigating IQ1 with c-keywords, then investigating IQ2.

The participants first practised the use of iLSB for 30 minutes. After that, each participant was required to investigate the corresponding initial question under each condition for 30 to 60 minutes in each session. In case of exercise question, the c-keywords were presented as a list on paper. The participants were required to use iLSB to investigate the initial question involving selection of the sub-questions from the c-keywords to create their learning scenario.

After the two sessions, the three authors manually assessed each question expansion in the scenarios created by the participants with three ratings of "appropriate", "weak appropriate" and "inappropriate" referring to reliable Web resources for the initial questions. The ratings were determined by a majority vote of the three authors. If the three ratings did not match, it was judged as "weak appropriate".

In order to ascertain the effectiveness of exercise questions, we compared the ratios of the ratings for question expansion (QE for short) with and without c-keywords. We also divided 7 participants for each initial question into high rating group (3 participants) and low rating group (4 participants) according to the ratio of "appropriate" ratings for QE without c-keywords, and analysed the effectiveness in detail. In total, there were six participants in the high rating group. The remaining eight participants were categorised in the low rating group.

In evaluating the effectiveness of the exercise questions, we set the following hypotheses:

H1: Exercise questions promote appropriate question expansion.

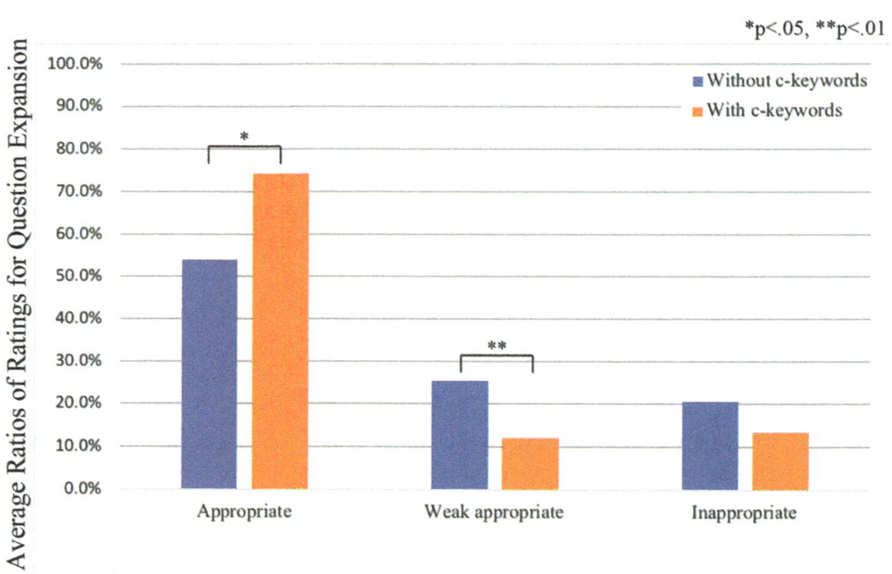

Fig. 14.7 Average ratios of ratings for question expansion in all participants

H2: Exercise questions particularly allow learners with lower skill to conduct question expansion.

14.4.2.2 Results

In this section, all t-test are one-sided and effect sizes reported for t-test are Cohen's d. Figure 14.7 shows the average ratios of three ratings for QE in all participants. In order to confirm the hypothesis H1, we compared the ratings in QE evaluation for learning scenarios created with and without c-keywords. The average ratio of "appropriate" rating for QE with c-keywords was significantly higher than the one without c-keywords ($t(26) = -2.34$, $p < 0.05$, $d = 0.88$). As for the "weak appropriate" and "inappropriate" ratings, the average ratios with c-keywords deceased compared to the ones without c-keywords. In particular, the average ratio of "weak appropriate" rating decreased significantly ($t(26) = 2.56$, $p < 0.01$, $d = 0.97$). This is caused by the significant difference in the "appropriate" rating. These results suggest that presenting c-keywords promotes appropriate question expansion and also prevents inappropriate expansion. This suggestion supports H1.

Figures 14.8 and 14.9 also show the average ratios of three ratings for QE in the high and low rating groups. As for the high rating group, the results of one-sided t-test showed no significant difference between every rating for QE with and without c-keywords. This suggests that the hypothesis H1 is not supported in the high rating group.

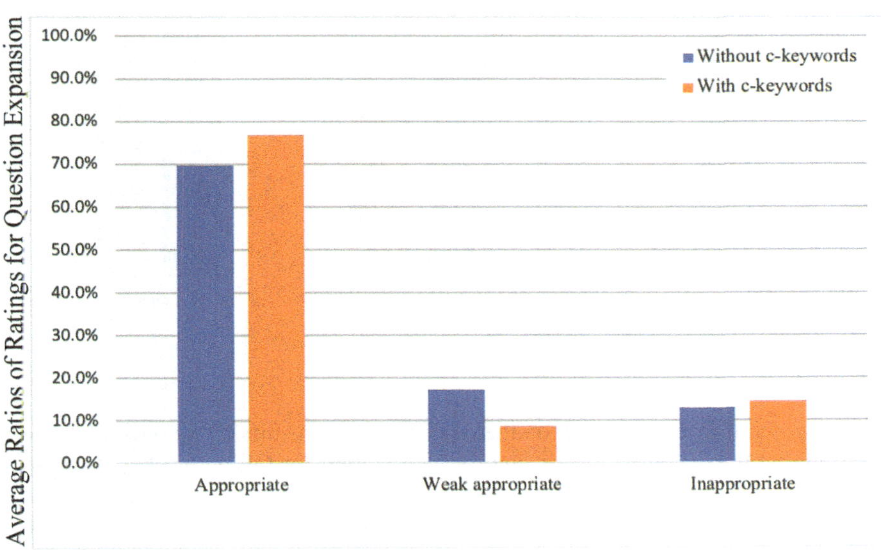

Fig. 14.8 Average ratios of ratings for question expansion in high rating group

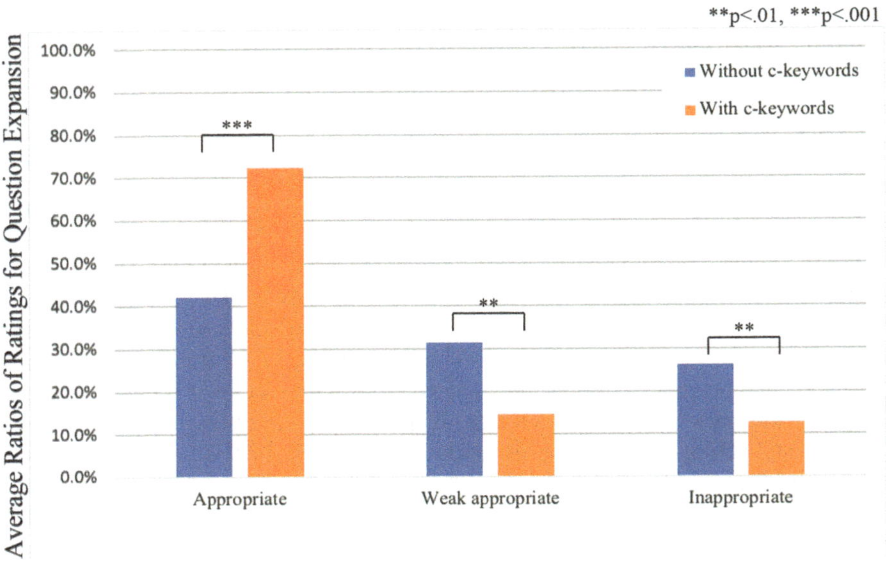

Fig. 14.9 Average ratios of ratings for question expansion in low rating group

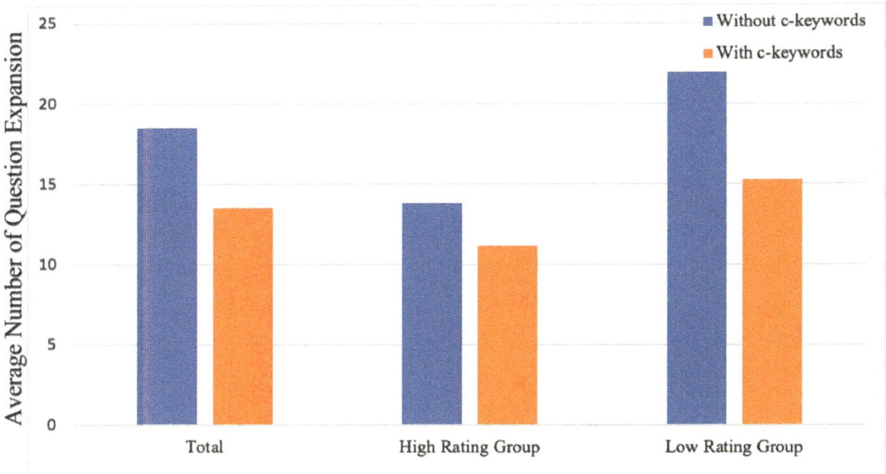

Fig. 14.10 Average Number of Question Expansion

As for the low rating group, on the other hand, the results of the one-sided t-test showed that the average ratio of "appropriate" rating for QE with c-keywords was significantly higher than the one without c-keywords ($t(14) = -4.39$, $p < 0.001$, $d = 2.20$). The average ratios of "weak appropriate" and "inappropriate" ratings for QE with c-keywords were also significantly lower (Weak appropriate: $t(14) = 2.75$,

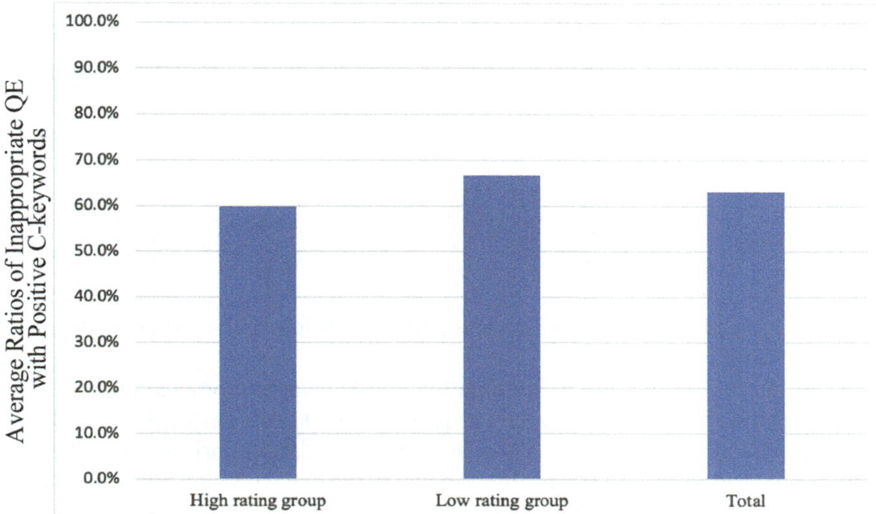

Fig. 14.11 Average ratios of inappropriate QE with positive c-keywords

$p < 0.01$, $d = 1.37$; Inappropriate: $t(14) = 2.72$, $p < 0.01$, $d = 1.36$). These results support the hypothesis H2.

Figure 14.10 shows the average numbers of QE with and without c-keywords. As shown in this figure, there were no significant differences between the numbers of QE with and without c-keywords. Figure 14.11 shows the average ratios of inappropriate QE with positive c-keywords to all QE rated as "inappropriate" in each group. This result suggests about 60–65% of inappropriate QE are conducted with positive c-keywords.

14.4.2.3 Discussion

Let us first discuss question expansion conducted by the participants. As shown in Fig. 14.7, overall, presentation of c-keywords significantly contributed to the increase in "appropriate" rating for QE and to the decrease in "weak appropriate" rating, although there was no significant decrease in "inappropriate" rating. This suggests that c-keywords could promote appropriate question expansion.

As shown in Fig. 14.8, secondly, presentation of c-keywords was not effective in question expansion conducted by the high rating group. Even in investigative learning without c-keywords, they could conduct appropriate question expansion as shown in this figure. This would be the main reason why c-keywords did not work for the high rating group. In addition, there was a slight increase in "inappropriate" rating for QE with c-keywords (see Fig. 14.8). As shown in Fig. 14.11, there is the possibility that the positive c-keywords prevented proper question expansion. We

Table 14.1 Correlation between data about question expansion

*p<.05, **p<.01

	Number of Appropriate QE	Number of Weak Appropriate QE	Number of Inappropriate QE	Number of QE	Number of Positive c-keywords used for QE	Ratios of Appropriate and Weak Appropriate QE	Ratios of Inappropriate QE
Number of Appropriate QE	–						
Number of Weak Appropriate QE	0.266	–					
Number of Inappropriate QE	-0.073	0.268	–				
Number of QE	0.820 **	0.664 **	0.398	–			
Number of Positive c-keywords used for QE	0.830 **	0.629 *	0.280	0.952 **	–		
Ratios of Appropriate and Weak Appropriate QE	0.381	-0.009	-0.917 **	-0.043	0.050	–	
Ratios of Inappropriate QE	-0.381	0.009	0.917 **	0.043	-0.050	-1.000	–

accordingly need to improve the accuracy of extracting positive c-keywords using LOD and Word2vec.

As for the low rating group, presentation of c-keywords promoted proper question expansion, which suggests exercise questions could be an effective scaffolding for novice learners (see Fig. 14.9). The average ratio of "appropriate" rating for QE with c-keywords in Fig. 14.9 was almost equal to the one without c-keywords in the high rating group in Fig. 14.8. This suggests that c-keywords allows less skillful learners to conduct appropriate QE equivalent to the one conducted without c-keywords by skillful learners.

Finally, we analysed the correlation between data related to QE as shown in Table 14.1. The results showed that the correlation between the number of appropriate QE and the number of positive c-keywords used for QE was relatively high. From this result, we think the participants could select positive c-keywords to expand question into appropriate sub-questions. This suggests that positive c-keywords make high contribution to question expansion.

As shown in Table 14.1, on the other hand, there was no correlation between the number of QE and the average ratios of "appropriate" and "weak appropriate" ratings for QE. From this, we think the participants who could conduct appropriate QE did not necessarily expand a question into plenty of sub-questions in a wider and deeper way. Although exercise question works effectively for promoting appropriate question expansion, we would need another method that interactively helps learners find out more sub-questions appropriate for investigating an initial question.

14.5 Conclusion

In this paper, we have addressed the issue how to generate exercise questions for self-directed investigative learning on the Web. Towards this issue, we have proposed a method to extract and select keywords related to an initial question as sub-question candidates by using LOD and Word2vec, and present them as an exercise question.

We have also reported the case studies whose purposes were to ascertain the appropriateness and effectiveness of the proposed method. The results suggest that the proposed method provides appropriate exercise questions, and particularly allows novice learners to conduct appropriate question expansion. On the other hand, we have confirmed the need to improve the accuracy of extracting sub-question candidates with LOD.

As for limitations of the work presented in this paper, the proposed method could not generate exercise questions adaptive to learners' skills in question expansion. In addition, exercise questions generated could improve the appropriateness of question expansion, but there is a need to help learners conduct more question expansion for better understanding of initial question.

In future, we will address the adaptation issue how to adjust the difficulty of exercise questions by varying the number and ratio of positive and negative c-keywords according to learners' skills in question expansion.

Acknowledgements This work is supported in part by JSPS KAKENHI grant number 17H01992.

References

DBpedia Japanese. (2016). Retrieved from http://ja.dbpedia.org/

Edwards, L. D. (2012). *Computers and exploratory learning* (vol. 146). Springer Science & Business Media.

Henze, N., & Nejdl, W. (2001). Adaptation in open corpus hypermedia. *International Journal of Artificial Intelligence in Education, 12*(4), 325–350.

Hill, J. R., & Hannafin, M. J. (1997). Cognitive strategies and learning from the world wide web. *Educational Technology Research and Development, 45*(4), 37–64.

Jonassen, D. H. (2000). *Computers as Mindtools for schools. Engaging critical thinking* (2nd ed.). Upper Saddle River, NJ: Prentice-Hall.

Kashihara, A., & Akiyama, N. (2013, July). Learner-created scenario for investigative learning with web resources. In *International conference on artificial intelligence in education* (pp. 700–703). Berlin: Springer.

Kashihara, A., & Akiyama, N. (2016a). Widening and deepening questions in web-based investigative learning. In *The 13th international conference on cognition and exploratory learning in digital age* (CELDA 2016) (pp. 227–234).

Kashihara, A., & Akiyama, N. (2016b). Learning scenario creation for promoting investigative learning on the web. *The Journal of Information and Systems in Education, 15*(1), 62–72.

Kinoshita, K., & Kashihara, A. (2013, October). Scaffolding learning scenario building with Web resources. In *2013 12th international conference on information technology based higher education and training (ITHET)* (pp. 1–6). IEEE.

Kudo, T. (2013). Retrieved from https://taku910.github.io/mecab/

Land, S. M. (2000). Cognitive requirements for learning with open-ended learning environments. *Educational Technology Research and Development, 48*(3), 61–78.

Ling, W., Dyer, C., Black, A. W., & Trancoso, I. (2015). Two/too simple adaptations of word2vec for syntax problems. In *Proceedings of the 2015 conference of the North American chapter of the Association for Computational Linguistics: Human language technologies* (pp. 1299–1304).

Raij, K. (2007). Learning by developing. Laurea Julkaisut A58.

Saito, R., Sato, Y., Hagiwara, M., Ota, K., & Kashihara, K. (2019). Towards generating exercise questions with LOD for web-based investigative learning. In *The 16th international conference on cognition and exploratory learning in digital age* (CELDA 2019) (pp. 117–124).

Sato, Y., Kashihara, A., Hasegawa, S., Ota, K., & Takaoka, R. (2019). Diagnosis with linked open data for question decomposition in web-based investigative learning. In *Foundations and trends in smart learning* (pp. 103–112). Singapore: Springer.
Sellwood, P. (1991). The investigative learning process. *Design & Technology Teaching, 24*(1), 4–12.
Zumbach, J., & Mohraz, M. (2008). Cognitive load in hypermedia reading comprehension: Influence of text type and linearity. *Computers in Human Behavior, 24*(3), 875–887.

Chapter 15
Collaborative Learning: Collegiate Pedagogy Utilizing Web Conferencing

Joan Ann Swanson, Susan L. Renes, and Anthony T. Strange

15.1 Introduction

We exist in places. It's where we live, work, learn, create, travel to, and imagine. While we often think of place in terms of a physical context, Hutchison (2004) notes that places can be a socially constructed reality. He remarks that the "significance of place is often enhanced by the personalities and idiosyncrasies of the individuals who populate a place" (p. 11). The place in which learning occurs for today's college students is evolving in large part due to technological advances. The traditional classroom has morphed into any setting in which a screen can capture synchronous or asynchronous instructional activity. This online place is both a locality and special representation of community building and learning. Bringing together individuals in an online setting allows for a reimagining of the learning place. Technology such as web conferencing allows for a mode of social construction which has the ability to move beyond the constructs of one geographic setting. Advances in technology now enable synchronous e-learning through online conferencing, however, empirical studies involving web conferencing for collegiate instruction are still limited (Kang & Shin, 2015). This study chapter seeks to provide an analysis of how

J. A. Swanson (✉)
Franklin Pierce University, Rindge, NH, USA
e-mail: Swansonj@franklinpierce.edu

S. L. Renes
University of Alaska Fairbanks, Fairbanks, AK, USA
e-mail: slrenes@alaska.edu

A. T. Strange
Wayland Baptist University, Plainview, TX, USA
e-mail: stranget@wbu.edu

D. Ifenthaler et al. (eds.), *Balancing the Tension between Digital Technologies and Learning Sciences*, Cognition and Exploratory Learning in the Digital Age,
https://doi.org/10.1007/978-3-030-65657-7_15

the movement toward web conferencing for educational purposes has impacted the place of collegiate academia.

15.2 Web Conferencing in the Collegiate Setting

Educators now have increased opportunities for synchronous, multimodal communication within academic settings as technological advancements provide web-based communication learning tools. In the 1990s, new technologies surfaced on college campuses which allowed for a limited number of users to have real-time communication and collaboration via the internet (Business Matters, 2015). With the introduction of internet services like Skype and iChat, telecommunications became available for free in the 2000s, and by 2010 video conferencing was put to the Cloud and freely available via mobile devices (ezTalks, 2017). The use of video and web conferencing has gained in popularity in both the traditional and online classrooms as an e-learning tool. Additionally, in 2019 with the Covid-19 pandemic, academic institutions were forced to move in the direction of online education if they were to continue instruction. Web conferencing involves the use of real-time video conferencing software that enables individuals to interact virtually and can be accomplished using any technological device that provides a screen enabling the sight of others and the sound to hear them. The scope of web conferencing ranges from being the vehicle for a guest speaker, a virtual field trip, group collaboration space, online synchronous instruction time or even virtual office hours. Video and web conferencing can be supplemental for face-to-face courses, or the vehicle to deliver a blended portion of a course or an entire online course.

15.2.1 Video Conferencing Versus Web Conferencing

Confusing video and web conferencing is understandable as they deliver similar opportunities. While both involve real-time communication between two or more parties from laptops, desktops, or mobile devices, there are subtle differences (see Table 15.1). Video conferencing has traditionally required special equipment for two-way, high quality video and audio sharing with limited file sharing for specific, controlled audiences. Web conferencing allows for live feed through a web browser that is not limited to a geographical location or number of participants. Web conferencing tends to be less expensive and offers collaborative communication, polls, surveys, whiteboard features and media streaming (Eli, 2017; Erwin, 2019). The course needs will dictate the type of conferencing chosen. For the purposes of this chapter and study, web conferencing will be the focus as it has the advantages of being used globally, is less expensive, and offers more academic pedagogical tools.

Table 15.1 Web and video conferencing contrast

	Web conferencing	Video conferencing
Purpose	Content sharing; ideal for face-to-face, impromptu meetings and delivering large amounts of information	Communication; ideal at face-to-face communication for groups and collaboration, provides real-time
Technical quality	Internet connection is needed and works across many geographic locations, sometimes has frozen images and pixilation due to low bandwidth hence limiting interaction	High quality images and sound, most similar to face-to-face, broadcast from one stationary location on a room-based system, specialized equipment needed
Ease of use	Authorization not required, only a browser needed, cheaper than video conferencing, webcast (one-way, non-interactive) or webinar conference, allows collaborative communication, polls, surveys, and whiteboard features	Created by an administrator who creates user accounts for every user and issues personal credentials for all users, limits potential number of users
Tools needed	Any device connecting to the web such as personal computer, laptop, tablet, mobile phone	Requires a PC, camera, and a microphone
Users	Unlimited users	Limited number of viewers by conferencing server's capacity
Application	Presentations, online trainings, distance learning	Regular meetings, discussions

Ch (2019), Eli (2017), Erwin (2019), ezTalks (2017)

15.2.2 Web-Based Pedagogy

Web-based pedagogy involves extending and enriching learning communities beyond the traditional setting. Stevenson and Hedberg (2011) acknowledge tremendous growth in internet-informed pedagogies due to the expansion of web tool availability and cloud computing. Wang, Jaeger, Liu, Guo, and Xie (2013) suggest that synchronous interaction in the collegiate academic arena aids in bridging both geographic and cultural gaps. Similarly, Mupepi (2014) contends web-based technologies are now allowing for expanded world-wide multicultural exchanges and education.

15.3 Statement of the Problem: Challenges in Higher Education

In an era of globalization and the Covid-19 pandemic, higher education is facing many challenges to stay current technologically and keep students engaged and motivated. "Bridging the geographical divide between on campus, off campus, rural, and remote learners has been an ongoing challenge for many universities often resulting in a different learning experience based on the mode of study"

(Martin & Broadley, 2018, p. 55). Knight, Dixon, Norton, and Bentley (2004) caution that the application of banking-model pedagogy and literacy in the online setting could be limiting and detrimental to culturally diverse groups. Martin and Broadley (2018), however propose that the use of distributed learning supported through video and web conferencing will assist in "real time student-centered learning experiences with diverse student perspectives" (p. 55). The increased ability for access, alone opens the door for increased diversity in course participants. Additionally, Stevenson and Hedberg (2011) remark that "despite the inherent challenges for adoption within the institution, Web 2.0 technologies do more to unify people across divides—generational, economic, geopolitical and digital—than they do to separate them" (p. 324).

While web conferencing has so much potential, another challenge within higher education is to assist instructors and students to develop the competencies needed to successfully engage in such synchronous activities. Bower (2011) notes synchronous collaboration competencies need to be developed for multimodal education which includes operational, interactional, managerial, and design abilities. Effectively managing groups and interactions as well as the technology necessitates some level of skill development by all parties. The instructor must then design the learning environment and activities which will support optimal learning. Bower (2011) illustrates that teaching effectively in web conferencing environments does have challenges that can potentially lead to misunderstandings or misuses if institutions do not prepare faculty and students in developing necessary competencies.

The reality of not having sufficient professional development in academia related to virtual innovations became quite evident when institutions suddenly had to move to virtual education due to the Covid-19 pandemic in the spring of 2020. While higher-education institutions were more ready to move to online than K-12 systems (Kennedy, 2020), there still were significant hurdles including instructors unprepared for the leap to online. The new normal suddenly included web conferencing tools such as Google Meet and Zoom, as well as increased dependency on virtual course management systems (Moodle, Canvas, Blackboard, etc.). Torres, Buck, and Gouldin (2020) from Wheaton College's Center for Collaborative Teaching and Learning note, "we have lost our ability to ignore how much we need to learn about technology and the changing world, but most importantly, to learn about one another. Reframing **educators** as learners central to our mission was no longer a difficult sell." After having made the sudden leap to online education, many instructors surveyed now report feeling underprepared to deliver online learning and assessments (Watermeyer, Crick, Knight, & Goodall, 2020). The forced shift to online pedagogies may now serve to break through hesitancy to learn more about effective digital pedagogies and practice.

The purpose of this study is to understand more fully the place web conferencing currently has pedagogically in the collegiate setting through systematic literature review and a case study utilizing web conferencing. The study was guided by the following research questions:

1. What is the scope of empirical studies about web conferencing in the collegiate instructional setting between 2000 and May 2019?
2. What lessons can be gleaned from a collaborative class case study experience in which web conferencing was utilized?

15.4 Methods

This study sought to investigate the use of web conferencing for collegiate instruction. The research was conducted in phases using two methodologies: systematic literature review and a collegiate case study. The systematic literature review approach was useful in identifying, selecting, and critically analyzing previous empirical studies concerned with the use of web conferencing for instructional purposes (Grant & Booth, 2009). The second methodology utilized a study approach (Creswell, 2007) and included data collection using course documentation, archival records, interviews, direct observation, participant observation and physical artifacts (Stake, 1995).

15.4.1 Phase 1: Systematic Literature Review

Data related to web conferencing for collegiate instruction was collected using a contemporary systematic literature review ranging from the year 2000, when video and web conferencing began to be utilized in higher education through the present, 2019. The Ebsco database was used and specific searches were completed utilizing Academic Search Complete, Education Search Complete, ERIC, and PsycINFO. The selection process was guided with the following criteria: papers published between January 2000 and May 2019, papers published in peer-reviewed journals, papers published in English; and papers including the key words web conferencing and college. Relevant, empirical publications were chosen resulting for analysis ($N = 76$).

15.4.2 Phase 2: Case Study and Procedures

Three collegiate instructors, representing three higher-education institutions collaborated for a cross-institution experience in an effort to establish a learning community place that stretched beyond the borders of one institution in one locale, to multiple regions and individual representations. These institutions and their respective students were different in both the type of institution and the diversity of individuals who attend these institutions. The first institution is a private liberal arts college offering undergraduate degrees and the other two were public and enrolling students in undergraduate through doctoral programs. Students attending the private institution which is located in the north eastern United States included students from across the United States as well as a few international students, however a large number of these students were from urban and suburban locals. These students additionally were primarily engaged in courses that met face-to-face. The second institution was a public university located in the southern portion of the United States and included students who experienced courses in both blended and fully online formats, and were working toward bachelor's degrees. Students in the third institution were in the north western region of the United States and attending a large public institution and working toward a master's degree. Additionally, students in this institution constitute both local and a large number of international students. Large segments of this student population are also from rural regions and access their courses remotely.

While there was a diverse representation in the constituents of the collaborative class, there was a central focus for the collaborative class learning experience. This collegiate learning community adopted the guiding framework provided by West and Williams (2017) which includes access, relationships, visions, and functions. Access would include a common meeting place (digitally through the web conferencing), relationships (sense of belonging to the learning cohort), vision (sharing the same purpose of learning), and function (shared practice and course materials). Despite their various locals, across the nation, these courses navigated differing time zones and combined to provide a group of students an experience of a shared cross-institution course.

Each of the participants were registered for a course at their home institution and then met three times throughout the semester for a combined synchronous web conference course session between the three institutions. Students were assigned an introductory activity, pre-class session readings, and activities including post-class reflections where students were able to respond to each other. Each of the three synchronous sessions was led by one of the lead faculty members from the different institutions and conducted through the Zoom platform. Each participant was enrolled in a private WordPress (WordPress.com) website which hosted all shared course related materials including required media, assignment descriptions and readings, as well as a depository for assignments.

The web conferencing tool Zoom (https://zoom.us/) was used to facilitate the collaborative classes. Students across all three geographic locations joined online at the appointed times, having completed the pre-class session work which served as a basis for the discussions and activities commenced in the hour-long synchronous time. Ground rules were reviewed for discussion times and for ease of facilitating discussion. When a speaker was vocalizing, the Zoom web conferencing program automatically enlarged their image in the center of the screen and all others were minimized in a frame around the perimeter of the screen. At different points in the course, for the purpose of the discussion, screen sharing allowed for documents and images to be viewed as well. All synchronous web conferenced class sessions were also recorded for students and instructors to review.

15.5 Results and Analysis

15.5.1 Phase 1: Results

To explore the ways in which web conferencing was being utilized for collegiate instruction, a systematic literature review of educational databases was conducted searching for empirical studies about collegiate web conferencing, which yielded ($N = 76$) articles with only four of the articles having lead authors publishing more than one publication. These results first indicated a limited number of research specialists focusing in on the use of web conferencing. The methodology utilized was mainly qualitative, followed by mixed methods and then quantitative (see Fig. 15.1).

While some studies were mixed methods, using more than one type of method to collect data, the largest amount of studies involved questionnaires or surveys, followed by case studies and interviews (see Fig. 15.2). Additionally, other methods noted included observations, correlational studies, experimental studies and narrative or systematic literature reviews. The qualitative studies reflected largely on detailed observations and descriptions of data relating to web conferencing in collegiate settings. The data complexities analyzed within the represented systematic literature review were found to be especially useful to describe the human experience in these online environments.

The content area in which the reported studies were associated varied a great deal. While most studies focused upon one course, there were 18% that were multi-discipline focused, looking globally across a campus. The content areas most represented in web conferencing studies over the past 19 years were Education (16%), Medical/Health (12%), Business (8%) and Library related (6%). Other content areas in which web conferencing was studied included: counseling, engineering, geography, history, chemistry, club events, communication, computer programming, criminology, environmental science, math, music, psychology, social work, tutoring, study abroad, and writing.

The most common web conferencing platform or tool reported was Adobe Connect (33%), followed by Blackboard Collaborative (16%), Skype (7%), and Elluminate Live! (5%). A total of 17 other tools were also reported however each

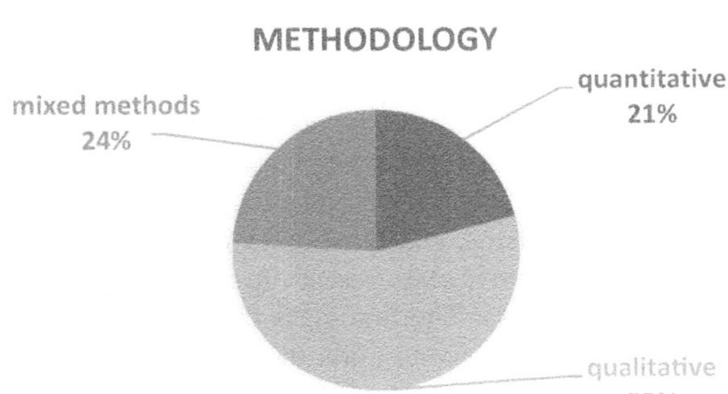

Fig. 15.1 Percentages of methodology most utilized when researching web conferencing as a collegiate pedagogy

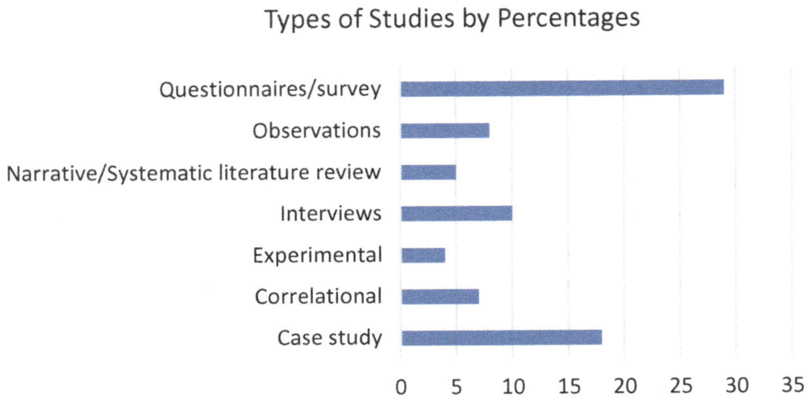

Fig. 15.2 Percentages for the types of studies utilized to research web conferencing for collegiate instruction

was only used in one study. The variety of tools utilized indicates there was no singular consensus on the best platform for web conferencing. This is reflective of the proprietary nature of web conferencing tools and that only two seem to have broken into the collegiate technology market in a stronger manner. Since most educational institutions have an established Learning Management System, it would further enrich their products with web conferencing tools, similar to the Blackboard Collaborate extension.

Many of the studies described within this literature review did not report a subject size (50% of studies reported an $N = x$), however, a majority of those who did report the number of participants dealt with smaller subject sizes, which is common among case studies, also strongly represented in this review (see Fig. 15.3). A total of 62% of studies reported under 100 subjects, and 44% had 36 and under for the number of subjects. Considering most college courses contain 40 students or less, the smaller number of subjects in studies involving college courses is commensurate with the typical population.

15.5.2 *Phase 2: Results*

Data analysis in phase two involved the researchers reviewing the recordings of the collaborative classes and reviewing the written submissions of students and faculty. Additionally, analysis was conducted within the theoretical frameworks of access, relationships, vision, and function (West & Williams, 2017).

The collaborative class was set up to deliberately concentrate on topics which are difficult to discuss. The first class theme focused upon the awareness of personal perspectives. The second class theme was about resolving conflict. The last class theme was cultural sensitivity. Students were assigned pre-class work involving

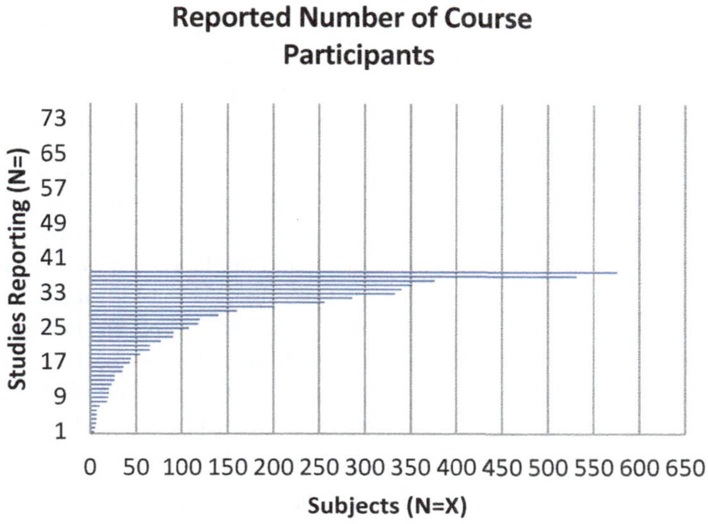

Fig. 15.3 This figure represents the number of subjects in the studies that included a report of participants

watching a media presentation, readings, and completing experiential activities related to the topics. Students were also to come to class having prepared responses to discussion questions, and following the class were required to post reflection pieces. These reflections were then available to be commented upon by other members of class, thus extending interactions. Examples of post-class reflections included the following comments which revealed both vulnerability and growth (see Fig. 15.4).

15.5.2.1 Access

On the outset of the course, each member of the community was required to present an "Introduction Selfie" on the course website. This could include their image or anything they felt was representative of them. Additionally, they were to note their home institution, major and why they chose that particular image. This served to humanize and acquaint all course members. Each member of the community had access to each other via course communications. There was also a common meeting place for collaborative sessions, discussions, and access to materials. Regardless of the location of the student, access and opportunity for participation was equal.

The course served to break barriers through web conferencing. For example, in rural areas, web conferencing increases access to higher education. In Alaska, where the communities and the demographics and culture of those communities are so varied, classes such as this increase a student's ability to work following graduation, as they have a better idea of other situations.

Student 1: I don't think differences should separate and divide people. I think acceptance is a large trend in the world right now and there are a lot of movements trying to increase cohesion between different groups of people. I believe a lot of great opportunities are lost when people decide to divide themselves based on ethnic, cultural, socioeconomic, religious, mental or physical differences.
Student 2: I think that this impacts my work as a teacher because I cannot simply ignore my students' identities. I should not lump all Asian students under one moniker just like I should not lump all people who identify as queer under the idea of gay or being gay. The population of non-cis-straight people is much larger than gay and lesbian, similar to how other populations have branches and subcultures that should be respected. I recognize that I cannot know the entire history of every individual culture that my students may have, but I can put the effort into discussing their cultures and identities with them and researching a little about each of their cultures in order to make sure I am not just assuming they experience life in a certain way or that they identify in a certain way.
Student 3: This experience, the combination of the classroom experiment and my thought processing exercise, evoked feelings of helplessness, a sense of frustration, and feelings of being "outside" the group, or set apart by a difference. It also evoked feelings of empathy as I considered how being blind might impact my own daily life and change the way I live. This experience reminded me of the importance of mindfulness in listening to clients, considering the daily obstacles and roadblocks they may be facing that might be overlooked or missed by anyone not having experience with a disability. As a society, we tend to "other" anyone with a disability. The label of disabled has been used to justify exclusion and derision, in the same way that race, and gender were used previously. …We view this other able-ness as negative, a weakness, instead of recognizing there is more than one way to navigate the world and experience life that is neither wrong nor less valuable.
Student 4: Like many of the people Chimamanda Ngozi Adichie has encountered in her lifetime, I too was told a single story about other places, whether they were about different continents or different states. I grew up in New York City and as a child I identified anywhere outside of the city as the "countryside." Even suburban areas I dismissed as rural. The term "redneck" was often used in my environment to describe and stereotype people who lived in the southern states. Many people in the city can be very elitist, and I was definitely guilty of this while I was living there. But after high school I was able to travel and go outside of my little city bubble and humanize the places I use to dismiss or overlook. I finally saw that people are more similar than different, but for some reason we like to focus on the differences, maybe to feel better about ourselves and to put ourselves on a pedestal.

Fig. 15.4 Student reflection samples

15.5.2.2 Relationships

Each person regardless of their locale was an accepted member of the course community and had opportunity to share knowledge, dialogue, and learning experiences. It was established that this was a "safe" community in which members were respectful of each other's differences—especially as topics were broached which challenged personal perspectives and experiences.

15.5.2.3 Vision

The vision for the collaborative learning community was to learn and grow both individually and collectively while examining many topics from multiple understandings, perspectives, and backgrounds. These were hard topics to discuss but enriching to watch the engagement of students across a wide geographic and cultural span.

15.5.2.4 Function

The function of the collaborative class was to socially and digitally create a space in which students could share projects and assignments while also breaching distance and cultural barriers. A Faculty Expectations handout was distributed at the individual institutions that stated the instructor's expectations regarding the class in general, and the interactions between students in the room, as well as distance students. The following ground rules were set up for discussion times: listen with focus and attention; speak without interruption; refrain from giving unsolicited advice or commentary; use I statements; avoid generalizing about people or groups; assume good intentions; and respect difference.

The instructor had to be vigilant about including all of the students in the conversations and activities. If something happened in the classroom that distance students could not see or experience, the instructor explained what just happened so they were not left to feel outside the classroom. The instructors also allowed the distance students plenty of time to talk about their location and how the various concepts being studied might differ elsewhere.

15.6 Discussion

Digital technologies, especially web conferencing, allow students to participate in classes in a highly interactive manner in real-time, using a variety of mobile devices as well as traditional lap top or personal computers. Incorporating these web-based technologies encourages collaboration and communication not only between students and teachers but student-to-student. Additionally, communication and collaboration as noted in this case study potentially spans across campuses as well. Web conferencing as described in both the systematic literature review and case study serves to increase accessibility and opportunity for diversity in learning experiences.

The web conferencing experience in the collaborative class served to enrich students' perspectives and expose students to a diverse range of fellow students and situations. One instructor noted, "I think in Alaska students get a much richer experience than they would otherwise; learning from others who are in remote locations, they come to appreciate the differences in access to health care and limitations in travel, etc. as well as the particular benefits and problems of living remotely in Alaska." Another instructor noted a student from New York City commented on how they realized their perspectives of other students prior to the collaborative class discussions were inaccurate.

The nature of a course being hosted through web conferencing, such as the collaborative class case study, ultimately adds pedagogical and methodological variety to instruction. The collaborative class case study illustrated how faculty were able to focus more on student-centered and active learning activities in which collaborative learning could occur within groups and across groups, locally and remotely. Web-based instruction has the potential for more authentic tasks and problem-solving via web technology connections which also promote the development of twenty-first century skills for a more global society.

In addition to the benefits of collaboration, the collaborative web conferencing class was financially feasible for each institution with no extra cost for individual students. Since all interactions were web-based, using Zoom and a WordPress website, there were no extra costs and all tuition remained at the home institution for each school. In an era when cutting cost is associated with institutional viability, this is significant.

The faculty who were involved in the collaborative course experience which utilized web conferencing were able to tackle this new experience together, strategizing procedures and protocols, sharing plans, concepts, and sharing the instructional presentation load. Based upon the experience of creating and instituting the collaborative class, the following recommendations have been suggested:

- *Key Recommendation 1: Clear Goals and Objectives*

- When establishing web conferencing and collaborative experiences, instructor should first be clear with their goal or objective—beyond just an opportunity to use new technology. In this case study, there were many logistical hurdles that were conquered, but it took time and organization centered upon the purpose for the experience.
- *Key Recommendation 2: Practice with the Technological Tools Prior to Implementation*
- Since the technological tools will vary across institutions, the faculties involved in this study recommend that instructors practice prior to the first instructional session. Step-by-step instructions on how to access the tools and materials should be provided.

- *Key Recommendation 3: Establish Course Expectations and Protocols*
- On the outset of the course, instructors should establish expectations and protocol for how the class will function and ground rules for interactions. Clear expectations might include expectations for when cameras should be on, mics muted (such as during entry), and participation requirements. Students will have varied experience with expectations concerning when and how they should "attend" and what participation will encompass.
- *Key Recommendation 4: Consistent and Clear Access to Course Materials and Repository/Submission Procedures*
- The organization of course materials should follow a consistent pattern.. Instructors often assume their students have a certain level of technological savviness which can lead to significant problems.
- Just as there is a wide range of instructional approaches instructors may draw from, there is also a wide range of preferences instructors may default to when organizing a course with an online component. It is essential that instructors walk their students through their organization of course materials within the online site utilized for the course. This communication is ideally accomplished in a document distributed at the outset of the course and explained in-person as they course begins virtually. Clear step-by-step directions should be also be provided which detail how to access the course materials. Communication should point to consistent, well-labeled places where students can access lectures, readings, activities, etc., in addition to clear instructions on where to upload assessment material and the preferred format for that material (i.e., Word document, Excel file, presentation slide format).
- *Key Recommendation 5: Trouble Shooting Assistance*

- Students commonly have technical difficulties thus it is essential that instructors provide a trouble shooting resource and technical support person for them to contact for assistance. A commonly asked questions sheet may be of help, but there are times in which a student's issue is unique and calls for specificity in response.

15.7 Conclusion

The systematic literature review revealed many holes in the knowledge base about web conferencing as a collegiate pedagogical tool. Even though it has been actively used since the year 2000, the documentation about the foundation for pedagogical practices is fragmented. Empirical studies are quite limited over the 19-year span, with very few quantitative studies. Additionally, there has been little analysis of the pedagogical effectiveness of web conferencing in singular disciplines. While case studies and singular uses within disciplines can serve to provide individual in-depth knowledge about experiences using web conferencing, these results are more difficult to generalize across other collegiate situations. This study served to expose the need for more sophisticated, generalizable studies regarding web conferencing, especially for collegiate instruction.

One limitation of this study could be the exclusion of literature review related specifically to video conferencing. The researchers focused upon web conferencing as a more amenable vehicle for synchronous pedagogy. However, future studies will encompass video conferencing as well, especially in light of immense changes in how instruction changed in recent times. One significant result of the global pandemic is that there has been a movement in education toward increased online learning. Further study is needed post-pandemic times to determine instructional lessons learned from recent instruction involving web and/or video conferencing.

An instructor, teaching in two formats (face-to-face and simultaneously web conferencing) must constantly pay attention to the local and remote students. Instructors and higher-education institutions have to ask themselves why they are instituting web conferencing. Why are they seeking to offer this option? It's optimal to be able to offer options that allow students to take classes the way they learn the best. Distance or web conferencing pedagogy adds a new dimension to education, one that increases access and therefore opportunities to many students who might not otherwise be able to take advantage of education past the secondary level. Web conferencing and collaborative classes also allow for the opportunity for students across geographic and cultural spans to interact, thus promoting and extending diversity acceptance. The goal of web conferencing and collaborative classes is construct a new place of social reality which increases learning and moves far beyond the physical constructs of one geographic setting and any singular pedagogy. Berens (2012) concludes, "It doesn't matter to me if my classroom is a little rectangle in a building or a little rectangle above my keyboard. Doors are rectangles; rectangles are portals. We walk through." Collegiate educators are challenged to venture into a new place for learning experiences, including web conferencing and collaborative teaching situations.

References

Berens, K. I. (2012, December 3). *The new learning is ancient.* New Media Curious. Retrieved from http://kathiiberens.com/2012/12/03/ancient/

Bower, M. (2011). Synchronous collaboration competencies in web-conferencing environments—their impact on the learning process. *Distance Education, 32*(1), 63–83.

Business Matters. (2015). *The history of video conferencing*. Retrieved from https://www.bmmagazine.co.uk/tech/history-video-conferencing/

Ch, R. (2019). *What is the difference between web conferencing and video conferencing?* Retrieved from https://www.quora.com/What-is-the-difference-between-web-conferencing-and-video-conferencing

Creswell, J. W. (2007). *Qualitative inquiry and research design: Choosing among five approaches*. Thousand Oaks, CA: Sage.

Eli, B. (2017). *Understanding the difference between web conferencing and video conferencing*. Retrieved from https://www.eztalks.com/video-conference/difference-between-web-conferencing-and-video-conferencing.html

Erwin, J. (2019). *What is the difference between video and web conferencing?* Premiere Global Services. Retrieved from https://www.pgi.com/resources/articles/what-is-the-difference-between-video-and-web-conferencing/

ezTalks. (2017). *A brief history of video conferencing from 1964 to 2017*. Retrieved from https://www.eztalks.com/video-conference/history-of-video-conferencing.html

Grant, M. J., & Booth, A. (2009). A typology of reviews: An analysis of 14 review types and associated methodologies. *Health Information & Libraries Journal, 26*(2), 91–108.

Hutchison, D. (2004). *A natural history of place in education*. New York, NY: Teachers College Press.

Kang, M., & Shin, W. S. (2015). An empirical investigation of student acceptance of synchronous e-learning in an online university. *Journal of Educational Computing Research, 52*(4), 475–495.

Kennedy, M. (2020). Classes dismissed: The Covid-19 virus pandemic has shut down virtually the entire U.S. education system and disrupted the lives of millions of students and staff. *American School & University, 92*(6), 14–17.

Knight, M. G., Dixon, I. R., Norton, N. E. L., & Bentley, C. (2004). Extending learning communities: New technologies, multiple literacies, and cultural blind pedagogies. *The Urban Review, 36*(2), 101–118.

Martin, R., & Broadley, T. (2018). New generation distributed learning: Models of connecting students across distance and cultural boundaries. *Australian and International Journal of Rural Education, 28*(1), 55–72.

Mupepi, M. G. (2014). A schematic description of the nature of video-conferencing and internet exchange: Enhancing global understanding. *International Journal of Web-Based Learning and Teaching Technologies, 9*(1), 33–40.

Stake, R. (1995). *The art of case study research*. Thousand Oaks, CA: Sage.

Stevenson, M., & Hedberg, J. (2011). Head in the clouds: A review of current and future potential for cloud-enabled pedagogies. *Educational Media International, 48*(4), 321–333.

Torres, M. G., Buck, C., & Gouldin, C. (2020). Making the leap from the traditional to the virtual educational experience. *New England Journal of Higher Education*. Retrieved from https://nebhe.org/journal/making-the-leap-from-the-traditional-to-the-virtual-educational-experience/

Wang, C. X., Jaeger, D., Liu, J., Guo, X., & Xie, N. (2013). Using synchronous technology to enrich student learning. *TechTrends, 57*(1), 20–25.

Watermeyer, R., Crick, T., Knight, C. & Goodall, J. (2020). Forced shift to online teaching in coronavirus pandemic unleashes educators' deepest job fears: 'Culture-change moment' for higher education. Nature Index. Retrieved from https://www.natureindex.com/news-blog/forced-shift-to-online-teaching-in-coronavirus-pandemic-unleashes-educators-deepest-job-fears-

West, R. E., & Williams, G. S. (2017). "I don't think that word means what you think it means": A proposed framework for defining learning communities. *Educational Technology Research Development, 7*, 1–14. https://doi.org/10.1007/s11423-017-9535-0

Chapter 16
Interaction Effects of Teachers' Educational Policies for Seminars and Students' Learning Goal Orientation on Students' Learning-as-Duty Conception

Mai Yokoyama and Kazuhisa Miwa

16.1 Introduction

16.1.1 Learning-as-Duty Conception

Conception of learning is defined as students' ideas and beliefs about "what learning is." Conception of learning is very important, as it shapes the foundation of students' learning behaviors and can greatly influence their visions for life (Horino & Ichikawa, 1993). Studies have indicated differences in students' learning behaviors, depending on how they conceptualize learning. For instance, Van Rossum and Schenk (1984) conducted an empirical study on learning behaviors in relation to reading materials. Students who perceived learning as memorization adopted a superficial learning behavior in which they only read a summary, whereas students who perceived learning as the abstraction of meaning or an interpretative process aimed at understanding reality adopted a deep-learning behavior and read the sentences while processing the relationships between the paragraphs. Dart et al. (2000) suggested that students who had qualitative conceptions, such as personal fulfillment, and experiential conceptions, such as a process not bound by time, were more likely to use deep-learning approaches, compared with students who had quantitative conceptions, such as an increase in knowledge, who were more likely to rely on superficial approaches.

In the present study, we focused particularly on "learning-as-duty conception." Sugie (2011) stated students' poor academic performance is primarily caused by the learning they have experienced and the idea that learning is not something that students voluntarily work on but is taught by teachers and parents. Peterson, Brown,

M. Yokoyama (✉) · K. Miwa
Nagoya University, Nagoya, Japan
e-mail: yokoyama.mai@c.mbox.nagoya-u.ac.jp; miwa@is.nagoya-u.ac.jp

D. Ifenthaler et al. (eds.), *Balancing the Tension between Digital Technologies and Learning Sciences*, Cognition and Exploratory Learning in the Digital Age,
https://doi.org/10.1007/978-3-030-65657-7_16

and Irving (2010) argued in an empirical study that learning-as-duty conception has a negative effect on academic achievement. Additionally, Takayama (2002) and Yokoyama and Miwa (2018) indicated that learning-as-duty conception can suppress adaptive learning behaviors, such as deep learning or active learning. Furthermore, Suzuki showed that students who think that tests are administered to force them to study do not study effectively and perform adaptive learning behaviors (Suzuki, 2011a; Suzuki, 2011b). Thus, learning-as-duty conception has been shown to have a negative effect on learning behaviors and academic performance; however, no studies to date have examined the factors or methods that suppress learning-as-duty conception. Thus, the purpose of this study was to clarify basic knowledge that can prevent students from conceptualizing learning as a duty.

In this research, we focused on teachers' educational policies as an intervening factor in shaping how students conceptualize learning. One's conception of learning is considered to be influenced by factors such as individual learning experiences, parents, teachers, and friends (Horino & Ichikawa, 1993). Further, learning-as-duty conception may also change, depending on one's learning experiences. Many universities in Japan have seminars in which one or several teachers and a small number of students' study specialized content in groups. The content and methods of seminars are usually chosen at the teachers' discretion; thus, each class reflects the teachers' educational policy. Additionally, since seminars are conducted with a small number of students over 1 or 2 years, teacher–student relationships are expected to become deeper than in general classes, due to the relatively long term of seminars. These conditions imply that teachers' educational policies in seminars greatly affect students' learning-as-duty conception. Therefore, this study examined the effects of teachers' educational goals in seminars, and what they emphasize in achieving those goals (i.e., teachers' educational policies), on students' learning-as-duty conception.

16.1.2 Moderating Effect of Learning Goal Orientation

As an attribute that defines students' learning behaviors—an important factor along with conception of learning—is goal orientation, which can be defined as the individual characteristics of students' learning goals (Nicholls, Patashnick, & Nolen, 1985). Dweck (1986) proposed goal achievement theory, which posits that differences in learning behaviors depend on students' goals. According to this theory, students' goals are classified into two categories: learning and performance. The purpose of learning goals is to acquire new knowledge and skills through challenging activities, while the purpose of performance goals is to seek positive and avoid negative evaluations. Students oriented toward learning goals tend to select challenging tasks and remain motivated even when they encounter failure, regardless of their confidence in their own abilities. Performance goal-oriented students behave similarly to students with learning goal orientation provided they are confident in their abilities; however, if they lack confidence in their abilities, they are less likely remain motivated and will use passive strategies to complete tasks.

Elliott and Dweck (1988) examined differences in behaviors after failure due to differences in students' goals, and their findings supported goal achievement theory. Specifically, they provided instructions to set learning goal and performance goal groups. Participants were asked to recognize whether their ability to perform a task was high or low, based on differences in correct and incorrect feedback. As a result, in the performance goal group, students who recognized their ability was low were unable to cope with the task, and made many negative statements, such as remarks about anxiety or escape, compared with students who recognized their ability was high. Such variations in learning behaviors due to differences in ability recognition were not observed in the learning goal group.

Many studies have investigated the relationship between goal orientation measured by questionnaires and variables in academic achievement, including Ames and Archer (1987, 1988). Kaplan and Midgley (1997) observed that learning goal orientation has positive effects on adaptive learning behaviors, and performance goal orientation has no relationship to or negative effects on adaptive learning behaviors. Nolen and Haladyna (1990) demonstrated that learning goal orientation has positive effects on deep-processing behaviors, such as monitoring comprehension and memory and elaboration of ideas. Similar results were reported in Fenollar, Román, and Cuestas (2007) and Liem, Lau, and Nie (2008). Learning goal orientation has also been shown to predict motivational variables, such as intrinsic motivation (e.g., Heyman & Dweck, 1992; Kavussanu & Harnisch, 2000). In subsequent studies, learning goal orientation has been found to be positively related to adaptive learning, with the superiority of learning goal orientation consistently emphasized (e.g., Chea & Shumow, 2014; Hudaykulov, Hongyi, & Galib, 2015; Tercanlioglu & Demiröz, 2015).

Although performance goal orientation has been regarded as negatively affecting academic achievement (Ames, 1992), positive associations were also found with variables related to academic achievement, such as self-efficacy and adaptive learning strategies (e.g., Meece, Blumenfeld, & Hoyle, 1988; Pintrich & Garcia, 1991); thus, consistent results have not been obtained. Therefore, in this study, we focused on learning goal orientation, which has consistently been shown to lead to superior academic achievement.

Ames (1992) proposed teacher involvement from the three dimensions of "task," "authority," and "evaluation/cognition," to create a classroom environment that increases students' learning goal orientation. However, to date, no empirical study has investigated this proposal. Geitz, Joosten-Ten Brinke, and Kirschner (2015) intervened in students' learning goal orientation by using a method to increase student involvement in their feedback, and examined the effects on learning goal orientation and learning behaviors. However, the intervention did not directly influence learning goal orientation. Thus, an intervention method that increases learning goal orientation, which has been repeatedly shown to have a positive effect on academic achievement, is expected to be developed; however, an effective intervention method has not yet been established. One reason for this may be that goal orientation is a stable characteristic.

Nakayama (2005), Yamaguchi (2012), and Akamatu (2017) examined the relationship between goal orientation and conception of learning in English learning, demonstrating that goal orientation is a predictor of conception of learning in English learning. Yamamoto and Ueno (2015) and Yokoyama and Miwa (2018) demonstrated that goal orientation is a predictor of not only English learning but also of conception of learning that is not dependent on a specific subject. Furthermore, Yokoyama and Miwa (2018) showed that learning goal orientation has a negative effect on learning-as-duty conception.

Students oriented toward learning goals see challenging situations as opportunities for growth, and this positive thinking promotes effective and fulfilling learning experiences (Garcia, Restubog, Toledano, Tolentino, & Rafferty, 2011). Considering that learning-as-duty conception is influenced by learning experiences, even if teachers provide the same education according to the same policy, students' learning experiences may vary, depending on their learning goal orientation; therefore, the educational effect on learning-as-duty conception may also vary. Thus, it is expected that the effects of teachers' educational policy on students' learning-as-duty conception will vary, depending on students' learning goal orientation.

16.1.3 Purpose

Figure 16.1 shows the outline of this study. This study's purpose was to examine the effects of teachers' educational policies for seminars on students' learning-as-duty conception, focusing on the moderating effect of learning goal orientation. It was expected that the effects of teachers' educational policies on students' learning-as-duty conception would vary, depending on the students' learning goal orientation. In the following, "teachers' educational polices for seminars" is abbreviated as "teachers' educational policies."

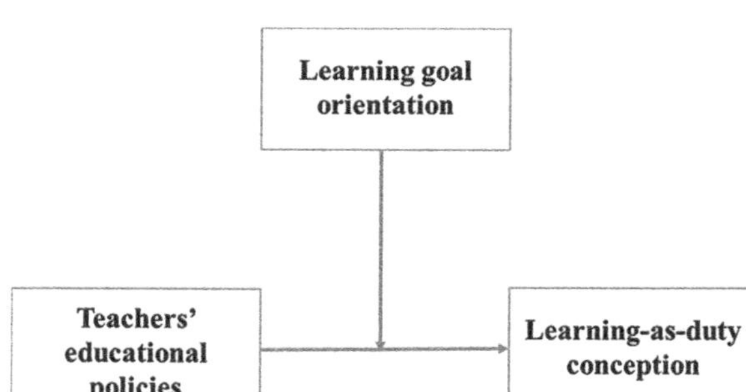

Fig. 16.1 Outline of the study

16.2 Method

16.2.1 Survey Method

The class practice was conducted in the School of Integrated Arts and Sciences of a Japanese public university. A questionnaire survey was conducted with 50 university teachers who taught seminars for 2 years, and 189 undergraduate students who participated in the seminars during their third and fourth years of university from 2017 to 2018. The seminars were compulsory subjects, and took more than 180 minutes a week. The survey was conducted from February to March 2019.

16.2.2 Instrument

16.2.2.1 For Teachers

Educational policies: 48 items developed by Fushikida, Kitamura, and Yamauchi (2014) were used in a comprehensive questionnaire to measure teachers' educational policies for running seminars. Each item was rated on a 5-point scale. The specific items used in the analysis are described in Sect. 16.3.1.

16.2.2.2 For Students

Learning-as-duty conception: Five items (Table 16.1) from the "Duty and Memorizing" factor in Yokoyama and Miwa (2018) were used, modified from Takayama's (2002) learning conception scale. Each item was rated on a 5-point scale.

Table 16.1 Items of learning-as-duty conception

Learning-as-duty conception
Learning is being forced without freedom
Learning is being forced by parents or teachers
Learning is being forced to do things that you do not want to do
Learning is accurately memorizing the content of materials
Learning is memorizing textbook content at a desk

Table 16.2 Items of learning goal orientation

Learning goal orientation
I hope my knowledge is broader and deeper when I am done with classes
I want to learn as much as possible from class
I prefer course material that arouses my curiosity, even if it is difficult to learn
I prefer course material that really challenges me, so I can learn new things

Learning goal orientation: Four items (Table 16.2) from the "Learning Goal" factor in Yokoyama and Miwa (2018), translated from Elliot and Church's (1997) Achievement Goal Scale, were used. Each item was rated on a 5-point scale.

16.2.3 Analysis Method

Hierarchical linear model analysis was performed, using learning-as-duty conception as a dependent variable, and learning goal orientation, teachers' educational policies, and interaction terms of learning goal orientation and teachers' educational policies as independent variables.

Student data were hierarchical data provided by multiple students from one seminar. If there are in-group similarities, hierarchical data should be divided into a group-level variable and an individual-level variable. We calculated the intra-class correlation coefficient (ICC) of learning goal orientation and learning-as-duty conception to evaluate intra-group similarities. If the value of the ICC was 0.10 or more, it was judged that the answers among students in the same seminar were consistent. If the learning goal orientation variable showed similarities within the group, the average value for each seminar was defined as the learning goal orientation at the seminar level, and the group-mean-centering variable was defined as the learning goal orientation at the individual level. The group-mean-centering variable was calculated by subtracting the mean value of each group from the individual-level variable, in order to remove the group-level effect from the individual-level score.

To evaluate intra-group similarities, two or more responses from students belonging to the same seminar were needed. After performing list-wise deletion, we identified 43 seminars for which there were responses from two or more students. Thus, data from 43 seminars, including 43 teachers and 179 students, were analyzed.

16.3 Scale Structure

16.3.1 Teachers' Educational Policies

A ceiling effect was judged to be present if the average + 1SD value exceeded the upper limit of 5.0 of a 5-point scale. In Fushikida et al. (2014), 22 items showed ceiling effects. Although it was possible to exclude these items from the survey, we used all 48 items, because different teachers answered the questions. As a result, ceiling effects were found for 23 items, including 19 items in which ceiling effects were found in Fushikida et al. (2014). The applicable 23 items were about learning motivation (e.g., "Improve students' motivation to learn") and standard educational methods when running a seminar (e.g., "Students give a presentation on what they have investigated for a subject"). These are items about standard educational methods for seminars; thus, they were excluded, because they were too ordinary for conducting a seminar. We performed exploratory factor analysis (maximum likelihood method with promax rotation) using the remaining 25 items. Items that were loaded at 0.35 or less, and items that were loaded at 0.35 or more on two or more factors, were excluded. The third eigenvalue was sufficiently larger than the fourth and subsequent eigenvalues, and a three-factor solution was considered appropriate. As a result, 7 items were excluded and 18 items were adopted. Factor categories were decided from the items with high factor loadings. Details for each item and analysis results are presented in Table 16.3. Cronbach's α coefficient was calculated for each factor. The first factor showed high internal consistency at $\alpha = 0.88$, and the second factor showed relatively high internal consistency at $\alpha = 0.76$. The third factor showed low internal consistency at $\alpha = 0.68$; however, we judged that it could be used for analysis. An average value of the items was regarded as the respective value of each factor.

16.3.2 Students' Learning-as-Duty Conception and Learning Goal Orientation

Learning-as-duty conception: The α coefficients of five items from the "Duty and Memorizing" factor in Yokoyama and Miwa (2018) were calculated, and a high internal consistency was shown at $\alpha = 0.84$. An average value of the five items was regarded as the learning-as-duty conception variable. The average value was $M = 2.36$, and the standard deviation was $SD = 0.70$.

Learning goal orientation: The α coefficients of four items from the "Learning-Goal Orientation" factor in Yokoyama and Miwa (2018) were calculated, and a relatively high internal consistency was shown at $\alpha = 0.74$. An average value of the four items was regarded as the learning goal orientation variable. The average value was $M = 3.79$, and the standard deviation was $SD = 0.63$.

Table 16.3 Teachers' educational policy items and factor loadings

Factors and items	I	II	III
Factor I. Focus on collaborative learning ($M = 2.62$, $SD = 1.22$, $\alpha = 0.88$)			
Students do fieldwork (observation, survey, etc.) in groups	**0.96**	−0.05	0.03
Students carry out joint projects with outside parties (other universities, companies, etc.)	**0.89**	−0.12	0.06
Students plan events in groups	**0.76**	0.01	0.05
A person outside the university gives a lecture at the teacher's request	**0.70**	−0.13	0.10
Students create documents (reports, resumes, etc.) in groups	**0.58**	0.23	−0.16
Factor II. Focus on relationships with society ($M = 3.81$, $SD = 0.59$, $\alpha = 0.76$)			
Students are aware of relationships with society	0.08	**0.77**	−0.02
Teachers teach the significance of learning through relationships with nature and society	0.10	**0.68**	−0.23
Students' job-hunting status	−0.14	**0.60**	0.17
Students deepen their systematic understanding of learning content	−0.19	**0.52**	0.08
Students acquire generic skills beyond their major	0.13	**0.49**	−0.14
Students collaborate with others	0.24	**0.47**	0.12
Teachers reliably teach the basics of learning	−0.07	**0.40**	0.12
Factor III. Focus on specialized learning ($M = 4.04$, $SD = 0.49$, $\alpha = 0.68$)			
Students have access to the latest research results	0.05	−0.09	**0.79**
Students acquire specialized skills in their major academic field	−0.01	0.29	**0.67**
Strengthen the relationships between teachers and students, and among students	0.20	0.08	**0.50**
Teachers introduce their specialized field and research	0.00	−0.14	**0.43**
Students acquire evolving knowledge in their major academic field	−0.15	0.27	**0.37**
Cognitive proficiency of students	0.09	−0.01	**0.35**

16.4 Results

The ICC for learning goal orientation was 0.12 ($p < 0.01$), and the ICC for learning-as-duty conception was 0.18 ($p < 0.01$), showing that responses among students in the same seminar were consistent. As shown in Sect. 16.2.3, learning goal orientation variables were calculated at the seminar and individual levels, respectively. Hierarchical linear model analysis was performed. Learning-as-duty conception was utilized as the dependent variable, and learning goal orientation at the seminar level, learning goal orientation at the individual level, the three variables of teachers' educational policies, and the interaction terms between each level of learning goal orientation and each variable of teachers' educational policies were utilized as independent variables. When the interaction terms were included in the same model, the variance inflation factor (VIF) of the interaction terms exceeded 2.0, and it was considered that multicollinearity existed. Thus, the interaction terms were put into different models, and tested by dividing them into three models. A model with the interaction terms of two levels of learning goal orientation and "Focus on collaborative learning" was defined Model 1, a model with the interaction terms of two levels of learning goal orientation and "Focus on relationships with society" was defined Model 2, and a model with the interaction terms of two levels of learning goal orientation and "Focus on specialized learning" was defined Model 3. As a result of performing the likelihood ratio test, the deviance of all three models was significantly lower than that of the NULL model (Table 16.4); thus, these models were considered acceptable. There were significant interactions between learning goal orientation at the seminar level and "Focus on collaborative learning," and between learning goal orientation at the seminar level and "Focus on specialized learning" (Table 16.5).

Simple slope tests were conducted to examine the two interactions (Fig. 16.2). The regression line of teachers' educational policies to learning-as-duty conception was calculated as the average of learning goal orientation at the seminar level ± 1 SD.

In the relationship between focus on collaborative learning and learning-as-duty conception (Fig. 16.2a), in the seminar where the students had low learning goal orientation (-1SD), the slope was not significant ($\beta = 0.11$, n.s); however, in the seminar where the students had high learning goal orientation (+1SD), the slope was significantly negative ($\beta = -0.17$, $p < 0.05$). This showed that in the seminar where students had high learning goal orientation, the more the teacher emphasized collaborative learning, the lower the students' learning-as-duty conception became.

In the relationship between focus on specialized learning and learning-as-duty conception (Fig. 16.2b), in the seminar where the students had low learning goal orientation (−1SD), the slope was a significantly positive ($\beta = 0.27$, $p < 0.01$); however, in the seminar where the students had high learning goal orientation (+1SD), the slope was not significant ($\beta = -0.16$, n.s). This showed that in the seminar where students had low learning goal orientation, the more the teacher emphasized specialized learning, the higher the students' learning-as-duty conception became.

16.5 Discussion

In this study, we examined the effects of teachers' educational policies for seminars on students' learning-as-duty conception, focusing on the moderating effect of students' learning goal orientation. As a result of examining the interaction between the three variables of teachers' educational policies and two levels of students'

Table 16.4 Likelihood ratio test results

	Degree of deviation	χ^2 (*df*)	*p*-value
NULL model	381.27		—
Model 1	347.45	33.82 (8)	0.00
Model 2	355.59	25.68 (8)	0.00
Model 3	352.91	28.36 (8)	0.00

Table 16.5 Hierarchical linear model analysis results

Model 1		Model 2		Model 3	
Independent variable	β	Independent variable	β	Independent variable	β
Focus on collaborative learning (FCL)	−0.05	Focus on collaborative learning (FCL)	0.01	Focus on collaborative learning (FCL)	−0.05
Focus on relationships with society (FRS)	0.06	Focus on relationships with society (FRS)	0.02	Focus on relationships with society (FRS)	−0.00
Focus on specialized learning (FSL)	−0.01	Focus on specialized learning (FSL)	0.08	Focus on specialized learning (FSL)	0.04
Seminar-level learning goal (SLG)	−0.38**	Seminar-level learning goal (SLG)	−0.35**	Seminar-level learning goal (SLG)	−0.35**
Individual-level learning goal (ILG)	0.02	Individual-level learning goal (ILG)	0.01	Individual-level learning goal (ILG)	0.02
SLG × FCL	−0.26**	SLG × FRS	−0.00	SLG × FSL	−0.16*
ILG × FCL	−0.00	ILG × FRS	−0.07	ILG × FSL	−0.06
R^2	0.17***	R^2	0.12**	R^2	0.14**

***$p < 0.001$, **$p < 0.01$, *$p < 0.05$

learning goal orientation, two interactions showed significant values. The results of simple slope tests suggested that the effect of teachers' educational policies on students' learning-as-duty conception may change, depending on students' learning goal orientation at the seminar level. Below, the effects of teachers' educational policies for seminars on students' learning-as-duty conception are specifically discussed.

16.5.1 Effects of Educational Policies that Emphasize Collaborative Learning on Learning-as-Duty Conception

The findings indicated that, in the seminar consisting of students who had high learning goal orientation, the more the teacher emphasized collaborative learning, the lower the students' learning-as-duty conception became. Conversely, in the seminar consisting of students who had low learning goal orientation, there was no effect of teachers' educational polices that emphasized collaborative learning on students' learning-as-duty conception.

In collaborative learning, teachers are only facilitators, and students take the initiative in class and learn from each other. Students who have high learning goal orientation tend to seek challenges in learning (Elliott & Dweck, 1988; Pintrich,

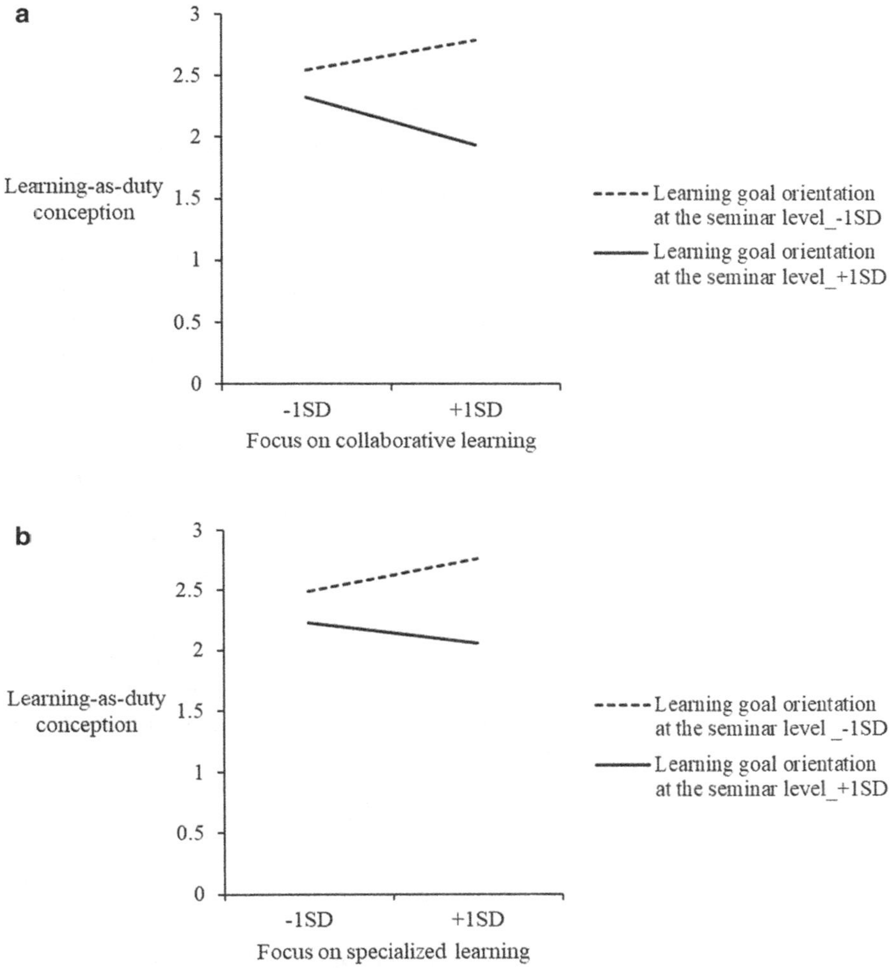

Fig. 16.2 (**a**) Interaction of focus on collaborative learning and learning goal orientation at the seminar level in predicting learning-as-duty conception. (**b**) Interaction of focus on specialized learning and learning goal orientation at the seminar level in predicting learning-as-duty conception

2000), and are actively involved in learning (Yokoyama & Miwa, 2018). It is assumed that, in a seminar consisting of students with such characteristics, each student would actively participate in learning, learn from each other, and recognize themselves as learners. As a result, their learning-as-duty conception would be suppressed. However, in a seminar consisting of students who have low learning goal orientation, even if there are many opportunities for collaborative learning, students would not actively participate in learning or learn experientially. As a result, the effects of educational policies that emphasize collaborative learning on students' learning-as-duty conception would not be seen.

16.5.2 Effects of Educational Policies that Emphasize Specialized Learning on Learning-as-Duty Conception

The findings indicated that, in the seminar consisting of students who had low learning goal orientation, the more the teacher emphasized specialized learning, the higher students' learning-as-duty conception became. However, in the seminar consisting of students who had high learning goal orientation, there was no effect of teachers' educational policy that emphasizes specialized learning on students' learning-as-duty conception.

There is a large difference in knowledge between teachers who are professionals in their fields and students; thus, when teaching specialized content in seminars, teachers often unilaterally transfer their knowledge to students. Students who have low learning goal orientation tend not to be interested in learning content (Pintrich, 2000), and will not show active learning behaviors (Yokoyama & Miwa, 2018). In a seminar consisting of students with such characteristics, even if there are many opportunities for specialized learning, the students will not be interested in the content, will not make statements or ask questions, and will passively listen to the lecture. As a result, students' learning-as-duty conception may increase. However, for students who have high learning goal orientation, learning specialized content from teachers is viewed as an opportunity to acquire knowledge and skills. Thus, in a seminar consisting of students who have high learning goal orientation, there is no positive or negative effect of an educational policy that emphasizes specialized learning on students' learning-as-duty conception.

16.5.3 Student' Learning Goal Orientation at the Seminar and Individual Levels

Since learning goal orientation data provided by students showed within-group similarities, we analyzed learning goal orientation by dividing it into two levels: the seminar level and individual level. As a result, it was suggested that students' learning goal orientation at the seminar level may moderate the effect of teachers' educational policies on students' learning-as-duty conception. Notably, however, that there was no moderating effect of students' learning goal orientation at the individual level. One possible reason for this is that conception of learning is influenced by learning experience (Horino & Ichikawa, 1993), and learning experience is affected by the member characteristics of the group to which a student belongs

(Adachi & Nakao, 2000). Thus, depending on whether the seminar consisted of students with high or low learning goal orientation, students' learning experiences would vary. As a result, the moderating effect of students' learning goal orientation at the seminar level became significant. Okubo and Kurosawa (2003) pointed out the need to consider students' learning experience together with, rather than separate from, the learning environment. Additionally, Ichikawa (1995) showed that students are highly influenced by learning groups, such as school classes, and it is important to consider the social environment, including individual students. Considering the suggestions of these previous studies, the results of this study supported the importance of examining the characteristics of seminar members as one environmental factor that shapes students' learning-as-duty conception.

16.5.4 Examination of Causal Results

Generally, what seminar a student belongs to is determined by the student's request, and the same applies to the faculty. Therefore, students' characteristics and beliefs, such as learning goal orientation and conception of learning, can be factors that affect selecting a seminar with a specific educational policy. In the following, the possibility of a causal relationship opposite to the assumptions of this study is examined.

The faculty surveyed were from a wide range of specialized fields, such as biology, engineering, mathematical sciences, social sciences, and humanities. In such a condition, the specialized field of the seminars to which students belonged in their third year was directly related to the theme of the graduation research students conducted during their fourth year. Therefore, it is highly possible that students selected a seminar based on research theme, rather than a teachers' educational policy. This is also presumed based on the fact that the faculty's explanatory materials for seminars only describe the subject and research content, not the teachers' educational policy or method.

Furthermore, path analysis verified whether students' learning goal orientation and learning-as-duty conception are the determinants of selecting a seminar with a specific educational policy. As a result of a path analysis using the three variables of teachers' educational policies as dependent variables, and learning-as-duty conception and learning goal orientation at the seminar level as explanatory variables, the goodness of fit was CFI = 0.661, RMSEA = 0.159, and the model was not established. Moreover, neither path showed a significant value.

Based on the above two points, it is unlikely that students' characteristics and beliefs, such as learning goal orientation and leaning-as-duty conception, are the factors that determine selecting a seminar with a specific educational policy; thus, the assumptions and results of this study are considered valid.

16.6 Conclusion

In order to clarify the basic factors that suppress learning-as-duty conception, we examined the effects of teachers' educational policies for seminars on students' learning-as-duty conception, focusing on the moderating effect of students' learning goal orientation. In the two analyzes in which the teachers' educational policy variables were "Focus on collaborative learning" and "Focus on specialized learning," the interaction between teachers' educational policies and students' learning goal orientation at the seminar level was significant. In order to examine the two interactions, simple slope tests were conducted. As a result, the following three suggestions were made: (1) the relationship between teachers' educational policies and students' learning-as-duty conception may change depending on students' learning goal orientation at the seminar level; (2) education which focuses on collaborative learning would be effective in suppressing students' learning-as-duty conception in seminars consisting of students who have high learning goal orientation; and (3) education which focuses on specialized learning can be a factor that increases students' learning-as-duty conception in seminars consisting of students who have low learning goal orientation. This study is significant, in that it was able to provide basic knowledge regarding suppression of students' learning-as-duty conception in general university seminars, which had not been examined to date.

This study has some limitations, which can be addressed in future research. First, we did not take into consideration students' tendencies in the initial stage, before the seminars started. It is not clear whether the responses among students in an individual seminar were consistent because the same type of students gathered in the same seminar, or because they were educated by the same teacher. Perhaps both factors are involved. In the future, it will be necessary to conduct a longitudinal study before students are assigned to seminars, and conduct an analysis taking into account student characteristics and environmental factors. Second is the elucidation of the process of how students' learning-as-duty conception is suppressed or increased. This study revealed a phenomenon in which students' learning-as-duty conception is suppressed or increased, and the elucidation of the process and detailed factors behind it require further investigation.

Acknowledgments This work was partially supported by JSPS KAKENHI grant number 18H05320.

References

Adachi, K., & Nakao, S. (2000). Relationship between group-based activity process, results and members' characteristics in CSCW. *Japan Journal of Educational Technology, 24*, 73–78. https://doi.org/10.15077/jmet.24.suppl_73

Akamatu, D. (2017). Relation between high school students' beliefs and learning strategies, and their academic achievement in learning English. *Japanese Journal of Educational Psychology, 65*(2), 265–280. https://doi.org/10.5926/jjep.65.265

Ames, C. (1992). Classrooms: Goals, structures, and student motivation. *Journal of Educational Psychology, 84*(3), 261–271. https://doi.org/10.1037/0022-0663.84.3.261

Ames, C., & Archer, J. (1987). Mothers' beliefs about the role of ability and effort in school learning. *Journal of Educational Psychology, 79*(4), 409–414. https://doi.org/10.1037/0022-0663.79.4.409

Ames, C., & Archer, J. (1988). Achievement goals in the classroom: Students' learning strategies and motivation processes. *Journal of Educational Psychology, 80*(3), 260–267. https://doi.org/10.1037/0022-0663.80.3.260

Chea, S., & Shumow, L. (2014). The relationships among writing self-efficacy, writing goal orientation, and writing achievement. *Language Education in Asia, 5*(2), 253–269. https://doi.org/10.5746/LEiA/14/V5/I2/A07/Chea_Shumow

Dart, B. C., Burnett, P. C., Purdie, N., Boulton-Lewis, G., Campbell, J., & Smith, D. (2000). Students' conceptions of learning, the classroom environment and approaches to learning. *The Journal of Educational Research, 93*(4), 262–270. https://doi.org/10.1080/00220670009598715

Dweck, C. S. (1986). Motivational processes affecting learning. *American Psychologist, 41*(10), 1040–1048. https://doi.org/10.1037/0003-066X.41.10.1040

Elliot, A. J., & Church, M. A. (1997). A hierarchical model of approach and avoidance achievement motivation. *Journal of Personality and Social Psychology, 72*(1), 218–232. https://doi.org/10.1037/0022-3514.72.1.218

Elliott, E. S., & Dweck, C. S. (1988). Goals: An approach to motivation and achievement. *Journal of Personality and Social Psychology, 54*(1), 5–12. https://doi.org/10.1037/0022-3514.54.1.5

Fenollar, P., Román, S., & Cuestas, P. J. (2007). University students' academic performance: An integrative conceptual framework and empirical analysis. *British Journal of Educational Psychology, 77*(4), 873–891. https://doi.org/10.1348/000709907X189118

Fushikida, W., Kitamura, S., & Yamauchi, Y. (2014). The effects of lesson structures in undergraduate seminars on growth in generic skills studies. *Japan Journal of Educational Technology, 37*(4), 419–433. https://doi.org/10.15077/jjet.KJ00009296326

Garcia, P. R. J. M., Restubog, S. L. D., Toledano, L. S., Tolentino, L., & Rafferty, A. (2011). Differential moderating effects of student- and parent-rated support in the relationship between learning goal orientation and career decision-making self-efficacy. *Journal of Career Assessment, 20*(1), 22–33. https://doi.org/10.1177/1069072711417162

Geitz, G., Joosten-Ten Brinke, D., & Kirschner, P. (2015). Goal orientation, deep learning, and sustainable feedback in higher business education. *Journal of Teaching in International Business, 26*(4), 273–292. https://doi.org/10.1080/08975930.2015.1128375

Heyman, G. D., & Dweck, C. S. (1992). Achievement goals and intrinsic motivation: Their relation and their role in adaptive motivation. *Motivation and Emotion, 16*(3), 231–247. https://doi.org/10.1007/BF00991653

Horino, M., & Ichikawa, S. (1993). Fundamental concepts of learning in university students. *Japan Society of Educational Information, 8*(3), 3–10. https://doi.org/10.20694/jjsei.8.3_3

Hudaykulov, A., Hongyi, X., & Galib, M. A. (2015). Impact of goal orientation theory on social capital: The implications for effective team cooperation in Uzbekistan textile industry. *International Journal of Management Science and Business Administration, 1*(6), 58–71. https://doi.org/10.18775/ijmsba.1849-5664-5419.2014.16.1005

Ichikawa, S. (1995). *Psychology of learning and education*. Tokyo: Iwanami Shoten.

Kaplan, A., & Midgley, C. (1997). The effect of achievement goals: Does level of perceived academic-competence make a difference? *Contemporary Educational Psychology, 22*(4), 415–435. https://doi.org/10.1006/ceps.1997.0943

Kavussanu, M., & Harnisch, D. L. (2000). Self-esteem in children: Do goal orientations matter? *British Journal of Educational Psychology, 70*(2), 229–242. https://doi.org/10.1348/000709900158074

Liem, A. D., Lau, S., & Nie, Y. (2008). The role of self-efficacy, task value, and achievement goals in predicting learning strategies, task disengagement, peer relationship, and achievement outcome. *Contemporary Educational Psychology, 33*(4), 486–512. https://doi.org/10.1016/j.cedpsych.2007.08.001

Meece, J., Blumenfeld, P., & Hoyle, R. (1988). Students' goal orientation and cognitive engagement in classroom activities. *Journal of Educational Psychology, 80*(4), 514–523. https://doi.org/10.1037/0022-0663.80.4.514

Nakayama, N. (2005). Testing a hypothesized model of English language learning: Japanese university students' goal orientation, beliefs, and learning strategies. *Japanese Journal of Educational Psychology, 53*(3), 320–330. https://doi.org/10.5926/jjep1953.53.3_320

Nicholls, J. G., Patashnick, M., & Nolen, S. B. (1985). Adolescents' theories of education. *Journal of Educational Psychology, 77*(6), 683–692. https://doi.org/10.1037/0022-0663.77.6.683

Nolen, S. B., & Haladyna, T. M. (1990). Motivation and studying in high school science. *Journal of Research in Science Teaching, 27*(2), 115–126. https://doi.org/10.1002/tea.3660270204

Okubo, T., & Kurosawa, K. (2003). A relational theory approach to social motivation. *Japanese Psychological Review, 46*(1), 12–23. https://doi.org/10.24602/sjpr.46.1_12

Peterson, E. R., Brown, G. T. L., & Irving, S. E. (2010). Secondary school students' conceptions of learning and their relationship to achievement. *Learning and Individual Differences, 20*(3), 167–176. https://doi.org/10.1016/j.lindif.2009.12.004

Pintrich, P. R. (2000). Multiple goals, multiple pathways: The role of goal orientation in learning and achievement. *Journal of Educational Psychology, 92*(3), 544–555. https://doi.org/10.1037/0022-0663.92.3.544

Pintrich, P. R., & Garcia, T. (1991). Student goal orientation and self-regulation in the college classroom. In M. L. Maehr & P. R. Pintrich (Eds.), *Advances in motivation and achievement: Goals and self-regulatory processes* (Vol. 7, pp. 371–402). Greenwich, CT: JAI Press.

Sugie, S. (2011). *An invitation to cooperative learning*. Kyoto: Nakanishiya Shuppan.

Suzuki, M. (2011a). How learning strategies are affected by the attitude toward tests: Using competence as a moderator. *Japanese Journal for Research on Testing, 7*(1), 52–65.

Suzuki, M. (2011b). Effects of a rubric: Values of a test, motivation for learning and learning strategies. *Japanese Journal of Educational Psychology, 59*(2), 131–143. https://doi.org/10.5926/jjep.59.131

Takayama, S. (2002). The relationship between learning conception, it's determinants and learning strategies. *Memoirs of the Faculty of Education, Shimane University, 37*, 19–26.

Tercanlioglu, L., & Demiröz, H. (2015). Goal orientation and reading strategy use of Turkish students of an English language teaching department. *The Qualitative Report, 20*(3), 286–311.

Van Rossum, E. J., & Schenk, S. M. (1984). The relationship between learning conception, study strategy and learning outcome. *British Journal of Educational Psychology, 54*(1), 73–83. https://doi.org/10.1111/j.2044-8279.1984.tb00846.x

Yamaguchi, T. (2012). Learning strategy use and cognitive and motivational factors in high school students: Individual differences in predicted test scores with a focus on effects of perceived utility. *Japanese Journal of Educational Psychology, 60*(4), 380–391. https://doi.org/10.5926/jjep.60.380

Yamamoto, M., & Ueno, M. (2015). Analysis of the effects of using rubric in constructivist learning: Focusing on goal orientation, conception of learning, motivation, learning strategy, and learning task performance. *Japan Journal of Educational Technology, 39*(2), 67–81. https://doi.org/10.15077/jjet.39007

Yokoyama, M., & Miwa, K. (2018). Relationship between goal orientation, conception of learning and learning behavior. In *Proceedings of 15th international conference of cognition and exploratory learning in digital age (CELDA 2018)* (pp. 233–240).

Index

A
Academic self-regulation, 129
ADDIE model, 80, 92
Adoption, 97, 98, 107
Affordances, 199–201
  communication, 201
  content, 201
  digital learning environment, 201
  organisational affordances, 204
  processes, 201
  research, 201
  social interaction, 205
  social software tools, 200
  sources, 202
  technology, 202
  virtual gaming, 202
American Civil War, 105
Artificial intelligence, 167
Augmented reality (AR), 185, 188–192, 194, 195
Average achieving (AA), 3

B
Banking-model pedagogy, 256
Behavioral changes, 145
Bipartite networks, 59
Bipolar construct, 36
BookRoll database, 139
Brain injury, 3

C
Cave Automatic Virtual Environment (CAVE), 170
Chess training
  analysis of variance, 27
  assessment tools, 24–26
  benefits, 27
  children's mathematical abilities, 28
  cognitive abilities, 19–21, 29
  cognitive skills, 29
  control groups, 28, 30
  data analysis, 26
  domain of mathematics, 28
  educational researchers, 27
  effects of training, 31
  groups of children, 28
  learning model, 30
  MANOVA, 26
  math and meta-cognitive abilities, 23
  math skills, 22, 23
  meta-cognitive abilities, 29
  meta-cognitive improvement, 29
  meta-cognitive teaching, 30
  participants, 24
  problem-solving skills, 28
  procedure, 24
  QAS test, 29
  school curriculum, 30
  school performance, 29
  significant differences, 28
  STEM, 27
  transfer process, 21, 22
  verbal skills, 28
Children's Embedded Figures Test (CEFT), 37
c-keywords, 241, 243, 247–250
  appropriateness, 244
  iLSB, 246
  positive and negative, 244

D. Ifenthaler et al. (eds.), *Balancing the Tension between Digital Technologies and Learning Sciences*, Cognition and Exploratory Learning in the Digital Age,
https://doi.org/10.1007/978-3-030-65657-7

Cognitive abilities, 19–21
Cognitive differences, 35, 44
Cognitive empowerment, 6
Cognitive functions, 6
Cognitive psychology, 21
Collaboration, 54, 55
Collaborative classes
  access, 261
  Alaska, 264
  conflict, 260
  cultural sensitivity, 260
  function, 263
  personal perspectives awareness, 260
  recommendations, 264, 265
  relationships, 263
  social reality, 266
  student reflection samples, 261, 262
  students' perspectives, 264
  vision, 263
  web-based instruction, 264
  web conferencing, 264
Collaborative course experience, 264
Collaborative Immersive Learning Virtual Reality Series (CILVRS), 171
Collaborative learning, 221, 277–279
Collaborative learning environments, 203
Collaborative problem-solving
  completion paths, 57, 63, 64, 72
  context, 57, 58
  data analytics, 58, 59
  evidence targets, 55, 56, 71
  evidence types, 56, 71
  feedback, 57, 65, 66, 72
  findings, 59, 60
  graph theory, 74
  learning assessment, 55
  learning outcomes, 57, 68, 70, 71, 73
  limitation, 73
  measures, 56, 59, 60, 71
  network analysis, 57, 71, 74
  participants, 57, 58
  personal learning, 55
  task participation, 57, 60–62, 72
  use of time, 57, 66–68, 72
Collegiate educators, 266
Collegiate instruction, 253, 257
Coloured Progressive Matrices (CPM), 7
Computational thinking
  algorithmic thinking, 34
  analysis of variance, 44
  children's descriptive statistics, 43
  cognitive skills, 45
  computer-based learning environments, 35
  computer science, 34
  control group, 44
  curriculum materials, 45
  definition, 33
  development, 33
  early elementary school, 35
  educational robotics, 35, 44
  educational settings, 33
  educational technology, 33
  effects of learning, 35
  FD learners, 44
  FDI, 34
  field of educational technology, 35
  gender, 35
  innovative programming environment, 35
  instructional tasks, 45
  mental model, 36
  method
    assessment of computational thinking, 42
    Bee-Bot, 38, 39
    CEFT, 37
    participants, 37
    problem-solving tasks, 40, 41
    research procedures, 41
    scaffolding, 39, 40
  problem-solving performance, 44
  research community, 34
  scaffolding, 34, 35
  scaffolding tools, 45
  school classroom, 45
  school curricula, 33
  teaching, 33
  technological tools, 35
  transfer of knowledge, 45
  visuo-spatial abilities, 36
Computer assisted training (CAT), 22
Computer-assisted learning programs, 79
Computer-assisted mathematical programme, 6
Computer-mediated communication systems, 205
Computer-supported collaborative learning (CSCL), 199–200
Conditional knowledge, 130
Control group, 7
Covid-19, 256
Creative Hybrid Environment for Robotic Programming (CHERP), 35

D

Data pre-processing, 156

Data processing module, 139
DBpedia Japanese, 240
Declarative knowledge, 130
Desktop-based virtual reality, 78
Diffusion of innovation, 98, 99, 106, 107
Digital learning environments, 54, 72
Digital technologies, 263
Dimensions of a Learning Organization Questionnaire (DLOQ), 112
Distance learning (DL), 183, 184, 186
  institutions, 184
  introduction, 184
  literature review, 185
  physical interaction, 184
  section, 188
Diversity, 153

E
Early numeracy learning
  adults' gender expectancies, 13
  analogies, 3
  arithmetical skills, 5
  Bonferroni's correction, 9
  coefficients, 8
  cognitive factor, 4
  daily life activities, 4
  demographic factors, 5
  dependency, 8
  development of numerical knowledge, 4
  developmental factors, 4
  domain-general functioning, 3
  effect of trainings, 8
  environmental factors, 3
  fixed effects, 8–10
  gender differences, 13
  general effects plots, 12
  groups, 8
  interventions, 4
  learning difficulties, 4
  learning disabilities, 3
  materials, 7, 8
  math and reading skills, 5
  math and visuospatial training, 12
  mathematical performance, 5
  mental rotation, 12
  non-pathological and non-innate condition, 3
  numerical skills, 4
  parameters, 8
  participants, 7, 8
  pencil-and-paper or computerized trainings, 6, 7
  post hoc comparisons, 11
  procedure, 7, 8
  psychoeducational intervention, 12
  skills, 12
  statistical analyses, 8
  stereotypes, 13
  variables, 4
  visuo-spatial and mathematical contents, 12
  visuo-spatial processing, 5
  working memory, 5
Ebsco database, 257
Educational institutions, 218
Educational policy, 272, 276, 280–282
Educational simulations, 200
Educational technology, 4
E-learning, 173, 176, 177
E-learning tools, 200
Exercise questions, 236
Experimental groups, 7
Experimental procedure, 155

F
Feedback, 65, 66, 72
Field Dependence (FD), 34
Field dependence/independence (FDI), 33–34, 36
Field Independence (FI), 34
Formal education, 6

G
Game Design Outline, 80
Generic learning skills, 21
Globalization, 162, 167
Graph theory, 54
Greenhouse gas, 239
Ground rules, 263

H
Head-mounted display (HMD), 78
Higher education, VR
  CAVE projects, 170, 176
  CILVRS, 171
  cost-effective smartphones, 179
  digital learning devices, 172
  free Wi-Fi, 177
  Geneva System, 171
  immersive learning experience, 170
  labs and infrastructure, 177
  learning platforms, off campus, 173

Higher education, VR *(cont)*
mixed-method survey, 171
mobile devices, 176
online learning content, 173, 177
qualitative and quantitative methods, 171
quality of life, 170
real-world environment, 170
remote learning experience, 179
self-administered questionnaire, 171
smartphones, 176
South Africa, 167–170, 176
student population, 173
student usage and knowledge, 173, 174, 177
students expectations, 174, 175, 177
student's perceived advantages, 175
theme identification, 172
training, 178

I
Ill-structured problems, 53
Immersive learning, 78–80
Immersive learning section, 193
Immersive technology, 178, 188
AR systems, 189
augmented reality, 189
cost, 192
education, 190
institutions, 189
learning outcomes, 190
realistic simulations, 189
virtual laboratories, 189
VR, 188
Information and communication technologies (ICT), 130, 187
Infrastructure readiness, 178
Interactive Learning Scenario Builder (iLSB), 236
c-keywords, 240
Firefox, 238
q-keywords, 239
scaffolds, 239
International Society for Technology in Education (ISTE), 95
Intra-class correlation coefficient (ICC), 274
Investigative learning, 235

K
Kindergarten, 4–6, 14
Knowledge construction, 237
Kyoto Protocol, 239, 243

L
Laboratory procedures, 190
Learner engagement, 191
Learning analytics (LA), 130, 132, 133
Learning analytics dashboard (LAD)
data sources, 133
functional design, 134
LA approach, 133
learning achievement, 134
metaboard (*see* Metaboard)
monitoring learning data, 133
online/blended learning settings, 133
students' online learning, 134
users, 134
visualization techniques, 133
Learning-as-duty conception
academic achievement, 270
academic performance, 270
collaborative learning, 277–279
content and methods, seminars, 270
definition, 269
hierarchical linear model analysis, 274, 276
ICC, 274
items, 273
learning behaviors, 270
learning experiences, 270
path analysis, 281
perceived learning, 269
specialized learning, 277, 280
students learning behaviors, 269
survey method, 273
teachers educational policies, 270, 272, 273, 275–277
Learning assessment, 55
Learning disabilities, 3
Learning experience, 280
Learning goal orientation
academic achievement, 271
categories, 270
collaborative learning, 277
conception of learning, 272
deep-processing behaviors, 271
definition, 270
feedback, 271
group-mean-centering variable, 274
ICC, 274, 276
individual level, 280, 281
items, 274
learning experiences, 272
learning-as-duty conception, 272
likelihood ratio test, 277
motivational variables, 271

path analysis, 281
performance, 270, 271
questionnaires *vs.* variables, academic achievement, 271
seminar level, 280
specialized learning, 277
students, 270
teacher involvement, 271
teachers educational policies, 276
Learning management system (LMS), 138
Learning organization model
assessment, 123
banking sector, 112
components, 111
definition, 111
dimensions, 118
education intensity, 119
education sector, 112, 119, 121
employees, 118
factors, 111
gender ratio, 120
human and structural level, 121
IT sector, 112, 117, 119, 121, 122
learning-related characteristics, 112
limitations, 122
method
design, 117
materials, 116
participants, 116
procedure, 117
questionnaire, 113
respondents' demographic profile, 117, 118
schools, 113, 115, 116
secondary vocational schools, 121
sectors, 123
self-education, 121
small and medium-sized enterprises, 114, 117, 120
systems level, 112
training of employees, 123
transformation, 112
type of school, 118
Learning outcomes, 190
Learning success, 82, 85–93
Linear mixed effects models (LME), 8
Linear regression, 157, 159, 160
Linked open data (LOD), 236, 239
c-keywords, 241
q-keyword, 240
RDF data, 240
structured data, 240
Word2vec, 240
Long-term working memory model, 20

M

Machine learning, 154, 157
data, 154
technology, 154
Machine learning methods, 161
Makerspace
artificial intelligence, 95
automated machinery, 95
conventional tools, 96
designing, 96, 97
diffusion of innovation, 98, 99
digital and physical materials, 97
digital/physical objects, 97
environment, 96
implication, 95, 106, 107
innovation, 96
ISTE, 96
learners, 95, 97
limitations, 107
materials, 96
mental representation models, 97
method
graphic representation, 104
influence, 105
instruction, 101, 102
instruments, 100, 101
materials, 103
PowerPoint, 103
pre-instructional and post-instructional self-assessments, 103
research participants, 100, 103, 105
tools, 103
type of tools, 103
multiple tools, 97
physical space, 97
pre-service teachers, 96
prototype, 97
research questions, 99
self-efficacy, 108
skills, 95
society, 95
students, 97
teachers, 95
technology acceptance model, 98, 99
technology integration, 97
tools, 96, 97
Math skills, 22, 23
Mathematical accuracy, 79
Mathematical learning disabilities (MLD), 3
Mathematical Problem-Solving (SPM), 25

Metaboard
behavioral changes, 145
BookRoll system, 135, 140
categories of elements, 145
comparison functions, 144
data processing module, 139
detailed view, 138
digital learning activities, 135
digital learning materials, 145
factors, 143
LAD success questionnaire, 140–143
learners' self-monitoring activities, 135
learning behaviors, 135
MAI, 140–142
monitoring functions, 144
monitoring learning, 135
participants, 140
perceived usefulness, 141
pre-MAI assessment, 141
reading path, 135–137
self-evaluation process, SRL, 136
visual attraction, 141, 143
visualization techniques, 138
web-based visualization module, 139
Metacognition
LA approach, 132, 133
LAD (*see* Learning analytics dashboard (LAD))
SRL, 130–132
Metacognitive Awareness Inventory (MAI), 140–142
Meta-cognitive skills, 24
Minimal brain dysfunction, 3
Modes of Production, 79
Multivariate analyses of variance (MANOVAs), 26

N
Network analysis, 57, 74
Neural network, 158

O
Online collaborative learning, 60
Online education, 256
Online learning environments, 200
Online learning technologies, 168
Organisational affordances, 203, 205, 210
Organising teamwork, 208
Oulu SWF context, 225

P
Peer assessment, 223
Perceived learning, 269
Perceived usefulness, 141, 144
Perceptual disabilities, 3
Performance goal orientation, 271
Personal learning, 55, 61
Primary school, 20, 22–24, 27, 30
Problem solving, 55
Procedural knowledge, 130
Programming languages, 162
Project-based learning (PBL), 217, 219
Psychoeducational intervention, 12
Psychoeducational programmes, 6
Psychoeducational trainings, 6, 7
Psychological issue, 23

Q
Qualitative research methodology, 185
Questionnaire of Approach to Studying (QAS), 24
Questionnaire of Study Strategies 1 (QS1), 24
Questionnaire of Study Strategies 2 (QS2), 25

R
Reading path, 136, 137
Realistic simulations, 189
Reflective diaries, 223, 229
Research-based tasks, 68
Robot Operating System, 220
Robotics, 34, 35, 37, 44, 45
Rogers' Diffusion of Innovation framework, 104

S
Sample selection processes, 22
Scaffolding, 34–36, 39, 41, 44, 45
Science technology engineering math (STEM), 20, 27
career goals and motivation, 152
cross-cultural training, 153
intercultural competencies, 152
job advertisements, 156
language barriers, 152
literature, 152, 153
multi-cultural stakeholders, 153
occupations, 155
statistical and machine learning methods, 153
Secondary schools, 77, 82, 86
Self-regulated learning (SRL)
academic achievement, 129
cognitive learning strategies, 130, 145
LA approach, 130

metacognition, 129–132, 145
Self-regulating navigation, 236
Self-regulation, 130
Self-reported inventories, 132
Simulation-based learning environment, 206
Simulation game exercise, 207
Simulation games, 202
Small and medium-sized enterprises, 114
Social competencies, 152, 154, 156, 160
Social network analysis, 59
Socio-interactional affordances, 203, 205, 206, 211
Software engineering (SE), 217
Software factory (SWF), 218, 224
collaborative learning, 221
communication, 221, 222
competencies, 227
course, 219, 220
customers, 222
design, 218
distribution, 223
execution plan, 222
laboratory, 220
learning approach, 220
PBL, 219
peer assessment, 223
research, 219
teaching staff, 229
workshop, 222
SPARQL queries, 242
Specialized learning, 277, 280
Static problems, 54
Student perception survey, 225, 226
ACCC, 226
CLEI, 225, 226
teaching staff, 225
Student-facing LAD, 134
Systematic literature review, 186

T
Teacher education, 96
Teachers' educational policies, 275–277
Team formation and mentoring, 223
Technological affordances, 202–204, 208, 211
Technology Acceptance Model (TAM), 99, 106
Technology-based learning environment, 201
Technology Integration Matrix (TIM), 97, 108
Technology tools, 184, 188
Transfer process, 21, 22
t-test, 247

V
Video conferencing, 254, 266
Videogame-like activities, 6
Virtual laboratory, 191
Virtual learning environments (VLEs), 193, 195
Virtual reality (VR), 188, 189
ADDIE model, 80
base area and height, 81
computer system-generated simulation, 168
curriculum, 81
educational institutions, 77
experience- and action-oriented learning, 77
facilitators, 168
field experiment
conventional lesson, 88
design, 86, 87
evaluation results, 88–92
experimental *vs.* control group, 89
gender, 91
learning goals, 78, 87, 88
learning success, 90
student observations and feedbacks, 92
head-mounted displays, 77
higher education (*see* Higher education, VR)
immersive experience, 81
immersive learning, 78, 79
learning success, 78
mathematics, 79, 80
prototype, 78
secondary schools, 77
sizes, 81, 82
usability tests
learning environment, 83, 85
paper-based questionnaire, 83
participants, 83, 86
secondary students, 82
usability-related criteria, 78
Visualization techniques, 133, 138
Visuo-spatial processing, 5
Visuospatial skills, 22
Vocational Education and Training programme (VET), 57

W
Web-based investigative learning, 236, 238, 239
iLSB, 237
knowledge construction process, 237–238
model, 237
navigation and knowledge construction, 237
question tree, 238

Web-based investigative learning process, 235
Web-based pedagogy, 255
Web-based visualization module, 139
Web conferencing
academic institutions, 254
academic settings, 254
challenges, higher education, 255, 256
collaborative class, 264, 266 (*see also* Collaborative classes)
collegiate instruction
collaborative class, 258
college courses, 260
content area, 259
cross-institution experience, 257
data analysis, 260, 263
educational institutions, 260
institution, 257
methodology, 259
students, 258
subjects, 261
systematic literature review, 257, 259
tool, 259
types of studies, 259, 260
zoom, 258
collegiate pedagogical tool, 265
Covid-19 pandemic, 254
internet services, 254
mobile devices, 263
real-time video conferencing software, 254
singular disciplines, 265
social construction, 253
social reality, 266
tools, 256
use of video, 254
web-based pedagogy, 255
Web resources, 236
Well-structured problems, 53
Wi-Fi, 177
Word2vec, 240, 241
Working memory, 5
Workshops goal, 222
Writing reflective diaries, 228

Y
Young children, 35, 42, 45

Z
Zoom web conferencing program, 258

The manufacturer's authorised representative in the EU is Springer Nature Customer Service Centre GmbH, Europaplatz 3, 69115 Heidelberg, Germany. If you have any concerns regarding our products, please contact ProductSafety@springernature.com

Printed and bound by CPI Group (UK) Ltd, Croydon, CR0 4YY
28/09/2026
02224089-0012